High Performance Scientific Computing Using Distributed Infrastructures

Results and Scientific Applications Derived from the Italian PON ReCaS Project

High Performance Scientific Computing Using Distributed Infrastructures

Results and Scientific Applications Derived from the Italian PON ReCaS Project

Editors

Giuseppe Andronico
INFN-Catania, Italy

Roberto Bellotti
University of Bari Aldo Moro, Italy

Guglielmo De Nardo
University of Naples Federico II, Italy

Giuliano Laccetti
University of Naples Federico II, Italy

Giorgio Maggi
Polytechnic University of Bari, Italy

Leonardo Merola
University of Naples Federico II, Italy

Guido Russo
University of Naples Federico II, Italy

Lucia Silvestris
INFN-Bari, Italy

Sabina Tangaro
INFN-Bari, Italy

Enrico Tassi
University of Calabria, Italy

World Scientific

NEW JERSEY • LONDON • SINGAPORE • BEIJING • SHANGHAI • HONG KONG • TAIPEI • CHENNAI • TOKYO

Published by

World Scientific Publishing Co. Pte. Ltd.
5 Toh Tuck Link, Singapore 596224
USA office: 27 Warren Street, Suite 401-402, Hackensack, NJ 07601
UK office: 57 Shelton Street, Covent Garden, London WC2H 9HE

Library of Congress Cataloging-in-Publication Data
Names: Laccetti, Giuliano, author. | Programma operativo nazionale ricerca e competitività 2007–2013 (Italy)
Title: High performance scientific computing using distributed infrastructures : results and scientific applications derived from the Italian PON ReCaS Project / Giuliano Laccetti (University of Naples Federico II, Italy), Leonardo Merola (University of Naples Federico II, Italy), Roberto Bellotti (University of Bari Aldo Moro, Italy), Giuseppe Andronico (National Institute for Nuclear Physics (INFN)--Catania, Italy), Guglielmo de Nardo (University of Naples Federico II, Italy), Giorgio Maggi (Polytechnic University of Bari, Italy), Guido Russo (University of Naples Federico II, Italy), Lucia Silvestris (National Institute for Nuclear Physics (INFN)--Bari, Italy), Enrico Tassi (University of Calabria, Italy) & Sabina Tangaro (National Institute for Nuclear Physics (INFN)--Bari, Italy).
Description: New Jersey : World Scientific, 2016. | Includes bibliographical references.
Identifiers: LCCN 2016000016 | ISBN 9789814759700 (hc : alk. paper)
Subjects: LCSH: Science--Italy--Data processing. | Engineering--Italy--Data processing. | Re.Ca.S. : rete di calcolo per SuperB ed altre applicazioni (Project) | Re.Ca.S. (Computer network) | Electronic data processing--Distributed processing. | High performance computing. | Computer networks--Italy.
Classification: LCC Q183.9 .H576 2016 | DDC 502.85/436--dc23
LC record available at http://lccn.loc.gov/2016000016

British Library Cataloguing-in-Publication Data
A catalogue record for this book is available from the British Library.

Desk Editors: Kalpana Bharanikumar/Amanda Yun

Typeset by Stallion Press
Email: enquiries@stallionpress.com

Printed in Singapore

Contents

Foreword

Started in October 2011 and completed in summer 2015, the ReCaS project was funded by the Italian Ministry of Research and Education (MIUR) through the Italian national *Programma Operativo Nazionale* (National Operational Programme, PON) for *"Research and Competitiveness"* (PON R&C) 2007–2013 Call 254 Action I *"support for structural changes and scientific & technological improvement for a transition towards a knowledge economy"* for the development and enhancement of a distributed computing infrastructure of Grid/ Cloud type over the four EU "Convergence" regions in Southern Italy: Campania, Puglia, Sicily, and Calabria.

The acronym ReCaS stands for "*Rete di Calcolo per SuperB e altre applicazioni*" (Computing network for SuperB and other applications). The SuperB was a *Istituto Nazionale Fisica Nucleare* (INFN) project for a new accelerator in the Rome area that has been cancelled by the Italian government due to budget limitations. Now ReCaS is a computing network infrastructure in Southern Italy devoted to scientific and non-scientific applications within the vision of a common European infrastructure for computing, storage, and network.

The actuators are the University of Naples Federico II, the University of Bari Aldo Moro and the Italian National Institute of Nuclear Physics (INFN) with its "sections" in Naples, Bari, Catania, and Cosenza. In Catania and Cosenza sites, there are fundamental operational synergies with the University of Catania and the University of Calabria, respectively.

A lot of experience in the context of previous PON 2000–2006 program, INFN and UE grid projects, has been fully exploited, e.g., SCoPE project in Naples, PI2S2 in Catania.

The main objective of the project is a better integration of the various initiatives already present in Southern Italy, carried out by research and academic institutions, including INFN and Universities. High throughput computing (HTC) and high performance computing (HPC), big data preservation in the medium and long term, need a national infrastructure federated between the various actors in Italy in order to be internationally competitive and integrated in a European Grid/Cloud infrastructure.

As an outcome of the ReCaS project, the present book is published by the World Scientific Publishing and it includes contributions from all the people who used (or collaborate to its maintenance) the ReCaS computing and storage resources for scientific and technological research activities from different scientific domains and applications, e.g., Materials Science and Technology, Aerospace Science and Technology, Mathematics and Computer Science, Chemistry, Life Science, and so on.

All the contributions are organized in different Sections: the first Section collects contributions describing the infrastructure in all its parts (from the hardware description to the policies used for user recruitment and support); the second one collects the contributions which are an overview on the type of applications using the ReCaS infrastructure. Each of the remaining sections collects contributions from a specific scientific/technological domain: Physical Science, Life Science, Engineering, Earth Science, Computer and Computational Science. In the "Physical Science" section there are contributions from the High Energy Physics domain related with the ATLAS, ALICE and The Belle II experiments. In the same section, contributions related with astronomical studies (the KM3Net experiment) and research on non-commutative geometry are also presented. The "Life Science" section includes contributions related to a new approach for the Study of Melanogenesis, the study of the structure of chromosomes in the nucleus of cells and the use of neural network to model complex biological phenomena. In the "Engineering" section are presented contributions from different engineering fields of interest: the building responses to an earthquake event; the computational simulation of rheology processes; the "in silico" analysis of architectures for the naval and aerospace worlds. In the "Earth Science" section, two contributions are present: the first one discusses results about the numerical simulation of seismic ground motion during the 2009 L'Aquila earthquake, the second one describes the results of the first section related with a new numerical approach for simulation of subsurface hydrological processes. Finally, the "Computer and Computational Science" section collects a first contribution related with a Scalability Analysis of Variational Data Assimilation Algorithms on Hybrid Architectures. All the other contributions in this section are related with the development of tools for the management, monitoring and use of the ReCaS infrastructure resources.

All of the above papers address original research and design and maintenance in the broad field of the HPC systems and environments and several science fields. We collected them with the aim to prepare a volume that could serve as resource for education, information, and reference to professors, researchers, graduate students and the more general HPC community. The editors would like to express their gratitude to the referees for their dedication and expert work in reviewing these papers. Last but not least, we are grateful to all authors for their contributions. We hope that readers will enjoy this volume as much as we enjoyed preparing it.

Leonardo Merola, Roberto Bellotti, Giuseppe Andronico, Guglielmo De Nardo, Giuliano Laccetti, Giorgio Maggi, Guido Russo, Lucia Silvestris, Sabina Tangaro, Enrico Tassi

Naples
September 2016

List of Contributors

M. Alfano
National Institute of Nuclear Physics, Naples, Italy

N. Amoroso
University of Bari A. Moro and National Institute of Nuclear Physics, Bari, Italy

G. Andronico
National Institute of Nuclear Physics, Catania, Italy

C. Annunziatella
University of Naples Federico II, Naples, Italy

M. Antonacci
National Institute of Nuclear Physics, Bari, Italy

R. Arcucci
Imperial College, London, UK
and University of Naples Federico II and SPACI s.r.l., Naples, Italy

R. Barbera
Università di Catania, Catania, Italy

M. Barbieri
University of Naples Federico II, Naples, Italy

G.B. Barone
University of Naples Federico II, Naples, Italy

E. Begovic
University of Naples Federico II, Naples, Italy

R. Bellotti
National Institute of Nuclear Physics
and University of Bari A. Moro, Bari, Italy

P. Belluomo
National Institute of Nuclear Physics, Catania, Italy

S. Bianco
University of Naples Federico II, Naples, Italy

V. Boccia
National Institute of Nuclear Physics, Naples, Italy

D. Bottalico
University of Naples Federico II, Naples, Italy

C. Bozza
University of Salerno, Salerno and National Institute of Nuclear Physics, Naples, Italy

A. Brasiello
University of Salerno, Fisciano, Italy

R. Bruno
National Institute of Nuclear Physics, Catania, Italy

F. Cafagna
National Institute of Nuclear Physics, Bari, Italy

G. Calise
University of Naples Federico II, Naples, Italy

R. Campagna
University of Naples Federico II
and National Institute of Nuclear Physics, Naples, Italy

G. Carlino
National Institute of Nuclear Physics, Naples, Italy

L. Carracciuolo
National Research Council (CNR), Naples, Italy

P. Castellano
National Institute of Nuclear Physics, Naples, Italy

M. Castellano
Politecnico of Bari and Genesis Consulting s.r.l, Bari, Italy

R. Cevenini
National Institute of Nuclear Physics, Naples, Italy

A.M. Chiariello
University of Naples Federico II, Naples, Italy

D. Ciliberti
University of Naples Federico II, Naples, Italy

D.P. Coiro
University of Naples Federico II, Naples, Italy

F. Colamaria
University of Bari A. Moro
and National Institute of Nuclear Physics, Bari, Italy

D. Colella
University of Bari A. Moro
and National Institute of Nuclear Physics, Bari, Italy

S. Corcione
University of Naples Federico II, Naples, Italy

N. Corriero
Politecnico of Bari, Bari, Italy

O. Crescenzi
University of Naples Federico II, Naples, Italy

F. Cristaudo
National Institute of Nuclear Physics, Catania, Italy

S. Cuomo
University of Naples Federico II, Naples, Italy

L. D'Amore
University of Naples Federico II,
Naples and Euro Mediterranean Center on Climate Changes (CMCC), Lecce, Italy

G. D'Avino
University of Naples Federico II, Naples, Italy

A. d'Onofrio
University of Naples Federico II, Naples, Italy

A.H. Day
University of Strathclyde, Glasgow, UK

M. de Palma
National Institute of Nuclear Physics
and University of Bari A. Moro, Bari, Italy

G. De Nardo
University of Naples Federico II
and National Institute of Nuclear Physics, Naples, Italy

P. De Michele
University of Naples Federico II, Naples, Italy

G. De Monaco
University of Naples Federico II, Naples, Italy

M. De Corato
University of Naples Federico II, Naples, Italy

A. De Marco
University of Naples Federico II, Naples, Italy

F. De Luca
University of Naples Federico II, Naples, Italy

D. Del Prete
National Institute of Nuclear Physics, Naples, Italy

S. del Gaudio
University of Naples Federico II, Naples, Italy

T. Del Vino
Genesis Consulting s.r.l, Bari, Italy

P. Della Vecchia
University of Naples Federico II, Naples, Italy

D. Diacono
National Institute of Nuclear Physics, Bari, Italy

G. Donvito
National Institute of Nuclear Physics, Bari, Italy

A. Doria
National Institute of Nuclear Physics, Naples, Italy

D. Elia
National Institute of Nuclear Physics, Bari, Italy

R. Errico
University of Bari A. Moro, Bari, Italy

R. Esposito
National Institute of Nuclear Physics, Naples, Italy

L. Evangelista
National Research Council (CNR), Naples, Italy

M. Fargetta
National Institute of Nuclear Physics, Catania, Italy

G. Festa
University of Naples Federico II, Naples, Italy

A. Franco
National Institute of Nuclear Physics, Bari, Italy

M. Gaetani d'Aragona
University of Naples Federico II, Naples, Italy

A. Galletti
University of Naples Parthenope, Naples, Italy

R. Gervasoni
National Institute of Nuclear Physics, Bari, Italy

F. Giannuzzi
University of Bari A. Moro, Bari, Italy

E. Giorgi
National Institute of Nuclear Physics, Catania, Italy

F. Greco
National Research Council (CNR), Naples, Italy

N. Guarracino
University of Calabria and National Institute of Nuclear Physics, Arcavacata di Rende, Italy

P. Guida
National Institute of Nuclear Physics, Naples, Italy

G. Ianniruberto
University of Naples Federico II, Naples, Italy

I. Iervolino
University of Naples Federico II, Naples, Italy

A. Incecik
University of Strathclyde, Glasgow, UK

A. Italiano
National Institute of Nuclear Physics, Bari, Italy

G. La Rocca
National Institute of Nuclear Physics, Catania, Italy

G. Laccetti
University of Naples Federico II
and National Institute of Nuclear Physics, Naples, Italy

M. Lapegna
University of Naples Federico II, Naples, Italy

V. Lavorini
National Institute of Nuclear Physics,
Arcavacata di Rende, Italy

F. Lizzi
National Institute of Nuclear Physics
and University of Naples Federico II, Naples, Italy

P.L. Maffettone
University of Naples Federico II, Naples, Italy

G. Maggi
National Institute of Nuclear Physics,
University of Bari A. Moro and Politecnico of Bari, Italy

S. Mancini
University of Naples Federico II, Naples, Italy

G. Manfredi
University of Naples Federico II, Naples, Italy

G. Marrucci
University of Naples Federico II, Naples, Italy

B. Mele
University of Naples Federico II, Naples, Italy

L. Merola
University of Naples Federico II
and National Institute of Nuclear Physics, Naples, Italy

D. Michelino
National Institute of Nuclear Physics, Naples, Italy

P. Migliozzi
National Institute of Nuclear Physics, Naples, Italy

G. Miniello
National Institute of Nuclear Physics
and University of Bari A. Moro, Italy

S. Miranda
University of Naples Federico II, Naples, Italy

A. Monaco
National Institute of Nuclear Physics, Bari, Italy

S. Monforte
National Institute of Nuclear Physics, Catania, Italy

A. Murli
Euro Mediterranean Center on Climate Changes (CMCC), Lecce and SPACI s.r.l., Naples, Italy

S. Naddeo
National Institute of Nuclear Physics and University of Naples Federico II, Naples, Italy

M. Nicodemi
University of Naples Federico II, and National Institute of Nuclear Physics, Naples, Italy

F. Nicolosi
University of Naples Federico II, Naples, Italy

S. Nicotri
National Institute of Nuclear Physics, Bari, Italy

P. Notarangelo
National Institute of Nuclear Physics, Bari, Italy

S. Nuzzo
National Institute of Nuclear Physics and University of Bari A. Moro, Bari, Italy

M. Paone
National Institute of Nuclear Physics, Catania, Italy

S. Pardi
National Institute of Nuclear Physics, Naples, Italy

G. Park
University of Naples Federico II, Naples, Italy

G. Passaro
National Institute of Nuclear Physics, Catania, Italy

C. Pensa
University of Naples Federico II, Naples, Italy

A. Piccolo
University of Naples Federico II, Naples, Italy

G. Platania
National Institute of Nuclear Physics, Catania, Italy

M. Polese
University of Naples Federico II, Naples, Italy

A. Pombo
Max-Delbrück Centre for Molecular Medicine, Berlin, Germany

A. Prota
University of Naples Federico II, Naples, Italy

C. Rocca
National Institute of Nuclear Physics, Catania, Italy

G. Russo
University of Naples Federico II
and National Institute of Nuclear Physics, Naples, Italy

B. Santeramo
National Institute of Nuclear Physics, Bari, Italy

A. Santo
University of Naples Federico II, Naples, Italy

G. Sava
National Institute of Nuclear Physics, Catania, Italy

R. Scognamiglio
University of Naples Federico II, Naples, Italy

G. Scotti
University of Naples Federico II, Naples, Italy

G. Selvaggi
National Institute of Nuclear Physics
and University of Bari A. Moro, Bari, Italy

G. Severino
University of Naples Federico II, Naples, Italy

F. Silvestri
University of Naples Federico II, Naples, Italy

L. Silvestris
National Institute of Nuclear Physics, Bari, Italy

A. Solla
National Institute of Nuclear Physics, Naples, Italy

V. Spinoso
National Institute of Nuclear Physics, Bari, Italy

B. Spisso
National Institute of Nuclear Physics, Naples, Italy

G. Staiano
University of Naples Federico II, Naples, Italy

S. Tangaro
National Institute of Nuclear Physics, Bari, Italy

M.A. Tangaro
Politecnico of Bari, Bari, Italy

A. Tarasio
National Institute of Nuclear Physics, Arcavacata di Rende, Italy

E. Tassi
University of Calabria and National Institute of Nuclear Physics, Arcavacata di Rende, Italy

A. Tateo
University of Bari A. Moro, Bari, Italy

A. Tebano
National Institute of Nuclear Physics, Naples, Italy

E. Tinelli
University of Bari A. Moro, Bari, Italy

R. Tognaccini
University of Naples Federico II, Naples, Italy

M. Trofa
Center for Advanced Biomaterials for Health Care @CRIB,
Italian Institute of Technology (IIT)
and University of Naples Federico II, Naples, Italy

R. Valentini
National Institute of Nuclear Physics, Bari, Italy

R. Vela
National Institute of Nuclear Physics, Naples, Italy

F. Ventola
University of Bari A. Moro, Bari, Italy

M. Villari
University of Messina, Messina, Italy

M.M. Villone
Center for Advanced Biomaterials for Health Care @CRIB,
Italian Institute of Technology (IIT)
and University of Naples Federico II, Naples, Italy

G. Vino
University of Bari A. Moro and National Institute
of Nuclear Physics, Bari, Italy

G. Zaccheo
Genesis Consulting s.r.l, Bari, Italy

Chapter 1

The ReCaS Project

L. Merola

Università Federico II, Napoli, Italy
Istituto Nazionale Fisica Nucleare, Napoli, Italy
leonardo.merola@na.infn.it

This paper gives a general description of the *Rete di Calcolo per SuperB e altre* (ReCaS) project and of the infrastructure which has been realized in four data centers in Southern Italy, namely Napoli, Bari, Catania, and Cosenza, thanks to the Italian National Operational Programme for "Research and Competitiveness" 2007–2013 (PON for R&C) administered by the Italian Ministry of University and Research. At present, the ReCaS data centers are fully operational and integrated in the national and international Grid/Cloud infrastructure.

1. Introduction

Started in October 2011 and completed in summer 2015, the ReCaS project[1] was funded by the Italian Ministry of Research and Education (MIUR) through the program National Operational Programme for "Research and Competitiveness" (PON for R&C) 2007–2013 Call 254 Action I "support for structural changes and scientific & technological improvement for a transition towards a knowledge economy" for the development and enhancement of a distributed computing infrastructure of Grid/Cloud type over the four EU "Convergence" regions in Southern Italy: Campania, Puglia, Sicily, and Calabria.

The acronym stands for "*Rete di Calcolo per SuperB e altre applicazioni*" (Computing network for SuperB and other applications, ReCaS).[2] The SuperB was a *Istituto Nazionale Fisica Nucleare* (INFN) project for a new accelerator in the Rome area that has been cancelled by the Italian government due to budget limitations. Now ReCaS is a computing network infrastructure in Southern Italy devoted to scientific and non-scientific applications within the vision of a common European infrastructure for computing, storage, and network.

The actuators are the University of Naples *Federico II*, the University of Bari *Aldo Moro* and the National Institute of Nuclear Physics (INFN) with its sections in Naples, Bari, Catania, and Cosenza. In the two sites of Catania and Cosenza, there are fundamental operational synergies respectively with the University of Catania and the University of Calabria.

A lot of experience in the context of previous PON 2000–2006 program, INFN and UE grid projects, has been fully exploited, e.g., SCoPE project in Naples, PI2S2 in Catania.

The program is integrated in the 2007–2013 European Union strategy and takes benefit from the partnership between MIUR and MiSE which is the Italian Ministry for the Economic Development (Fig. 1). It is based on total investment of 13.7 million euros, respectively 6.9 M to INFN, 2.1 ME to University of Naples Federico II (UniNA), 4.7 ME to UniBA. The objective is to transform these regions into drivers of social and economic development.

Fig. 1. Partners and funding agencies.

The ReCaS project and infrastructure are managed by an Executive Committee chaired by L. Merola with representatives of the three institutions (INFN, UniNA, and UniBA).

2. The Objectives

The main objective of the project is a better integration of the various initiatives already present in Southern Italy, carried out by research and academic institutions, including INFN and Universities. The collaboration between INFN and University in scientific and technological research and development dates back to early '50 of the last century and has been crucial for the present project. High throughput computing (HTC) and high performance computing (HPC), big data preservation in the medium and long term, need a national infrastructure federated between the various actors in Italy in order to be internationally competitive and integrated in a European Grid/Cloud infrastructure.

The network will be open and accessible to all researchers, public and private, and will be characterized by unprecedented — in the academic context — computing power and storage capacity. Posted in the European Grid Infrastructure (EGI), ReCaS is also an opportunity to the countries of the Mediterranean area and extends the potential of the current network, composed of four major supercomputing infrastructure made by the PON 2000–2006. Thanks to numerous technological spin-offs of the territory, ReCaS fully responds to the need to promote the competitiveness of enterprises in Southern Italy.

The main scientific users for the development of the ReCaS infrastructure are the 4 major international collaborations (ALICE, ATLAS, CMS, LHCb) of the CERN Large Hadron Collider (LHC). In all four ReCaS sites, there are scientific communities who actively participate in the above experimentation.

Furthermore, the ReCaS computing and storage resources are also available to any interested public and private researchers coming from different scientific domains and applications, e.g., from the

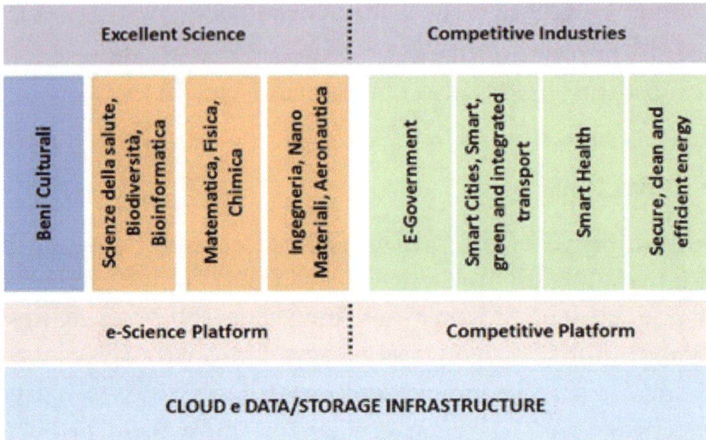

Fig. 2. Scientific and industrial users.

Science and Technology of Materials, Science and Technology of Aerospace, Mathematics and Computer Science, the Chemistry, Science of Life, Cultural Heritage, and so on (Fig. 2).

The ReCaS infrastructure is also supporting a number of projects and collaborations both of national and international interest as PRISMA-Smartcities, BioVeL, ELIXIR-ITA, KM3NeT, LCG-WLCG, Belle2 experiment (Japan).

It also aims at promoting collaborations with scientific and academic institutions in the Mediterranean area; recently an agreement has been signed with Egyptian universities.

ReCaS provides services based on open standards that allow interoperability between datacenters and data-based architectures which can implement all the elements of the model platform for the processing of big data and that has been recently defined by the European Commission for Horizon 2020 (Fig. 3).

Users can choose from an exhaustive set of application software (Fig. 4).

The project also included a training activity through the establishment of university post degree masters, in Naples and in Bari, followed by summer schools on Cloud Computing and Science Gateways and stages for the best students at the four ReCaS sites.

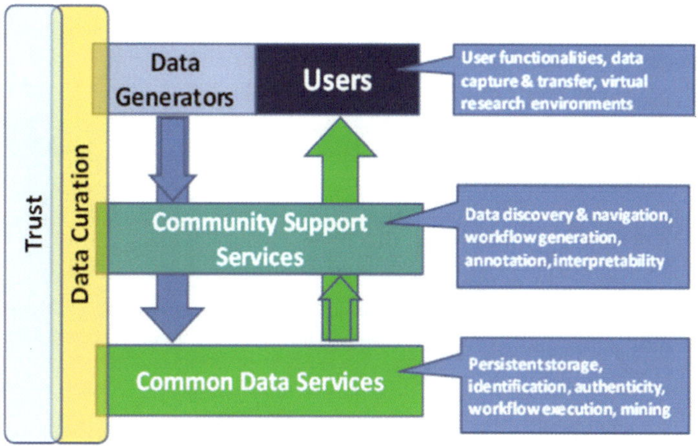

Fig. 3. Horizon 2020 model for big data processing.

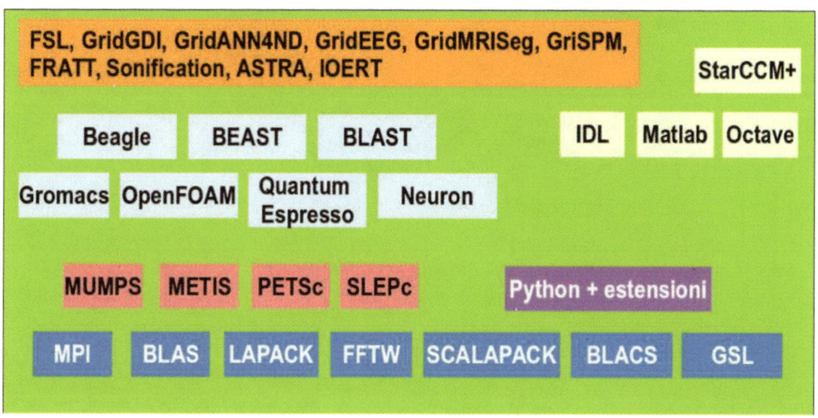

Fig. 4. Application software.

3. The Infrastructure

The four data centers are connected to each other through a high bandwidth network (Fig. 5) which is foreseen to be upgraded with links at 40/100 Gbps in the near future. They are also connected to the rest of the Italian GARR research and academic network and through it to CERN and to international links not shown in the figure.

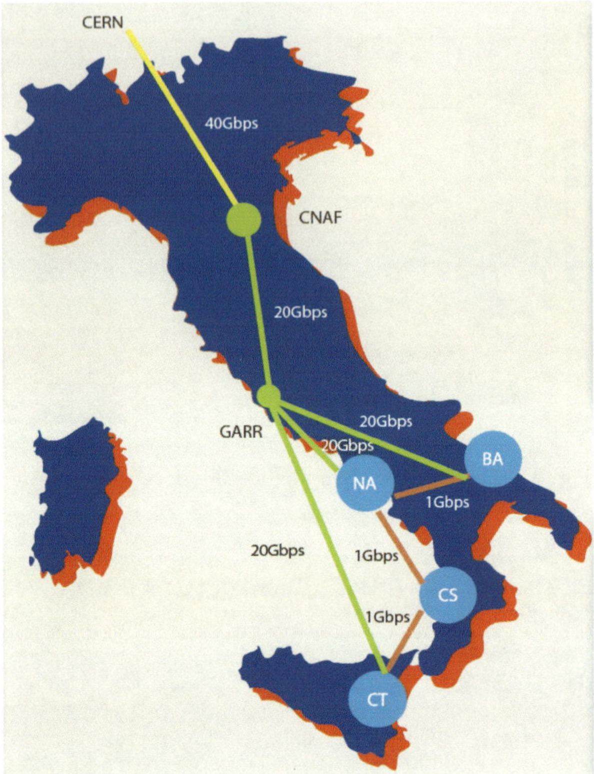

Fig. 5. The ReCaS network.

Fig. 6. The Naples data center. Computing resources: 132 servers, 4,956 cores. Storage capacity: 4,957 TB.

All the four data centers (Figs. 6, 7, 8 and 9) have realized HPC infrastructures which are used both by local users and by the ReCaS and the national/international community. The details of each configuration will be described elsewhere in this review volume.

Fig. 7. The Bari data center. Computing resources: 128 servers, 8,192 cores, HPC cluster with 20 GPU nodes for a total of additional 800 cores. Storage capacity: 3,552 TB on disk and 2,500 TB on tape (tape library).

Fig. 8. The Catania data center. Computing resources: 54 servers, 2,562 cores. Storage capacity: 324 TB.

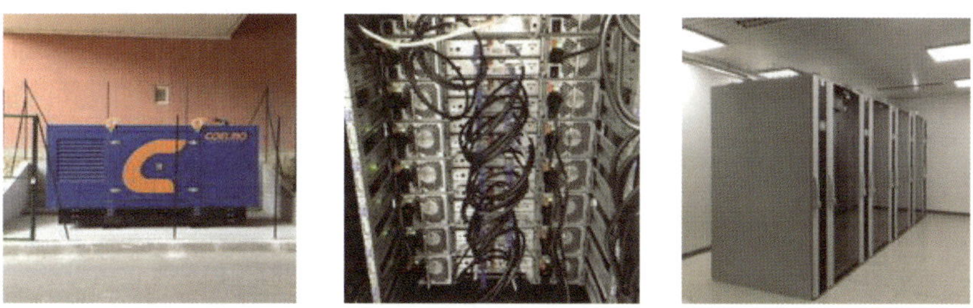

Fig. 9. The Cosenza data center. Computing resources: 90 servers, 3,500 cores. Storage capacity: 900 TB.

4. The Training Program

By the training program associated to the project, a total of 50 experts were formed, with two different profiles: (a) experts in the field of data analysis, with particular reference to the applications and the use of

HPC for physics research and applications; (b) developers of applications and/or managers of the infrastructure.

In particular, the objective has been achieved to qualify technical-scientific, management and business personnel in the ICT sector, particularly with regard to systems of HPC. The trainees have gained expertize in the development, use and management of ICT services and platforms for HPC to support different fields of interests, like Science, Industry, and Public Administration.

For this purpose, two University Masters have been set up, one on "Technologies for High Performance Scientific Computing," organized and managed by the University of Naples Federico II, for first level graduated students, the other, second level, on "Development and Management of DataCenter for High Performance Scientific Computing," organized and managed by the University of Bari for second level graduated students.

The ReCaS training program is managed by an Executive Committee chaired by R. Bellotti with representatives of the three institutions (INFN, UniNA, and UniBA).

References

1. Available at http://www.ponrec.it/en/.
2. Available at http://www.ponrec.it/en/open-data/projects/?Cerca=pona3_00052.

Chapter 2

The CASAP Project

R. Bellotti

Istituto Nazionale Fisica Nucleare
Sezione di Bari via E. Orabona 4, 70126 Bari, Italy

Dipartimento Interateneo di Fisica
Università degli Studi di Bari,
via E. Orabona 4, 70126 Bari, Italy
roberto.bellotti@ba.infn.it

Calcolo Scientifico ad Alte Prestazioni (CASAP or High Performance Scientific Computing) was the training program associated with the infrastructural strengthening program ReCaS. CASAP was organized in 2 academic masters, 5 Cloud Computing schools, and over 100 months of post-master's specialized training.

1. First Level Master — Napoli[1]

The First Level Academic Master (Tecnologie per il calcolo scientifico ad alte prestazioni) was designed for those who have a 3-year degree in any scientific or engineering subject of study. It lasted 1-year and had the objective to train a skilled professional in software and hardware technologies, particularly in Cloud and Parallel Computing.

The 10 best master students were given an internship in Research Laboratories and High Performance Computing Centers. Table 1 reports the students placements after the CASAP master and

Table 1. Students placements — Napoli.

Student	Thesis	After master placement
F. De Bonis	La valutazione delle prestazioni dei datacenter di nuova generazione	Studente Ingegneria Navale
M. Naviglio	Analisi delle problematiche relative all'I/O in ambienti virtualizzati	Studente Corso di Formazione Informatico
G. De Luca	Tecniche e metodologie di imputazione singola per dataset non normali	Studente corso di Formazione Biotecnologie
D. Guarino	Sviluppo del modello SVP su GPGPU per la classificazione di immagini 3D di risonanza magnetica	Studente di Laurea Magistrale Ingegneria Elettronica
A. Tebano	Sviluppo di tecnologie Saas e PaaS su piattaforma OpenStack	Accenture S.p.A.
F. Iannone	GaaS, un approccio per l'integrazione dei paradigma GRID/Cloud: i servizi di storage	Studente di Laurea Magistrale Ingegneria Informatica
C. Napolitano	GaaS, un approccio per l'integrazione dei paradigma GRID/Cloud: La gestione delle risorse e i servizi di calcolo	Web Agency & Marketing Digitale | 3d0 Digital Agency
A. Solla	GAPP: Una applicazione Android per la piattaforma GaaS	Accenture S.p.A.
S. Scamardella	Unina Data Center Portal: verso la realizzazione di un portale unico per l'accesso ai servizi Grid/Cloud	Azienda nel settore del networking
G. Riccio	Un web service per il workflow di simulazione di immagini astronomiche in ambienti di calcolo distribuito	Assegno di ricerca INAF
L. Perillo	Valutazione delle prestazioni di algoritmi paralleli per il calcolo scientifico in ambienti con CPU multi-core: caso di studio	Code This Lab s.r.l.
M. Garofalo	Classificazione del traffico di rete mediante tecniche di data mining su architetture parallele GPGPU	Studente di Ingegneria Industriale

internship. It can be seen that the reported post-master activities are of Research and Development type, and the students that had found a job remained in the computing world.

2. Second Level Master — Bari[2]

The Second Level Academic Master (Sviluppo e Gestione di Data Center per il Calcolo Scientifico ad Alte Prestazioni) was designed for those who have a 5-year degree in scientific or engineering subject of study. It lasted 1-year and had the objective to train an experienced researcher in Data Center Management and High Performance Scientific Computing. Table 2 reports the students placements after the master conclusion and the internship, and shows that many of them remained in the research field.

The second edition of this Masters has been funded by PON PRISMA, and some of the courses have been taught also in a third master, "Metodologie e tecnologie per lo sviluppo di infrastrutture digitali," that has been funded by the "Progress in Training GARR@ UNIBA" project.

3. Cloud and Grid Computing Schools

The CASAP project also founded 5 Grid and Computing schools; 3 of them was located in Bari,[3] 1 summer school was located in Cosenza[4] and 1 summer school was located in Catania.

The schools held in Bari were aimed at system administrators, technicians, technologists and researchers from any Italian region, the schools held in Cosenza and Catania were designed for young graduates of scientific faculty (MPN Science and Engineering) of the convergence regions, with basic training in the ICT sector or those, coming from different areas of research, intend to retrain themselves.

The three editions of Bari Cloud schools had 28, 25, and 22 students respectively, the Catania and Cosenza Cloud schools had 30 participants.

Table 2. Students placements — Bari.

Student	Thesis	After master placement
M. Antonacci	Realizzazione e verifica delle funzionalità e delle prestazioni di una infrastruttura di cloud computing di tipo IaaS basata su OpenStack	Tecnologo a tempo determinato INFN
P.R. Altieri	Sviluppo e test di codici di simulazione per adroterapia in ambiente Fluka/ Geant4 utilizzando l'infrastruttura Grid di RECAS	Collaboratore Tecnico Fisico a tempo indeterminato ARPA Puglia
F. Barile	Deuteron Production in Pb–Pb collisions at 2.76 TeV with the ALICE HMPID detector at the LHC using the Alice Grid Environment	Assegnista di ricerca INFN
C. Calculli	Sviluppo ed elaborazione di modelli gerarchici multi-variate per la mappatura di dati di qualità dell'aria a livello regionale tramite l'infrastruttura di calcolo avanzato del DataCenter ReCaS	Esperto statistico con contratto a progetto presso il dipartimento di Biologia — UNIBA
G. De Risi	Development of probes for monitoring and accounting of a large computing farm	Assegnista di ricerca press oil Politecnico di Bari
N. Firza	Business plan per un Data Center basato su sistemi Grid e Cloud	CoCoCo press oil Dipartimento di Fisica
P. Inglese	Learning methods for automatic hippocampal segmentation in a cluster computing environment	London Imperial College — Ph.D Student in Clinical Medicine Research
L.M. Minervini	Sviluppo di interfaccia utente per l'analisi dati in ambiente AliEn-Grid nell'esperimento Alice al CERN	Dottorando presso il Politecnico di Bari
G. Miniello	Adopting NoSQL Databases to manage Big Data end simplify monitoring application development	Assegnista di ricerca presso il Dipartimento di Fisica — UNIBA

(*Continued*)

Table 2. (*Continued*)

Student	Thesis	After master placement
A. Monaco	Studio ed implementazione di metodologie per il support al calcolo scientifico in ambiente distribuito	Tecnologo di III livello presso l'INFN (Art. 23)
S. Nicotri	Realizzazione e verifica delle funzionalità e delle prestazioni di un'infrastruttura di cloud computing di tipo Infrastructure as a Service basata sul software OpernStack	Tecnologo di III livello presso l'INFN (Art. 23)
G.C. Palestra	Servizio di Virtual Private Network dinamico e on- demand per piattaforme di Cloud Computing basate sul paradigma IaaS	Contratto a termine — Progetto di ricerca
E. Stasolla	Studio e analisi delle funzionalità di autenticazione e sicurezza del software open source di Cloud IaaS	Informatico per il Ministero dell'Interno presso la Prefettura di Sondrio
A. Tateo	Analisi su Farm per la selezione delle feature applicato ad un algoritmo di segmentazione dell'ippocampo	Co.Co.Co. presso il Dipartimento Interateneo di Fisica di Bari
M.T. Toritto	Test di accesso per la valutazione di database NoSQL finalizzati allo sviluppo di un monitoring di una farm di calcolo di grandi dimensioni	Contratto a termine con Servizio Protezione Civile — Regione Puglia
A. Turnone	Studio e utilizzo di un modello statistico spazio-temporale per il monitoraggio della qualità dell'aria nella Regione puglia avvalendosi dell'infrastruttura di calcolo avanzato ReCaS di UNIBA-INFN	Contratto a tempo determinato presso ARPA Puglia
M.L. Avelluto	Confronto di algoritmi di allineamento per dati generati con le nuove tecnologie di sequenziamento massivo	Costituendo Spin off di Biotecnologie del CNR
V. Finamore	Analisi, installazione e test di applicazioni per il desktop remoto autenticato su server Linux	Docente in scuola superiore

(*Continued*)

Table 2. (*Continued*)

Student	Thesis	After master placement
V. Fiorentino	Progettazione di un sistema di monitoring di una farm basata su grid: la scelta del DBMS	Dipendente del Centro Servizi Informatici presso la Università degli studi di Bari "Aldo Moro"
V.A. Lasorsa	Dipendente del Centro Servizi Informatici presso la Università degli studi di Bari "Aldo Moro"	Borsista presso il CEINGE di Napoli: Analisi di dati NGS in Cancer Genomics

4. Conclusion

The Training Project CASAP, connected to ReCaS, offered a real opportunity to acquire new skills on topics of high scientific and technological content for a wide range of different interested parties, technicians, technologists and researchers. With 2 Masters at different levels, and 5 Cloud Computing Schools, CASAP has taught more than 150 students, outperforming its initial objectives.

The post master activities of the students were mostly in the Research and Development field, with jobs in private enterprises or public administration, and strong links with other projects as PRISMA and Smart Health.

References

1. Available at http://www.pon-recas.it/web/napoli.
2. Available at http://recas.ba.infn.it.
3. Available at http://www.ba.infn.it/scuolacloud.
4. Available at https://agenda.infn.it/conferenceDisplay.py?confId=8219.

Infrastructure

Chapter 3

The ReCaS Project: The Bari Infrastructure

M. Antonacci*, R. Bellotti*,†, F. Cafagna*, M. de Palma*,†, D. Diacono*, G. Donvito*, A. Italiano*, R. Gervasoni*, G. Maggi*,‡,§, G. Miniello*,†, A. Monaco*, S. Nicotri*, S. Nuzzo*,†, P. Notarangelo*, B. Santeramo*, G. Selvaggi*,†, L. Silvestris*, V. Spinoso*, S. Tangaro*, E. Tinelli† and R. Valentini*

**INFN Sezione di Bari,*
Via Edoardo Orabona 4, Bari 70126, Italy
†*Dipartimento Interateneo di Fisica,*
Università degli Studi di Bari,
Via Edoardo Orabona 4, Bari 70126, Italy
‡*Dipartimento Interateneo di Fisica, Politecnico di Bari,*
Via Edoardo Orabona 4, Bari 70126, Italy
§*giorgio.maggi@ba.infn.it*

This document describes the infrastructure built by the Università degli Studi di Bari 'Aldo Moro' and the Istituto Nazionale Fisica Nucleare (INFN) Sezione di Bari, within the Rete di Calcolo per SuperB e altre applicazioni (ReCaS) project,[1] founded by Ministero dell'Istruzione, dell'Università e della Ricerca (MIUR) within the Programma Operativo Nazionale (PON) 2007–2013 funding program. The ReCaS project, whose objective was to upgrade the distributed computing infrastructure existing in Southern Italy, started in the year 2012 and was completed in July 2015. The ReCaS data center built in Bari adds, to the existing computer center, Bc^2S, 128 servers for a total of 8,192 core; 3500 TB of disk space; an HPC cluster consisting of 20 bi-processors systems, each one equipped with 40 cores, 256 GB of memory, a graphics accelerator NVIDIA Tesla K40 GPU model, two 10 Gbps ports and a 40 Gbps InfiniBand port; an IBM TS3500 tape library with a total capacity of 2.5 PB.

1. Introduction

The Bari infrastructure of the Rete di Calcolo per SuperB e altre applicazioni (Computing Network for SuperB and other applications, ReCaS) project represents a significant upgrade of the computing facility set up by Istituto Nazionale Fisica Nucleare (INFN) since the year 2000 in the Bari "Sezione INFN."

The "Bari Computer Centre for Science" (Bc^2S) was the result of more than 10 years of work and represented the contribution of Bari to the realization of the European Grid Infrastructure (EGI) and the Worldwide LHC Computing Grid. It came in operation in 2009 due to a joint effort of the INFN of Bari and the University of Bari, in particular of the Bari Physics Department. The Bc^2S infrastructure included 18 racks and 10 heat exchangers (water-cooled), arranged in 2 rows with the hot aisle in the middle.

The farm management was based on OPEN Source software. In particular the following tools were used:

- Torque-Maui for the Batch System,
- PXE, Kickstart for automatic installation of the Operating System,
- Lustre as distributed parallel file system,
- Nagios, Ganglia, and in-house developed tools for the monitoring,
- in-house developed tools for nodes configuration.

The Bc^2S infrastructure, before the ReCaS upgrade, consisted in about 250 computing nodes corresponding to a total of more than

4,000 cores, 2,000 TB of disk space organized as a single posix file-system shared between all nodes (Lustre), two servers with Tesla C2070 graphics accelerator cards. All the computer nodes could exploit, without any limitation, the full bandwidth of 1 Gbps. While the connection to the Italian research network (managed by GARR — ***Italian Research & Education Network (NREN)***) reached a bandwidth of 10 Gbps.

The main use of the Bc^{2}S was as TIER2 for the Large Hadron Collider (LHC) experiments ALICE and CMS, however it was conceived and developed as a single unit capable of serving multiple use cases. For this reason it attracted both users and IT resources of other INFN groups (astro-particle and theoretical groups) as well as groups of University of Bari and CNR, in research fields related to medical physics, remote sensing, materials chemistry, theoretical computer science, phylogenetic, biology, physics, etc.

2. The ReCaS DataCenter

The ReCaS DataCenter in Bari is a shared infrastructure set up by the University of Bari and INFN. It is housed in a two-story building, built specifically for the ReCaS project (Fig. 1), with an area of 430 square meters per floor, of which 180 square meters are occupied by services (electrical cabinets, UPS, ...) while more than 270 square meters are dedicated to house the computing resources (Fig. 2).

At its full capacity, the data center can accommodate up to 80 racks, with an average power consumption of 12.5 kW, for a total nominal power consumption of 1 MW. The racks are arranged in 4 islands each formed by 2 rows of 10 racks with a cold corridor in between. The cooling is obtained by means of a stream of cold air conveyed through the subfloor by heat exchangers (fed with chilled water coming from the chiller located on the roof) arranged along the walls of the Data Center. The number of heat exchangers and the corresponding chiller is such as to ensure a redundancy N + 2. At its entry in operation, the infrastructure will be limited to: the first two islands of racks, 6 out of 10 couple of chillers and computer room air conditioner (CRAC) and two 800 kW uninterruptible power supply (UPS). The racks of the Bc^{2}S, as soon as this farm is closed, will be used to

Fig. 1. The Bari ReCaS DataCenter building.

Fig. 2. On the ground floor of the Bari ReCaS DataCenter building there is the area with all the equipment of computing, storage and network, while the first floor are offices, the control room, and service facilities.

form the fourth island of racks, which will be used for future expansions. The third island hosts the tape library and the associated services. Figure 3 shows the layout of the ground floor of ReCaS farm.

Each rack is equipped with two power distribution units (PDUs) that are connected to two independent electrical distribution lines

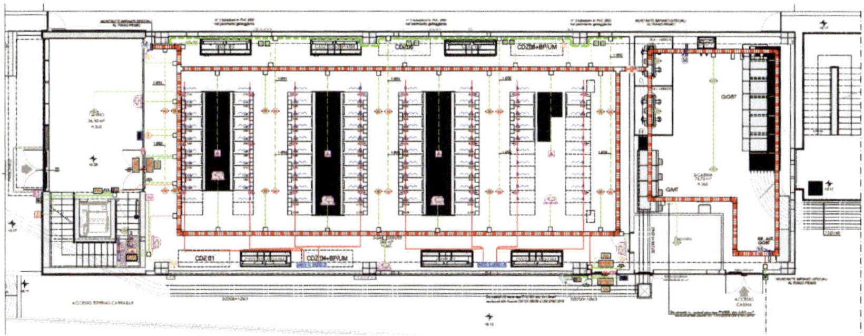

Fig. 3. Layout of the ground floor ReCaS DataCenter.

from two identical uninterruptible power suppliers (UPSs) The majority of the PDUs are the 22 kW Raritan metered PDUs. Some specific racks, which host critical services and which need to be remotely controllable, are equipped with 22 kW switched PDUs from Emerson.

The UPSs are of the type TRINERGY from Emerson. At the start up, each UPS can maintain 800 kW (expandable in the future up to 1200 kW) of load for 10 minutes.

In case of an electrical power blackout, the entire system can be powered by a 1600 kW diesel generator Green Power GP 1650.

All components of the ReCaS infrastructure are controlled by the "supervision system" (Fig. 4).

The following computing resources are part of the ReCaS DataCenter which add to the computing resources deployed into the Bc^{2}S:

- 128 servers for a total of 8,192 core,
- 3552 TB of disk space,
- a HPC cluster consists of 20 systems bi-processors. Each system has 40 cores, 256 GB of memory, 2-port 10 Gbps, and a port QSFP 40 Gbps InfiniBand, a graphics accelerator with NVIDIA Tesla K40 GPU model,
- an IBM TS3500 tape library model, equipped with 6 drives with Fiber Channel interface and an aggregate transfer rate of at least 3 TB/h, 1500 cassette slot and 1000 LTO6 tapes with a total of 2.5 PB of installed capacity.

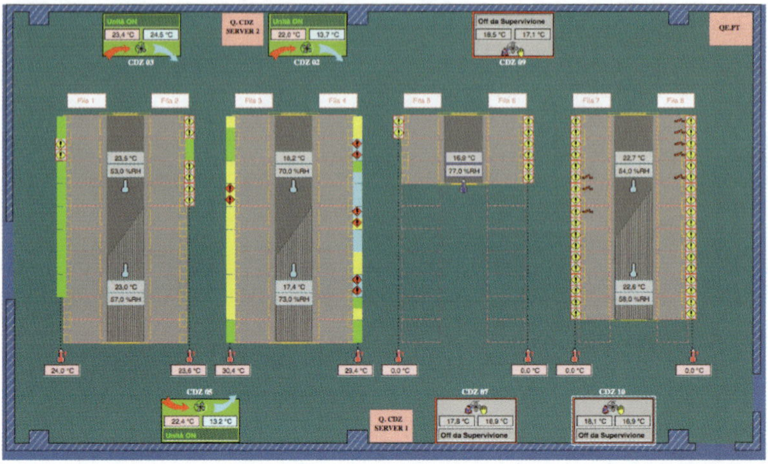

Fig. 4. A web image of the computer room with details about temperatures and status of the racks.

The LAN interconnecting the servers is based on a flat network matrix, based on a pair of identical HUAWEI Cloud Engine 12812 Core Switches, one of which is active and the second in stand-by mode, able to support a traffic of 10 Gbps between any two points of the matrix. The switch HUAWEI Cloud Engine 12812, which has a total capacity of 17820 Gbps of switching, is equipped with 480-ports 10 Gbps and can be expanded up to 576 ports. The Bari ReCaS DataCenter is connected to the research wide area network (GARR) with a band of 40 Gbps. Also the users connectivity to the Data Center has been improved. The entire campus, in fact, has been rewired with single-mode optical fibers so as to allow virtually unlimited speed between users and the computing resources.

3. The Management of the Bari ReCaS DataCenter

***Batch system*:** The Open Source and free version of Torque-Maui batch system, used in Bc^{2}S, does not scale to the size of the ReCaS DataCenter. Investigations have been performed on other Open source freely available products. After an initial evaluation of SLURM[2] which gave satisfactory results, an intensive test program has been set up to evaluate the performances of **HTCondor**[3] in order to choose between the 2 products. The final choice was HTCondor for two main reasons. The first one can be considered conceptual because, as the name suggests, HTCondor is a batch system oriented towards the High Throughput Computing (HTC) that generally means a lot of batch jobs waiting for a computing slot. On the other hand, the SLURM batch system is more High Parallel Computing oriented which generally means running a parallel batch job on the biggest number of slots available. This distinction makes the architectural, development and deployment details of both batch systems quite different. For instance, the underlying hardware of a HTC cluster can be a blend of different hardware resource type because the basic computation unit is the single core available on every processor while a HPC cluster is usually deployed using the same hardware in order to provide the same hardware features to the entire parallel job. Taking into account this clarification HTCondor is the more convenient choice because it fits the ReCaS DataCenter features and its main use case requirements. Anyway the strongest reason turns out to be the technical one. Actually we have tried to measure the batch system performance with different intensive stress tests and the detailed results are shown on a dedicated chapter of this book. What clearly stands out from the experience with HTCondor is the sustainability of this product under different aspects. First of all, it is the stability of batch system that is the main feature for these kind of products. The process, to accept a new batch job, allocate the right resource for it and let the batch job to start and finish on the allocated resource was found very smooth at the ReCaS DataCenter size. This is the key for a successful resource management because a CPU cycle lost is lost forever. Tests have proven the scalability of the product in terms of both resource and batch job managed. This means the choice that we have been taking can easily support an increase of the resources, and thus handle batch jobs, in the future. HTCondor is an open source product with a monthly release process that means that the code is well maintained and all of this for free.

***Storage*:** As storage solutions, along with the classic ones such as NFS, Bc^{2}S was using Lustre,[4] an open source cluster file system that scales to the order of tens of thousands of nodes and tens of petabytes of disk space without degradation of the performance while accessing files. It was decided to continue to use this product also in the ReCaS DataCenter.

For the most critical services requiring increased redundancy, we will exploit two different approaches depending on the context: in the legacy context, we will exploit DRBD[5] as software based replication between block devices, while in the Cloud environment we will use Ceph[6] as software replicated file-system.

***Installation and configuration*:** The topic concerning the server deployment has a lot of contributions. In particular every big player in the computing industry releases its own tool in order to install and configure a single server. What we have done was to select those that fit the ReCaS DataCenter requirements in order to test them. At the end, we have chosen Foreman[7] and Puppet.[8] In particular, Foreman will provide all the features to install a server while Puppet will be exploited in order to contextualize every server. The two products are developed and released separately, but Foreman in particular has been built on top of Puppet. They have been chosen for the following main reasons:

- Both are open source, which does not mean only free of charge, but it would mean to be able to modify the code in order to fit specific requirements.
- The server lifecycle management process must be a certainty. This means applying the same process to the same server several times so that its final status will be always the desired one. You can only achieve these requirements by centralizing the management because you can modify a server configuration only from that point. Actually Puppet provides a catalog which stores the server definition and it will be the same unless you modify the server configuration.

- A centralized management is supposed to provide a minor flexibility in the server definition but this is not the case. Actually both Foreman and Puppet provide the feature to be flexible as you want while defining the server configuration. You can define a configuration per host or group and while you are defining a configuration per group you always have the flexibility to customize a subset of the entire configuration of one host in the group.
- In spite of the fact that a centralized architecture could commonly lead to scaling issues, this kind of products and in particular Puppet can scale horizontally. We have also tested this feature: actually a single instance of a Puppet master can lead to scaling issues when the number of compiled catalogues rises to 30 per minute. It can easily be fixed by just deploying another Puppet master instance serving the same requests.
- The server configuration is always declared in a high level language. In this way, you can define the server configuration in a common language while you don't have to care about how the configuration will be applied on the server. This means you can define the same configuration using the same language for instance for both Linux and Windows operating systems.
- We are introducing a configuration language which is common for all the administrators. So everybody "speaks" the same configuration language avoiding personal customization and improving the collaboration: for instance all data center administrators can share their Puppet code and eventually use the code developed by others.

***Monitoring*:** After careful evaluation of several monitoring tools (Ganglia,[9] Nagios,[10] Monit[11]), the final choice was to base the monitoring on Zabbix,[12] since it integrates into a single tool all the desired features: alarm dispatching via IM, SMS, email; graphical representation of the monitored metrics; user friendliness also due the availability of most of the sensors to monitor the majority of metrics typical of a data center, easiness of installation, availability of the documentation and the excellent support system. The primary purpose of Zabbix is to represent the state of the data center in the ReCaS unique view (Dashboard) and also to be capable of cataloging the machines in groups. The entire IT infrastructure (machines, network, air conditioning, UPS) is scanned continuously by dedicated sensors, and in case of problems, site administrators are contacted according to the severity of the problem (SMS for serious problems, email for critical problems, instant messages for less severe problems). Another important feature is the ability to create graphics natively, the capacity to maintain the history even for years with the aid of down-sampling and housekeeping features.

Ticketing: The start of the operation of the Bari ReCaS DataCenter has experienced a growth in the number of users, also determined by activation of the Cloud infrastructure. So it becomes necessary to rationalize user support activities and to setup procedures and interfaces that can better meet the needs of both users and system administrators. To disseminate real-time news and activities related to the data center, Facebook (https://www.facebook.com/bari.recas) and Twitter (https://twitter.com/recasbari) accounts were created. To allow a direct contact of users with the site administrators a Skype account and an email list were set-up. Finally, Redmine,[13] the management tool used for managing the IT projects, planning the activities, sharing code, guide, wiki, is also used by the users as a ticketing system for reporting problems.

Inventory: The size of ReCaS-Bari DataCenter requires an efficient, reliable and scalable inventory system. For this purpose OCS-Inventory[14] and GLPI,[15] two open-source applications that respectively allow the automatic discovery of new hardware and software resources and their cataloging. Cataloging covers both the technology aspect (machines, switches, PDU, etc.) as well as the administrative part (invoices, delivery slips, etc.).

4. The Cloud Services

As already stated, a consistent part of the computing resources will be supplied to the users as cloud resources.

The cloud platform (Cloud@ReCaS) is built on top of the HW infrastructure of the Bari-ReCaS Computer Center and is based on a highly customized "(by the pon_smart_cities PRISMA[38] project) version of the middleware OpenStack,[16] integrating open-source components that represent the state-of-art virtualization technologies, storage, networking, etc.

Cloud@ReCaS is conceived as a modern data center and, therefore, implements advanced features such as high reliability of services, scalability, and disaster-recovery. The software is organized according to a plugin structure, in which each service is driven by a separate component.

The cloud platform delivers a wide range of services that span from the on-demand provisioning of compute and storage resources to high-level services that allow users to quickly and easily deploy complete and ready-to-run environments.

Three types of storage services are provided to the users that can select those that best fit with their needs:

- **Block Storage** service (based on OpenStack Cinder[17]) allows the on-demand provisioning of volumes (block device) directly usable by the virtual machines;
- **Object Storage** service (based on OpenStack Swift[18]) provides disk space that users can use for storing unstructured data (virtual images, backup, video files, web content, etc.);
- **File-sharing** service (based on OwnCloud[19]) provides personal storage supporting file synchronization and online collaboration.

Personal Storage as a Service and Object Storage are currently used by local academic users, as well as by public/private partnerships of universities and enterprises working on eLearning, to host personal and public data.

The backend storage solutions implementing these services have been configured for data replication among the storage nodes in order to ensure high-availability and fault tolerance.

After a preliminary phase of testing, Ceph has been selected among the available open-source distributed file-systems as storage backend for OpenStack main components. Ceph is a fully open source distributed object store, network block device and file system designed for reliability, performance and scalability from terabytes to exabytes. Ceph utilizes a novel placement algorithm (CRUSH[20]), active storage nodes, and peer-to-peer gossip protocols to avoid the scalability and reliability problems associated with centralized controllers and lookup tables. We have decided to use Ceph for:

- Hosting the disks of the running Virtual Machines, enabling also the live migration of the virtual instances among different hypervisors;
- Storing the virtual images that will be used to create virtual machines;
- Providing the block devices that are attached to the virtual machines.

Encryption and backup are advanced features that have been implemented on top of the basic storage services.

The ability to store data on encrypted volumes is critical in case of sensitive data such as those managed by e-health applications. The users of Cloud@ReCaS can easily and transparently encrypt their data without the need to know anything about data encryption.

Data backup is another very critical service: almost all our users require the possibility to automatically backup the data stored in the cloud. Unfortunately, this kind of service is currently not implemented by the Cloud Management Framework (i.e., OpenStack). Therefore, specific homemade tools have been implemented for:

- performing automatic incremental backups of the Ceph pool used by Cinder (using the export-diff/import-diff feature),
- performing automatic backups of the Virtual Machines and attached block devices according to a flexible schedule that foresees also the rolling of the older items.

Moreover, a tool for the automatic full backup of the data stored by database management system (DBMS) like MySQL[21] and MongoDB[22] has been developed: the files containing the backup are compressed, encrypted if requested, and stored on the Object Storage.

Cloud@ReCaS also provides a number of higher level services, most of which are offered within a Software as a Service (SaaS) paradigm:

- a complete Remote Desktop solution has been implemented based on Linux and X2Go,[23] an open source tool for remote desktop management that allows users to access their workstation from any thin client connected to Internet,
- a web-based Git repository hosting service has been deployed, based on the open source software Gitlab,[24] which provides web access to public and private repositories, with wiki and issue tracking features,
- a collaborative platform for LaTeX project has been implemented, based on the open source software ShareLaTeX,[25]
- advanced data analysis and programming environments, like Rstudio[26] Server and IPython Notebook[27] Server are provided.

The target users of these services are Postdocs, Researchers and Professors of the University of Bari, particularly in the Physics, Maths, and Computer Science Departments.

They take advantage of key features like elasticity, high-availability and automatic backup implemented by the infrastructure Cloud@ReCaS, focusing more on their research activities and less on the underlying computing platform.

The platform Cloud@Bc2S also supports the following advanced services:

(1) multiple authentication mechanisms: username/password, X.509 certificates, external authentication via IdP SAML2 and/or LDAP,
(2) multi-level monitoring services based on Zabbix,
(3) standard APIs like Open Cloud Computing Interface (OCCI)[28] and Cloud Data Management Interface (CDMI)[29] to access compute and storage resources,

(4) API compatible with Amazon AWS (EC2, S3),
(5) service orchestration for the automatic deployment of resources and IaaS applications,
(6) multiple configurations of the network, providing the user the ability to create private networks and use floating internet protocol (IP) and/or use public IP directly.

Many different scientific communities access the ReCaS cloud services: bioinformatics, genetics, financial data analysis, medical imaging and signal analysis, high-energy physics (like ALICE and CMS).

Some of these communities uses the cloud resources of the ReCaS DataCenter also through the EGI Federated Cloud,[30] which is a seamless grid of academic private clouds and virtualized resources, built around open standards and focusing on the requirements of the scientific community.

Currently, the supported EGI use-cases involve many fields: high energy physics experiments, mainly from CMS Virtual Organization; biodiversity research like the BioVeL project[31]; hydrometeorology, like the DRIHM project[32] that provides services for predicting risks related to extreme flooding events; execution of scientific code (R and Octave jobs) through the Chain-Reds Science Gateway.[33] A particular effort has been spent to support the CSC team from the Finnish Elixir node in order to enable the Chipster[34] suite in the Federated Cloud environment: Chipster contains over 340 analysis tools for next generation sequencing (NGS), microarray, proteomics and sequence data. Users can save and share automatic analysis workflows, and visualize data interactively using a built-in genome browser and many other visualization tools.

5. High-Level Services for Users: The Job Submission Tool (JST)

To simplify the submission of jobs, especially when there are so many, the ReCaS provides to its users a tool, JST, that can simplify this task.

JST has been improved recently with the addition of a web services interface technology with Simple Object Access Protocol

(SOAP) and Representational State Transfer (RESTful) interfaces. JST can therefore be used, using standard technologies, by different components such as client applications, browsers, workflow manager, etc. In addition to job submission to the GRID, JST is now able to submit jobs also to dedicated servers, to batch clusters as well as to the cloud infrastructure. Of particular interest is the possibility of using JST within workflow manager such as Taverna,[35] LONI,[36] and Galaxy,[37] tools particularly used in bioinformatics.

Conclusions

The ReCaS infrastructure is elastic, and many different applications from a wide panorama of scientific fields can be executed on it. The Data Center hosts a number of applications: starting with ALICE and CMS experiments belonging to high energy physics environment, but also astro-particles experiments as FERMI, Pamela, and T2K, bioinformatics applications for the ELIXIR, Lifewatch, and BioVeL communities. It can also provide a range of services to the highest level for users of EGI Federated cloud.

The final user can execute over 8,000 sequential jobs at the same time, use 800 cores interconnected with low latency network for parallel computations, exchange up to 10 Gbps of data per seconds between any computational node in the farm and the local storage systems, access grid resources distributed all over the world.

The deployment strategies adopted for the construction of the new ReCaS infrastructure in Bari have been focused on interoperability and integration of the different paradigms of parallel and distributed computing (HTC, Grid, and Cloud).

Acknowledgment

The authors would like to thank the people of the Centro Servizi Informatici (CSI) of the University of Bari responsible for the management of the University Campus network.

This work has been funded by PON 2007–2013, project code PONa3_00052.

References

1. Available at https://www.recas-bari.it.
2. Available at http://slurm.schedmd.com/.
3. Available at http://research.cs.wisc.edu/htcondor/.
4. Available at http://lustre.org/.
5. Available at http://drbd.linbit.com/.
6. Available at http://ceph.com/.
7. Available at http://theforeman.org.
8. Available at http://docs.puppetlabs.com/puppet/.
9. Available at http://ganglia.sourceforge.net/.
10. Available at https://www.nagios.org/.
11. Available at https://mmonit.com.
12. Available at http://www.zabbix.com/.
13. Available at http://www.redmine.org/.
14. Available at http://www.ocsinventory-ng.org/en/.
15. Available at http://www.glpi-project.org/spip.php?lang=en.
16. Available at https://www.openstack.org/.
17. Available at https://wiki.openstack.org/wiki/Cinder.
18. Available at http://docs.openstack.org/developer/swift/.
19. Available at https://owncloud.org/.
20. Available at http://ceph.com/papers/weil-crush-sc06.pdf.
21. Available at https://www.mysql.it/.
22. Available at https://www.mongodb.org/.
23. Available at http://wiki.x2go.org/doku.php.
24. Available at https://about.gitlab.com/.
25. Available at https://www.sharelatex.com.
26. Available at http://www.rstudio.com/.
27. Available at http://ipython.org/notebook.html.
28. OCCI, Open Cloud Computing Interface, Version 1.1, June 2011. Specification available at "http://occiwg.org/about/specification/.
29. CDMI, Cloud Data Management Interface, Version 1.0.2, June 2012. Specification available at <http://www.snia.org/cdmi>.
30. Available at https://www.egi.eu/infrastructure/cloud/.
31. Available at https://www.biovel.eu/.
32. Available at http://www.drihm.eu/.
33. Available at https://www.chain-project.eu/.
34. Available at http://chipster.csc.fi/.

35. Available at http://www.taverna.org.uk/.
36. Available at http://pipeline.loni.usc.edu/.
37. Available at https://galaxyproject.org/.
38. Available at http://www.ponsmartcities-prisma.it/

Chapter 4

The ReCaS Project Catania Infrastructure

G. Andronico*,‡, S. Monforte*, G. Sava*, F. Cristaudo*, G. Platania*, C. Rocca*, P. Belluomo*, G. Passaro* and R. Barbera†

**Istituto Nazionale di Fisica Nucleare, Sezione di Catania, Italy*
†*Dipartimento di Fisica e Astronomia, Università di Catania, Italy*
‡*giuseppe.andronico@ct.infn.it*

Hosted at the Catania division of *Istituto Nazionale di Fisica Nucleare* (INFN), inside the Physics and Astronomy Department of the local University, there is a data center developed along the last 15 years.

The genesis of this data center begun with the development of a middleware for the European Grid Computing, an effort where most of us had a part together with other people from other INFN divisions and from many other European partners.

This data center currently hosts the Tier 2 for Alice Large Hadron Collider (LHC) experiment and its resources were used within projects funded by regional, national, and European projects.

Recently we submitted, to answer a call for a national project, together with people from divisions of Naples, Bari, Cosenza and from Universities of Naples and Bari a project to upgrade our data centers.

In this paper, we describe the project we followed to upgrade the Catania data center.

1. Introduction

Around 15 years ago in Catania, lead by Roberto Barbera, we joined the effort made from INFN and from a number of other institutions and enterprises in Italy and in Europe to build a grid-computing infrastructure in Europe.

We worked both on the middleware and on the infrastructure, starting with prototypes to deploy and test the middleware we were contributing to develop.

This way we passed through several projects, some funded from Europe (European Data Grid, EGEE I, II and III, EGI, EMI) and others funded from *Regione Sicilia* (TRIGRID) and from Italy (PI2S2) by means of the funds for objective areas.

The activities developed in these projects, the expertize gained and the experience of our colleagues in other parts of Italy and Europe led us to develop the local data center enough to host a Tier 2 of the Large Hadron Collider (LHC) grid infrastructure, the network of data centers spread out all around the world to analyze, using the grid computing paradigm, the data produced from the experiments at the European Organization for Nuclear Research (CERN) looking for Higgs boson and for other predictions from the Standard Model.

In the next section, we will report about the architecture of the data center we developed.

In the year 2012, another call from the national program allowed us to propose an upgrade of our data centers in the objective area, then in Calabria, Campania, Puglia and Sicilia. The divisions present in this region joined and worked out a project, called *Rete di Calcolo per SuperB e altre applicazioni* (Computing network for SuperB and other applications, ReCaS) and described in another paper in this same book, that was accepted and funded. In the frame of this project, we developed a detailed project to upgrade the Catania data center that is reported in the following Section.

Our conclusion will end this work.

2. Old Catania Computing Room

The Catania data center is based on a small room for services to local people (mail, web, cloud storage, and more) and a big computing room we call *Sala Grid*. A schematic picture of the room is in Fig. 1.

Racks were grouped into two distinct areas. The former, actually the oldest, comprises of about 10 racks disposed in a row with Power Distribution Units (PDUs) and Uninterruptible Power Supplies (UPSs) disposed at the two ends of the row. Two cooling units (1) at the two ends, each of them with its external chilling unit (2), ensure cooling. They are actually two very big conditioners for computing rooms. The latter set of rack is more recent and it is based on the separation between cold and hot air. The 13 racks are disposed in two parallel rows in such a way that equipments (computers, storage units, switch, and more) get fresh air from outside, use it to cool and send the hot air in the space between the two rows. This space is closed, forming a hot aisle, and the hot air is confined. Cooling units amidst the racks get hot air from this and, using cold water provided from the external chiller (3), send out cold air.

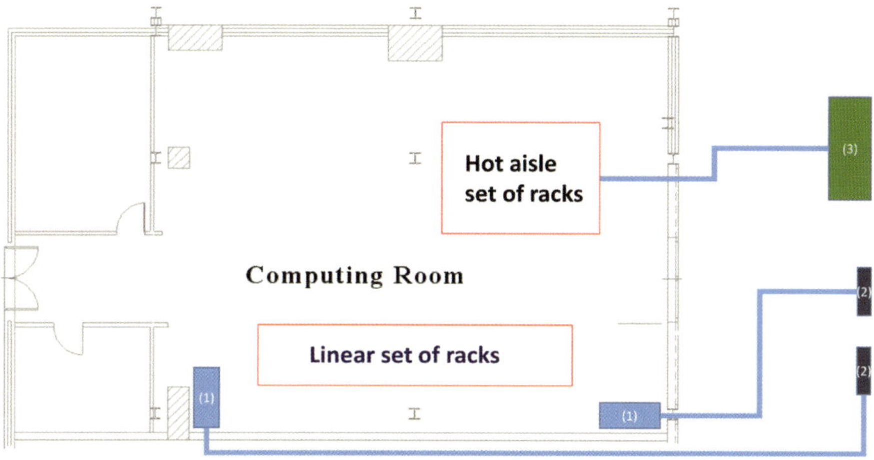

Fig. 1. The computing room before the ReCaS project.

External to this hot aisle set of racks, but inside at the computing room, there is a PDU and an UPS.

Both the set of racks are connected to the buildings electrical cabinet and to an external diesel generator. With this design in case of power failure, service continuity is ensured from UPSs till the start of the external diesel generator.

Network connection was assured from a devoted optical fiber until the next Point Of Presence (POP) of Group for Harmonisation of Research Networks (GARR), the Italian Research and Education Network, provided connection at 10 Gbps. The network is then distributed from the router to all the racks with Top of Rack topology and from the switches to the several equipment using 1 Gbps copper RJ45 cables.

3. Evolution of Sala Grid

The ReCaS project provided us with the resources to increase computing power, storage, and bandwidth of our data center.

However, increasing resources available in data center leads to bigger power consumption and bigger heat production. For every new resource, we would like to add at the data center that we have to check for available infrastructures (power distribution, heat dissipation, rack units available) or understand how to increase the infrastructures to accommodate the new resources. We needed to find an optimal point between desired resources and required infrastructure strengthening.

Another requirement was to have infrastructure able to host new resources for some time without the need to change the infrastructure again.

A period of studying and planning was needed, looking around not only for available solutions and technologies but also for future trends. From this period some conclusions arise:

1. How many new racks to buy to host ReCaS and future resources, their characteristics and their topology?
2. The maximal amount of heat to be dissipated from ReCaS and future resources.

3. The maximal power requirements from the computing room, needed for the present resources, for the resources we would be able to buy with the ReCaS project and for future resources.

All these numbers are strictly related. The number of resources we plan to host in the new infrastructure define the number of rack units required, the amount of power needed to operate them as well as the power needed to dissipate the amount of heat generated.

Also, the study we did clarify another point related to temperature in the computing room. In fact, the only element in the computing room with stringent limits for operational temperature appear to be the battery. It is usually suggested to maintain a temperature in the range 22–23°C, and it is reported that every degree more shortens significantly the battery lifetime. In Sicily, especially in summer, but also in spring and autumn, we reach easily external temperatures higher than 30°C with extraordinary peeks just minimally lower than 50°C. From these data, it was clear that to ensure a longer battery life and to contain cooling consumptions a practical solution was to realize a devoted room to host UPSs and batteries. It was possible to locate an appropriate space to realize such a room and this becomes part of the project.

To summarize the several points touched from the project we developed:

***Resources*:** We quantified the amount of resources we want to host in the new infrastructure, in term of rack units. At the end we converged on 8 racks, each with 42 rack units, deep 1200 mm instead of the existent deep 1000 mm. These racks are able to host longer units, are larger for easier maintenance and have four extra vertical rack units. They form a structure similar to the hot aisle set of racks, but this time the central aisle is cold (in Fig. 2). There is no strong technical motivation for this choice but a bit more interesting offer from producer.

***Cooling*:** We kept the existing cooling solutions, that worked quite nicely at the time of the project, and added a new chiller (4) devoted to the new set of racks by means of six cooling units. The chiller is able to dissipate the heat produced filling completely the eight racks.

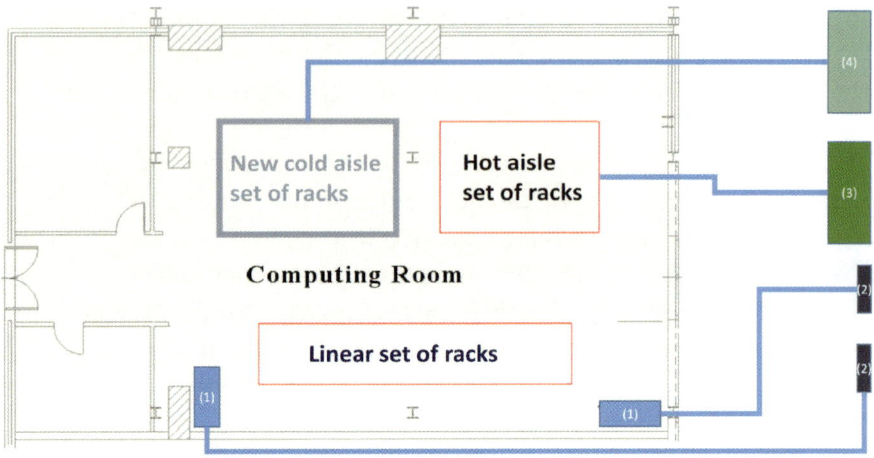

Fig. 2. Actual computing room.

New pipes were deployed to implement a completely new hydraulic system able to send cold water at the cooling units, retrieve the hot water and cool it. Moreover, a manual bridge between the new and old chiller was realized for the emergency case. In such a situation, the working chiller is able to treat, almost in part, the hot water coming from the cooling units and send the cold water.

Power: To be able to sustain all the equipment in the computing room, together with conditioners and chillers, we had to complement the existing electrical system with new power lines able to sustain the new infrastructure and the eight racks completely full of new equipment. The new UPS room is part of the new electrical system and the new power lines integrate the new UPS systems and extend to include the little data center devoted to local services. Finally, due to limited amount of space, the diesel generator was changed to support the maximum load of new and pre-existing systems.

Network: Together with the other ReCaS data centers, we agreed to switch from the Top of Rack local network topology to the Converged Core one, in which all the equipment are directly connected at only one big concentrator. To better exploit such a network topology, we included in the project the deployment of a new set of optical fibers from *Sala Grid* to the local GARR POP certified to support almost till 100 Gbps. The new concentrator was required to support also 100 Gbps connections. A little part of the new computing and storage infrastructure is based on Infiniband FDR with a limited number of computers and storage systems directly attached to a Mellanox Infiniband switch also connected at the concentrator with 4×40 Gbps links.

4. Conclusions

To conclude, some numbers. We added eight racks able to host longer units. Presently we have five racks available for future expansions in this new set of racks and two racks available in the old sets.

The improvement in the cooling power let us to have a bit more than 300 kW with a power consumption of a bit more than 100 kW.

The total power availability of the computing room increased from 340 kW to 604 kW, while the actual consumption is around 190 kW, providing large opportunities to grow.

The ReCaS project provided us the opportunity to technologically upgrade *Sala Grid* and let us to be ready for the next scientific computations.

Chapter 5

The ReCaS Project Cosenza Infrastructure

N. Guarracino*,†, V. Lavorini†, A. Tarasio† and E. Tassi*,†,‡

**Dipartimento di Fisica, Università della Calabria, Arcavacata di Rende, Italy*

†*Istituto Nazionale di Fisica Nucleare, Gruppo collegato di Cosenza, Arcavacata di Rende, Italy*

‡*enrico.tassi@fis.unical.it*

The new *Rete di Calcolo per SuperB e altre applicazioni* (ReCaS) Cosenza Data Center is briefly described. The new IT infrastructure reaches, within the limits of the allocated budget, the goals of scalability, redundancy and full efficiency. A unique IT infrastructure in the Calabria region, it will provide jointly with the other ReCaS sites and within grid and cloud paradigms, the computing and storage resources needed by diverse scientific and non-scientific communities.

1. Introduction

The *Rete di Calcolo per SuperB e altre applicazioni* (Computing network for SuperB and other applications, ReCaS) Cosenza Data Center is a newly realized IT infrastructure and part of the ReCaS project described in detail elsewhere[1] in this volume. Funded with a total budget of 1.3ME, the Cosenza Data Center represents a unique IT infrastructure in the Calabria region. In addition to providing computing and storage resources to the four international collaborations (ALICE, ATLAS, CMS, LHCb) taking data at the Large Hadron Collider (LHC) of the European Organization for Nuclear Research (CERN), the Data Center will support the computing needs of diverse scientific and non-scientific communities, so promoting and enhancing the competitiveness of applied research and enterprises in Southern Italy.

In this contribution, we briefly describe the characteristics of the Data Center focusing in particular on the realization of the support infrastructure (power and cooling systems), the description of the IT equipment and the monitoring system.

2. Support Infrastructure

The support infrastructure and its most critical components, power and cooling systems, has been designed to meet, within the constraints of the available budget the goals of efficiency, redundancy, and scalability. In this section, we briefly describe the solutions adopted for the power and cooling systems.

2.1. *Power System*

Electric power, up to 200 kW, is granted to the ReCaS Cosenza Data Center by a cabin (of total capacity of 1 MW) located in the underground floor of a building adjacent to the computer room. A diesel emergency generator set (COELMO Model. FDTC87-25),[2] with an onboard 400 liters fuel tank, supplies the emergency power of up to 250 kVA (see Fig. 1).

The energy back-up is provided by a transformerless uninterruptible power supply (UPS); the adopted solution is based on the RITTAL system UPS PMC 120[3] (see Fig. 2). The UPS system PMC 120 is designed for medium-sized data centers, and is scalable from 10 kW to 120 kW thanks to its modular structure based on (up to 6) 20 kW power modules that can be replaced during running operation

Fig. 1. The COELMO emergency generator set.

Fig. 2. (Left) The UPS PMC 120, (Right) the UPS assembled together with the rack hosting the batteries and the dedicated switchboard.

("safe swapping") without any switch-over to bypass being required. The UPS modules work on the double conversion principle and are classified to VFI-SS-111 (Voltage and Frequency Independent); efficient insulated-gate bipolar transistor (IGBT) power transistor technology achieves a high operating ratio of 95% even in part-load operation.

The UPS system of the Cosenza Data Center is presently configured with five power modules and a double set of batteries accommodated in a separate rack that guarantee an autonomy time of approximately 25 min. The UPS, as well as all the other critical components described in this section, are monitored by way of SNMP boards.

2.2. *Cooling System*

An energy efficient water cooling system has been adopted. The system is based on a set of 7 Liquid Cooling Package (LCP) units (one on each side of the rack) and a fully redundant two-chiller system with integrated free cooling. The LCP (RITTAL LCP CW-model 3311.230) draws in the air at the sides at the rear of the server enclosures, cools it using high-performance compact impellers, and blows the cooled air back into the front part of the server enclosure at the sides. The system offers optimum adaptability due to dynamic, continuous control of the cold-water volume flow; in addition by using high water inlet temperature, the proportion of indirect free cooling is increased, which in turn reduces operating costs. The two redundant chillers (RITTAL Model 3232.761), shown in Fig. 3, have each a power of 88 kW (thermal) and safety-relevant features such as redundant speed-controlled pumps and compressors.

As for the power system previously described, the cooling system (including the sizing of the hydraulic pipes) has been designed so to allow a factor 2 increase of the computing nodes (and racks) with no need for an upgrade of the components of the support infrastructure.

Fig. 3. The two-Chiller system installed at the ReCaS Cosenza Site.

2.3. *Additional Components*

Complete the support infrastructure with the following components:

(1) two electric switchboards (by Schneider Electric), one dedicated to the IT equipment, LCP and chillers and the other to the UPS system and generator set,
(2) five racks (Rittal TS IT) each equipped with two intelligent PDUs and sensors to monitor temperature, humidity and door opening,
(3) sensors for the detection of water leakages,
(4) a fire detection and extinguishing system localized to the racks,
(5) an air conditioning system,
(6) a system of controlled access to the Data Center area.

3. White Space

The white space (shown in Fig. 4) comprises two rooms. The first room, for which a raised floor has been realized, hosts a row of four racks with five LCPs. The fifth rack with two LCPs on its sides is in the second room that hosts also the ATLAS T3 pre-existent Grid cluster and is available for further expansions.

Fig. 4. The white space and the racks and LCPs hosting the IT equipment.

4. The IT Equipment and Deployment Strategies

The computing and storages systems have been chosen taking mostly into account the computing needs of the LHC experiments. For the servers, this require systems with a large number of cores per server and local disks with large capacity while the storage systems should guarantee high-throughput, efficiency and reliability. The deployment strategies adopted for the Cosenza ReCaS infrastructure have been chosen taking into account the issues of interoperability among the ReCaS DataCenters and the integration of the different computing paradigms: parallel and distributed, Grid and Cloud.

The IT equipment, consisting of computing and storage resources belonging to the most recent hardware generation, are shown in Fig. 5 after the final installation in the racks. A detailed description of the computing and storage systems is given in the following sub-sections.

4.1. *Compute Nodes*

The computing resources, keeping in mind the needs of the High Energy Physics (HEP) and High Performance Computing (HPC) users discussed above, have been acquired among the multi-node computer systems, with a large number of cores per node, available on the market. The bulk of the compute nodes at Cosenza Data Center is constituted by 50 SuperMicro Systems As-2042G,

Fig. 5. The IT equipment installed on the five racks.

Table 1. Characteristics of the compute nodes.

Model	**DELL C6145**	**DELL R515**	**DELL R820**	**SuperMicro As-2042G**
Number	2	2	2	50
CPU number and type	8 × AMD Opteron 6276 2.3 GHz	2 × AMD Opteron 4386 3.1 GHz	4 × Intel Xeon E5-4620 2.2 GHz	4 × AMD Opteron 6366 HE 3.1 GHz
Number of cores per CPU	16	8	16	16
RAM	512 Gb	64 Gb	384 Gb	256 Gb
Local storage	8 Tb	44 Tb	5 Tb	6 Tb
Connectivity	2 × 10 GbE, 2 × 1 GbE	2 × 10 GbE, 2 × 1 GbE	2 × 10 GbE, 2 × 1 GbE	2 × 10 GbE, 2 × 1 GbE

complemented by additional Dell systems (C6145, R515, and R820). The technical specifications of the systems are detailed in Table 1. The total computing power amounts to approximately 32 kHep-Spec06, for a total of 3,800 cores and more than 15 TB of memory (RAM).

4.2. *Storage Systems*

The main storage system adopted for the Data Center is based on two identical systems, each of which is composed of:

- two front-end nodes Dell PowerEdge R620,
- one Storage Unit providing 378 TB of disk space.

Each front-end node is directly attached via two channels — Fiber Channel (FC) at 8 Gbs to the storage unit that in turn is composed of:

- one controller tray Dell PowerEdge MD3600f equipped with
  - Double controller FC
  - 4 ports FC8 per controller
  - 4 optical transceivers SFP 8 Gb
  - 12 HDDs from 4 TB (NL_SAS 7.2 krpm)
- N. 7 Expansion Trays Dell PowerEdge MD1200, each with 12 HDDs from 4TB (NL_SAS 7.2 krpm).

5. The Network Infrastructure

The network infrastructure at the ReCaS site in Cosenza is based on a central core switch which directly connects each node via 10 GbE optical lines, with often two direct lines per node, thus guaranteeing a single hop model, reduced latencies and round trip times. The central switch is connected with a pair of optical fibers, via a Router Cisco ASR 1002, to the 10 GbE GARR PoP located in the University campus. A second connection to the GARR PoP, composed of 24 pairs of optical fibers, is presently being implemented and will soon become operational, increasing the network reliability and bandwidth.

The adopted core switch is the system HP10508 (see Fig. 6) characterized by a high-speed fully distributed next generation Clos architecture with up to 11.52 Tbps switching capacity and up to 13.72 Tbps switching fabric capacity to meet the demands of bandwidth-intensive applications.

With its ultra-high port density (achieved thanks to the possibility to use up to 12 interface modules) the HP10508 can provide up to 576 10 GbE ports with 3-microsecond latency. In its initial configuration, the switch adopted in Cosenza has been equipped with 96 GbE ports.

As far as the management network is concerned, all the connections have been implemented in copper. The copper connections

Fig. 6. The switch HP 10508.

internal to the rack are used for the servers as well as for the sensors (temperature, humidity, and door opening) and the PDUs. The servers are connected through a dedicated management port, with IPMI or iDRAC (for Dell hardware), while the environmental and power supply parameters are concentrated through a CMC unit. A network topology based on top-of-rack switches with 1 GbE ports and 10 GbE uplinks to a common concentrator has been adopted while different VLANs have been configured so to keep logically separated the different internal networks used for management purposes.

6. Grid and Cloud Platforms and Core Services

The services provided by the ReCaS infrastructure in Cosenza will be based on both the Grid and Cloud computing paradigms so to give to the end-users the ability to choose the most suitable resources to efficiently and effectively run their own applications. To this end, the virtualization platform Ovirt[4] and the Cloud operating system OpenStack[5] have been adopted. Ovirt has been installed in a

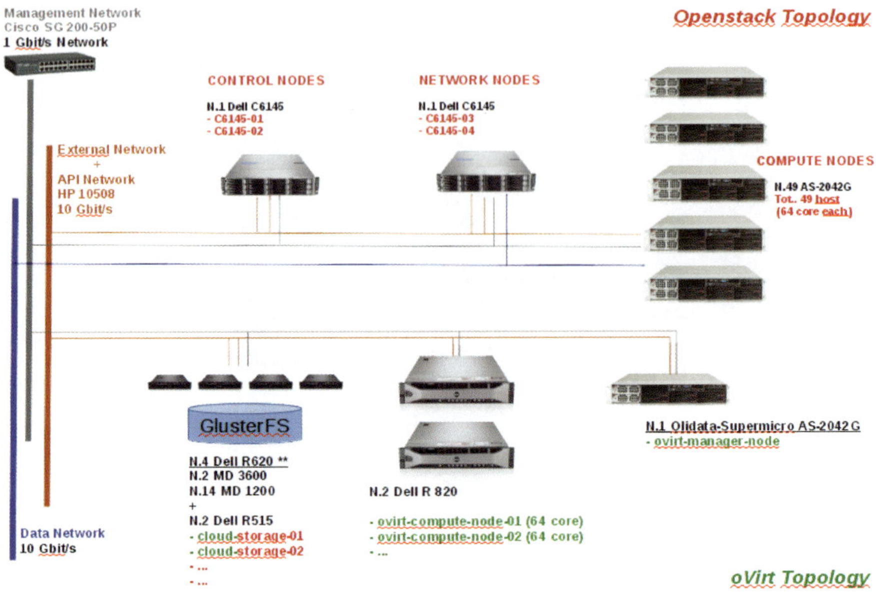

Fig. 7. A possible deployment schema of OpenStack and Ovirt.

high-availability multi-node configuration so to be able to rely on a robust virtualization platform on top of which to instantiate all the virtual hosts running key services such as the monitoring and installation servers as well as the Grid Computing Elements, User Interfaces, and the siteBDII.

OpenStack is a cloud operating system that controls large pools of computer, storage, and networking resources throughout a datacenter, all managed through a dashboard that gives administrators control while empowering their users to provision resources through a web interface. A scheme for a possible deployment of OpenStack with GlusterFS or Ceph together with Ovirt is shown in Fig. 7.

7. Monitoring

The ability to efficiently and continuously monitor the behavior of the Data Center and issue automatic alarms is clearly a critical component for a successful operation of the infrastructure. At the ReCaS Cosenza site, two independent monitoring systems have

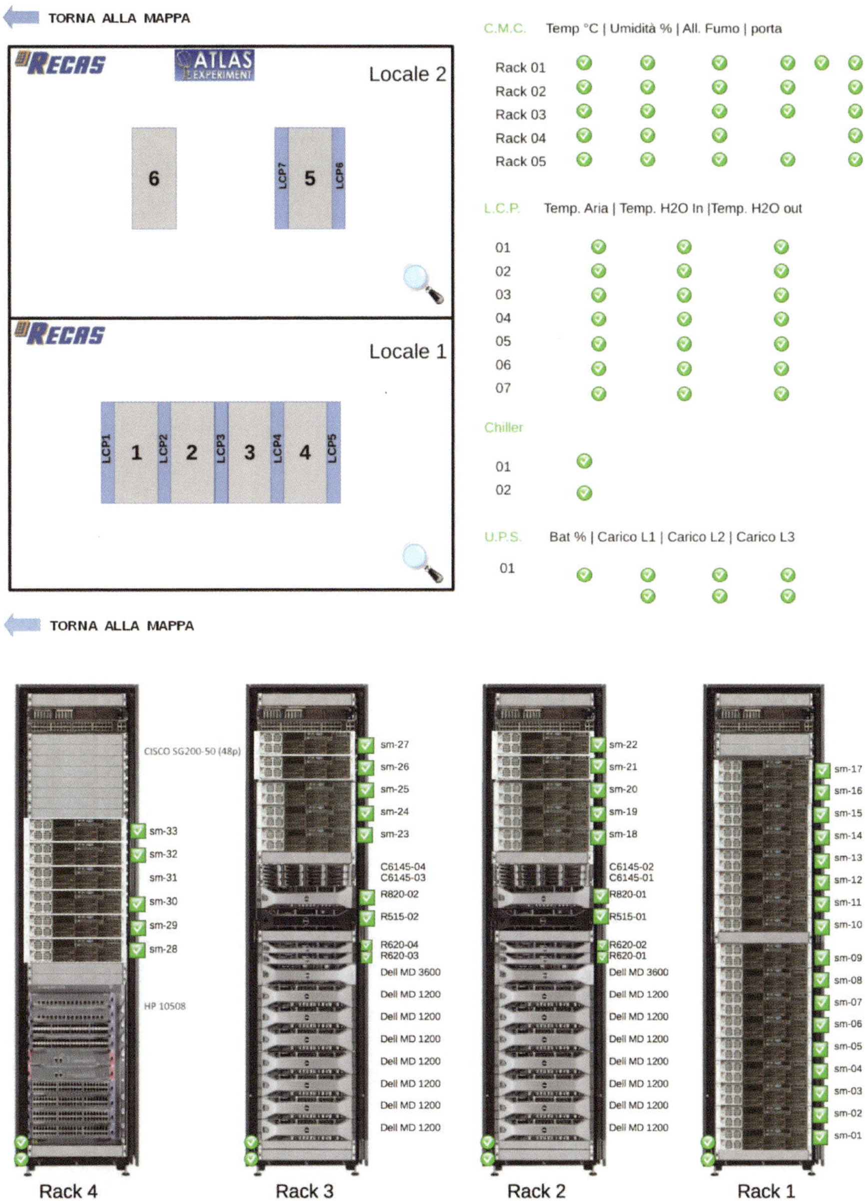

Fig. 8. Two screenshots of the Nagios–Ganglia based monitoring system.

been implemented to guarantee redundancy. A first monitoring system, described in detail elsewhere[6] in this volume, is implemented exploiting different tools like Nagios, Ganglia, customized plug-ins, Centreon and NagVis; it allows the efficient monitoring of all the critical components of the support infrastructure as well as of the computing resources via a simple web interface (see Fig. 8). The system also allows the implementation of management actions such as the clean shutdown of the IT equipment in case of exceptional circumstances. A second completely independent solution has been implemented by making use of the Rittal RiZone® Data Centre Infrastructure Management (DCIM) platform. RiZone® allows the efficient administration of the physical IT infrastructure from power supply and UPS, to cooling and rack monitoring, through the security system. Thanks to its Workflow Editor it is possible to define user-specific scenarios ("what happens if…") that help in implementing efficient control over the infrastructure. Complete the monitoring system with a video surveillance program for the Data Center area (see Fig. 9).

Fig. 9. A screenshot of the video surveillance system.

8. Conclusions

The ReCaS Cosenza Data Center represents a unique IT infrastructure in the Calabria region. It is a medium-sized Data Center that is characterized by an efficient, advanced and scalable support infrastructure that will guarantee high operational standards. Its present computing and storage resources will contribute in a substantial way, together with the other ReCaS DataCenters, to the computing needs of the experiments at the LHC as well as in supporting the needs of diverse scientific and non-scientific communities.

Acknowledgments

The authors would like to thank the Technical Division and Physics Department of the Università della Calabria for their support during the construction phase of the Data Center.

This work has been funded by PON 2007–2013, project code PONa3_00052.

References

1. L. Merola, *The ReCaS project,* this volume.
2. Technical specifications available at http://coelmo.net/.
3. Technical specifications available at http://www.rittal.com/.
4. The Ovirt platform is fully described at http://www.ovirt.org.
5. For additional information on OpenStack visit https: www.openstack.org/.
6. N. Guarracino *et al.*, *An integrated monitoring system, with Ganglia and Nagios, for the Cosenza ReCaS Site*, this volume.

Chapter 6

The ReCaS Project Naples Infrastructure

G. Russo*,[†],[‡], G. B. Barone*, G. Carlino[†] and G. Laccetti*,[†]

**Università Federico II, Napoli, Italy*

[†]*Istituto Nazionale Fisica Nucleare, Napoli, Italy*

[‡]*guido.russo@unina.it*

This document describes the infrastructure built in the University Federico II of Naples and in the *Istituto Nazionale Fisica Nucleare* (INFN) section of Naples, within the *Rete di Calcolo per SuperB e altre applicazioni* (ReCaS) project. This project started in the year 2012 with European funding (PON 2007–2013). It was completed in 2 years with the integration and upgrading of structures for distributed computing existing in Southern Italy. At the end of the work the data center in Naples has 132 hosts, 4,956 cores and 4960 TB of storage with an aggregate computing power of 45.4 kHepSpec.

1. Introduction

As detailed in a previous paper of this volume, the PON 2007–2013 as founded, under contract PONa3_00052 *Rete di Calcolo per SuperB e altre applicazioni* (Computing network for SuperB and other application, ReCaS), the realization of two datacenters in Naples, namely for *The Istituto Nazionale di Fisica Nucleare* (INFN or National Institute for Nuclear Physics) and University of Naples Federico II (UniNA). The infrastructure at the two sites, ReCaS-A for the University of Naples and ReCaS-B for INFN, is detailed in the following.

2. The Common Infrastructure

The distributed data center built in Naples, is based on pre-existing infrastructures inherited from the SCoPE project (UniNA) and from INFN computational resources placed in the Physics Department of the University of Naples, both located in the Monte S. Angelo University Campus, and are 100 m apart. We will refer to the two sites, respectively, hereafter as ReCaS-A and ReCaS-B.

Excellent network infrastructures with many kilometers of optical fibers, structured according to a multi-ring shape with multiple differentiated ways connect all the departments; each fiber operates up to 10 Gbps, with several fiber pairs on each connection.

As commonly happens in data centers, the two main problems are power and cooling, not the computing components. The two sites ReCaS-A and ReCaS-B share a single power supply system, with a cabin capable of 1 MW electrical power.

A diesel power generator (Fig. 2), with a 5,000 L tank for diesel fuel, supplies the emergency power, while four uninterruptible power supplies (UPS) provides stable power: 2 × 400 kW for ReCaS-A, 2 × 250 kW for ReCaS-B (Fig. 1).

Regarding cooling concerns, we have to consider that most of the energy which goes into the servers is dissipated into heat, therefore one needs the same quantity of energy to cool back the servers. To achieve an efficient cooling, we adopted a cooling system based on water: cool water enters an air–water exchanger, thus cooling the air inside the rack. The water is slightly heated and goes to an external chiller, which cools back the water, in a virtuous loop. In ReCaS-A, we have two redundant chillers, 2 × 400 kW (thermal), and three air–water exchangers for each rack, each capable of 7 kW cooling power.

In ReCaS-B, the power of the chillers is 150 kW. The whole cooling system has been realized by RITTAL.

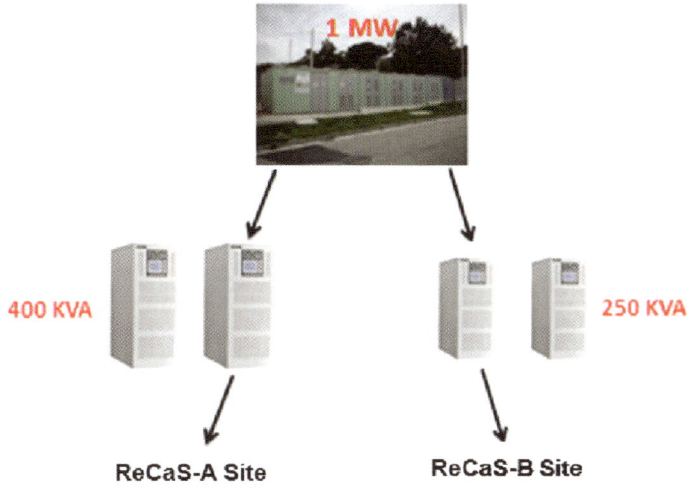

Fig. 1. Power distribution.

Fig. 2. Emergency power generator.

2.1. *The ReCaS-A Site*

The hardware inherited from the SCoPE project, consists of network, computing and storage devices.

The network is based on a pair of Cisco 6513 core switches, which connect, in a redundant way, all servers, thus providing a single switching matrix where each server can interact with any other server with a single hop. The two core switches connect each server with 2 × 1 Gbps lines, but are connected to each other with 4 × 10 Gbps lines.

In addition to the above Ethernet-based infrastructure, a Cisco SFS7024D Infiniband switch provides low latency connection to 280 blades dedicated to MPI-based applications, while a Cisco MDS9509 switch provides FC connectivity for accessing the storage, for 24 blades dedicated to data intensive applications that require frequent I/O phases toward the Storage Area Network (SAN).

The blades, in total 304, are arranged in groups of 16 blades per chassis; each blade has 2 quad core Intel CPUs and 8 or 16 Gb of memory, which is 2 Gb per core.

The storage system is based on EMC[2] hardware, a SAN system, with a FC infrastructure. In addition, there are two EqualLogic Network Attached Storage (NAS), based entirely on the iSCSI protocol. While the SAN storage provides direct connection to all data intensive applications through the MDS9509 switch, the NAS storage

Fig. 3. The SCoPE/ReCaS DataCenter (ReCaS-A). This data center hosts 33 racks.

provides all other nodes with block I/O access to storage, through the 6513 switches.

A detailed description of the SCoPE Data Center is in Ref. 1.

Summarizing, the resources inherited by the SCoPE project are:

- 304 compute nodes (DELL M600 blade) with 2 quad-core processor (organized in 19 Chassis DELL M1000) of which:
  - (i) 280 with Inter-node Connectivity-based on Infiniband for parallel MPI-based applications,
  - (ii) with FC Connectivity for applications with intensive I/O towards the SAN,
  - (iii) service nodes (DELL PowerEdge 2950) with two quad-core processor and FC connectivity to the SAN.
- a SAN EMC[2] (CX3-40) with the storage capacity of 130 TB;
- two Network Attached Storage (EqualLogic PS5000XV) with the total storage capacity of 32 TB.

All the above-described resources are placed into racks from 1 to 12, while the racks from 26 to 33 hosts computing and storage resources belonging to the Tier 2 of the ATLAS project.[2]

Thanks to the funding of the ReCaS project, the remaining part of the racks (from 13 to 25) have been filled with:

- a new network infrastructure described in detail in the next section,
- storage and computing resources belonging to the most recent hardware generation, described in the remaining part of this section.

The deployment strategies adopted for the construction of the new ReCaS infrastructure have been focused on the issues of interoperability of the infrastructure among the project partners and the integration of the different paradigms of parallel and distributed computing (HTC, Grid, and Cloud).

The infrastructure includes distributed computing resources interconnected by high-performance networks (10 Gbps Ethernet) and equipped with high-performance distributed file systems accessible through protocols of different levels.

Furthermore, in order to efficiently and effectively support users, both traditional high energy physics (HEP) and high performance computing (HPC) groups, the infrastructure includes multi-node computer systems with a high number of cores per node and, in some cases, also equipped with graphics processing unit (GPU) accelerators, following with the guidelines of other international projects for the design and implementation of HPC computer systems for the next 10 years (e.g., the International Exascale Software Project (IESP)).

This features will make the ReCaS infrastructure particularly suited to host applications based on different paradigms of parallelization (Message Passing, Multi-thread and GPU-based) possibly combined.

The presence of heterogeneous computational and storage resources can be managed at the level of the middleware in order to ensure end-users the ability to choose the most suitable resources to efficiently and effectively run their own applications.

The computing systems have been planned to suite all the needs of the resources, therefore based on very diverse needs of users of the University was chosen a general purpose approach implementing different computational paradigm and parallel architecture. Besides, the INFN users are more oriented to specific experiments, which require a large number of cores per server and local disks with larger capacity. In general, the storage systems are oriented to the scalability efficiency and reliability using redundancy systems. The new ReCaS computational resources accommodate both the requirements; in fact, the following nodes compose them.

The compute nodes are particularly suitable for data intensive applications and multi-core applications, besides the "accelerated" nodes offer a higher parallel environment via GPGPU and/or MIC resource.

The storage system (see Fig. 6) includes three new SAN Dell Power Vault 3660f for the aggregated capacity of 288 TB and one new Storage Area Network Dell Power Vault 3660f for the aggregated capacity of about 150 TB, bringing the total capacity to about 1,300 TB.

Model	DELL R270	SuperMicro SYS-72TRF +
Number	3	1
CPU number and type	2 × Intel Xeon E5-2680 2.7 GHz	2 × Intel Xeon E5-2603 1.8 GHz
Nr. cores per CPU	8	8
RAM	128 GB	128 GB
GPGPU model	Nvidia Tesla K20	Nvidia Tesla 2090
GPU number	5	1
CUDAcore number	2956	512
RAM per GPU	5 GB	6 GB
Other accelerator (ACC)		MIC Intel Xeon Phi Coprocessor 3120P
Specifications (ACC)		6 GB, 1.100 GHz, 57 × 86-cores
Local storage	4 TB	4 TB
Connectivity	2 × 10 GbE, 2 × 1 × GbE	2 × 10 GbE, 2 × 1 GbE

Fig. 4. Nodes with GPGPU and/or MIC.

Model	DELL M915	DELL M620	DELL R720XD	DELL R815	DELL R510	SuperMicco As-2042G	HP DL560G8
Number	8	8	15	3	3	11	10
CPU number and type	4 × AMD Opteron 6376 2.3 GHz	2 × Intel Xeon E5-2600 2.4 GHz	2 × Intel Xeon E5-2680 2.7 GHz	4 × AMD Opteron} 6376 2.3 GHz	2 × Intel Xeon E5-2600 2.4 GHz	4 × AMD Opteron 6366 HE 3.1GHz	2 × Intel Xeon E5-4600 3.3 GHz
Cores per CPU	16	12	8	16	6	16	8
RAM	256 GB	192 GB	64 GB	512 GB	32 GB	256 GB	256 GB
Local storage	900 GB	900 GB	44 TB	6 TB	12 TB	24 TB	3 TB
Connectivity	2 × 10 GbE, 2 × 1GbE	2 × 10 GbE, 2 × 1 GbE	2 × 10 GbE, 2 × 1 GbE	2 × 10 GbE, 2 × 1 GbE	2 × 10 GbE, 2x 1 GbE	2 × 10 GbE, 2x 1 GbE	2 × 10 GbE, 2 × 1 GbE

Fig. 5. Computational nodes.

2.2. *The ReCaS-B Site*

The pre-existing INFN site, in order to accommodate part of the new ReCaS infrastructure, went under an upgrade of the basic infrastructure: nine new racks were added whose cooling system is based on the same conceptual design used for SCoPE.

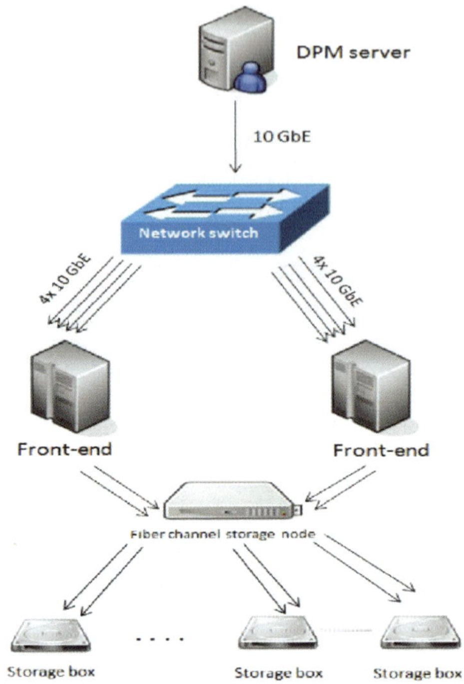

Fig. 6. Storage system architecture.

Fig. 7. The INFN/ReCaS DataCenter (ReCaS-B).

The cooling system has been improved with a new chiller placed to support the existing one and monitored via network. The new site layout is reported in Fig. 7.

3. The Networking Infrastructure

As with the inherited SCoPE data center, the network infrastructure at both ReCaS sites in Naples is based on a central core switch, which directly connects each node via 10 GbE lines, often with more than one line per node, thus guaranteeing a single hop model, to reduce latency and round trip times.

For ReCaS-A site, the core switch is a Juniper EX9214; all line cards have SFP+ connectors, and we use optical transceivers (short and long range) on each line, thus connecting all nodes via optical fibers, eliminating cross talk and the noise problems due to the crowded cabling of data centers.

For ReCaS-B site, the core switch is an HP10508, with similar characteristics and the same optical connections. In this site, we could not afford redundancy of the core switch, while in the ReCaS-A site the redundancy was achieved by a separated and dedicated backup network, based on top-of-the-rack (ToR) switches with 1 GbE ports and 10 GbE link to a concentrator. The 1 GbE links, ToR architecture, are in copper, but cabling is limited to the inside of the rack; the uplink connections are in fiber. A more detailed representation of the network is given in Fig. 8.

3.1. *The Backup Network*

Particular attention has been dedicated to the management system. The idea behind it is that the two sites are unattended, but operate 24 h/day, 365 days/year.

Therefore a good monitoring of the system, described in a separated paper in this volume, is necessary, with automatic alarms and with a remote user interface accessible from anywhere through a web interface. Two control rooms have also been set up in Monte S. Angelo Campus, and from each of them we can monitor the entire Naples infrastructure. The first element of a management network is a dedicated network, which is a complete out-of-band management with a ToR architecture. Each rack has a management switch, 48 ports at 1 GbE, which connects all internal devices; all of

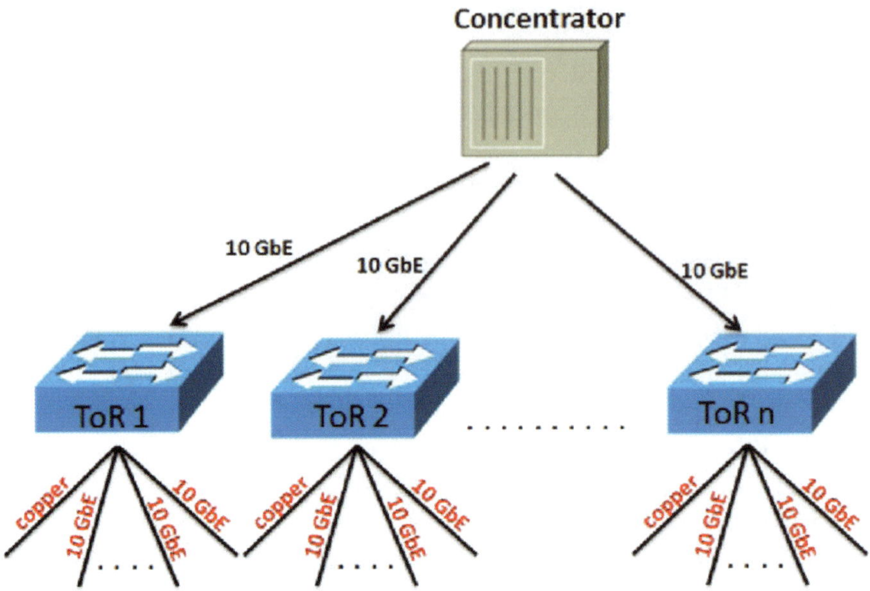

Fig. 8. Backup network.

these switches are connected to a concentrator switch as in the schema below:

All connections are in copper. For ReCaS-A, this network is not coincident with the backup network. In this case, each rack has therefore two ToR switches, one for backup and one for management (see Fig. 9).

In ReCaS-B, there is no backup network and hence one ToR switch. The copper connection internal to the rack is used for all devices: servers are connected through a dedicated management port, with IPMI protocol (or iDRAC protocols for the Dell hardware) while the environmental and power supply parameters are concentrated through a CMC unit, which delivers to the network the information coming from the probes (temperature, humidity, door opening, ...) and from the PDU's to monitor energy consumptions at each plug. The concentrator has a redundant connection to the general campus network to guarantee reachability at any time.

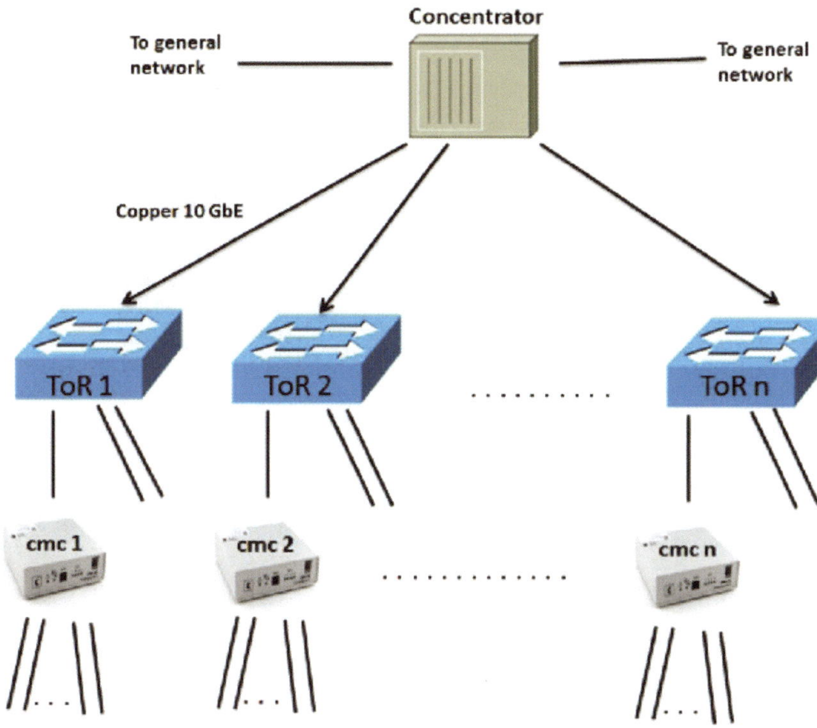

Fig. 9. Management network.

3.2. *The Inter-Site Network*

The external connections of the two ReCaS sites are quite complex, as they have to cope with the many local constraints and many users' needs. Before giving the geographical scheme, we will now point out the characteristics; we also anticipate that for each connection we always have put 12 pairs of fiber, even if we will not use all of them; this is because we want to leave space for future needs.

First, the two sites ReCaS-A and ReCaS-B are connected with 4 × 10 GbE links, in bonding from the HP 10508 to the Juniper EX9214 (shown in Fig. 10).

Each site has also its own link to the Garr-X PoP, also located in the Monte S. Angelo Campus. These links are at 4 × 10 GbE (ReCaS-B) and at 1 × 10 GbE (ReCaS-A). This asymmetry is due to

Fig. 10. The Juniper EX9214 switch (left) and the HP 10508 switch (right).

the larger amount of data that reach ReCaS-B every day, during LHC operation at CERN (from 15 to 20 TB per day, continuously for several months).

An additional 4 × 10 GbE link between ReCaS-A and ReCaS-B is dedicated to the internal traffic between the BelleII experiment nodes, located in both sites.

Sixteen lines at 10 GbE are used to connect the ATLAS experiment nodes and the storage present in ReCaS-A to the HP switch in ReCaS-B to maintain the single-hop paradigm among all nodes of the ATLAS experiment, whose main infrastructure are located in the ReCaS-B site.

A backup line to GARR is provided at 10 GbE by a Cisco 3570 switch connected to the HP switch in the ReCaS-B site; this backup line follows a different route through the campus, for reliability.

A similar backup link is made for ReCaS-A site; this is given by a fiber link at 2 × 1 GbE from the two Cisco 6513 switch in the site. These two switches are connected to the EX9214 and the backup concentrator at 10 GbE. The schematic diagram is given in Fig. 11.

3.3. *The Storage*

The storage architecture is neither a pure SAN nor a pure NAS system. It is rather a "storage service": a front end system node (paired for

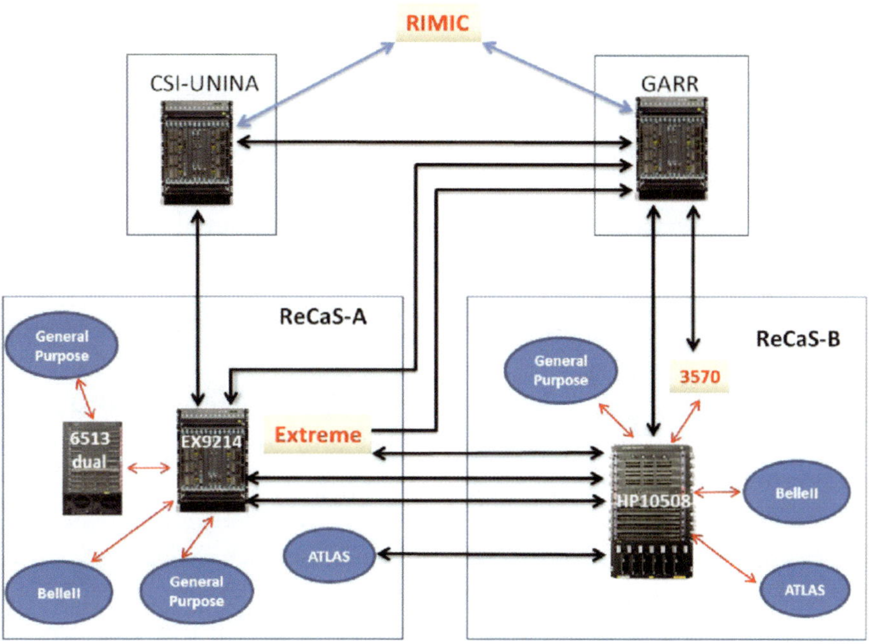

Fig. 11. The ReCaS connectivity architecture.

redundancy) guarantees access to the data, which are stored on a FC box, which in turn collects data from several SAS storage boxes; a further node runs the DPM software which maintains location and filename for all datasets. The scheme below represents our implementation, where, the front end are Dell R620 nodes connected at 4×10 GbE to the core switches. The FC storage boxes are MD3600f/MD3800f boxes and the storage boxes are MD1200 boxes. Each of these systems, from the front-end to the storage box, is an autonomous storage system of roughly 400 TB raw; these autonomous storage systems are replicated to reach the desired total storage capacity of about 3 PByte in each site (ReCaS-A and ReCaS-B).

4. Core Services

The services provided by the ReCaS infrastructure in Naples are based mainly on a Grid computing paradigm.

Among them there are:

- an information system on the computing power and the amount of storage available (implemented by means of a siteBDII),
- a computing service realized by means of a Computing Element with a local resource manager and a wide set of Worker Nodes,
- a storage service implemented by a DPM head node based on MySQL and a set of DPM disk nodes.

The middleware used is EMI: in particular the version 3 for all services. The operating system is SL6.

To provide a unified vision of the total computing and storage power belonging both to the University and to the INFN, we proposed the deployment schema shown in Fig. 12.

This deployment, besides, optimizes the management of sites increasing reliability and enables "autonomous" management of the resources belonging to the two partners.

In thc dcscribcd scenario, users can receive from the unique information system all the information they need about the total amount of disk space and cores. Moreover, on the worker nodes, a conspicuous set of libraries and scientific packages have been made available and published on the information system.

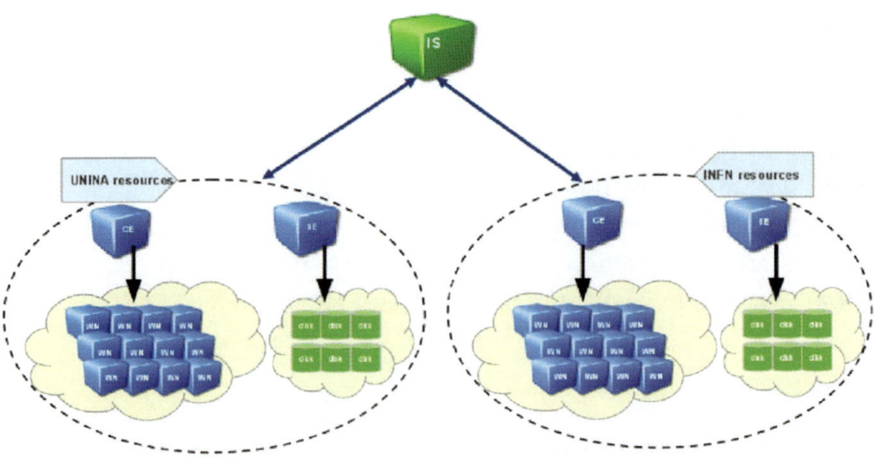

Fig. 12. The ReCaS services architecture.

Finally, other general-purpose services are available and are, instead, based on a Cloud Computing paradigm.

5. Conclusions

The realized infrastructure is general purpose because it is able to run and satisfy very different applications, also belonging to very different scientific fields.

The whole set of resources comprises servers with different equipment in terms of RAM, local disk capacity, number of Cores/CPUs that can be used in a generalistic configuration or, if necessary, redeployed into *ad hoc* application-specific shape.

From the final user point of view, the available resources provide the chance to execute on 5,000 cores (or equivalently job slots) up to 5,000 sequential jobs at the same time, to transfer up to 20 Gb of data per seconds on the local storage systems.

Just as an example, it is possible to run 30,000 sequential (or concurrent) jobs in a day having 4 hours of duration.

The deployment strategies adopted for the construction of the new ReCaS infrastructure have been focused on the issues of interoperability of the infrastructure among the project partners and the integration of the different paradigms of parallel and distributed computing (HTC, Grid, and Cloud).

Acknowledgment

The authors would like to thank the multi-disciplinary group (GTT), responsible of the management of the datacenters located at Site A and B. This work has been funded by PON 2007–2013, project code PONa3_00052.

References

1. L. Merola, The ScoPE Project, in Proc. of the Final Workshop of Grid Project *PON RICERCA 2000–2006, Avviso 1575*, pp. 18–35, 2009.
2. The ATLAS Collaboration, The ATLAS Experiment at the CERN Large Hadron Collider, *JINST* 3 S08003, 2008.

Chapter 7

Monitoring the ReCaS Project Infrastructure

S. Pardi*,‡, D. Del Prete*, S. Naddeo* and G. Scotti†

**INFN, Naples, Italy*

†*University of Naples Federico II, Naples, Italy*

‡*spardi@na.infn.it*

The complexity of a geographically distributed computing center requires a very efficient monitoring system at various operating levels. Must be ensured of many requirements i.e., reliability, efficiency in information management, user interface usability, etc. The most important one is certainly the scalability of the system; the monitoring system must itself be controlled on a distributed network. The aim of the ReCaS monitoring project is to provide a scalable and easy to use tool, allowing the monitoring (and in some cases to control) of all the services of the computing infrastructure systems including network services.

1. Introduction

Monitoring the infrastructure is a dirty work: one has to do it, but there is no public evidence of the work done. We are not talking about jobs, queues, space used, quotas, users, but of voltage, amperes, temperatures, humidity, and so on. In Naples, monitoring the infrastructure has always been considered important, almost a prerequisite before opening the infrastructure to the user. In this paper, we first describe the various parameters that we have considered for real time monitoring and we will describe the tools used.

2. The Parameters Being Monitored

The very first level of infrastructure monitoring is related to the power supply, as we know that power and cooling are the main systems in a data center. Monitoring the power supply means first of all to collect data about the quality of the main supplied by the energy company: in the *Rete di Calcolo per SuperB e altre applicazioni* (Computing network for SuperB and other applications, ReCaS) data center in Napoli, the energy company provides a line at 10,000 V, which we transform to 380 V, tri-phase. Through a switchboard, power goes to charge four uninterruptible power suppliers (UPSs), while emergency power coming from a diesel power generator gives power to the switchboard, when necessary. The switches on the switchboard are remotely controllable, and control is assured by the University campus technical offices, who maintain a guard office 24 hours a day, for the whole campus. The UPSs are equipped with a management board, SNMP-capable, and data on voltage level and battery status are acquired via an Ethernet network and moved to the database, described later on in this paper. The UPS status is displayed via a HTTP service as shown in Fig. 1.

Within the data center, data acquisition is performed by hardware provided by the company who manufactured the rack, namely RITTAL. In each rack there is a controller, SNMP-capable, which collects data from the sensors and, via the management network, sends them to the monitoring system. The main advantages of the CMC system are: reduced cabling and installation; standardization of the sensors; reduced space in the rack. The main component of the CMC is the Processing.

Unit/Compact is the central unit of the CMC III monitoring system. Up to 32/4 external sensors/CAN bus connection units may be connected to the integral sensors. The units may be connected to the data network via Ethernet, configured via Web/USB, can send alarms via an e-mail server and connect to the Network Management

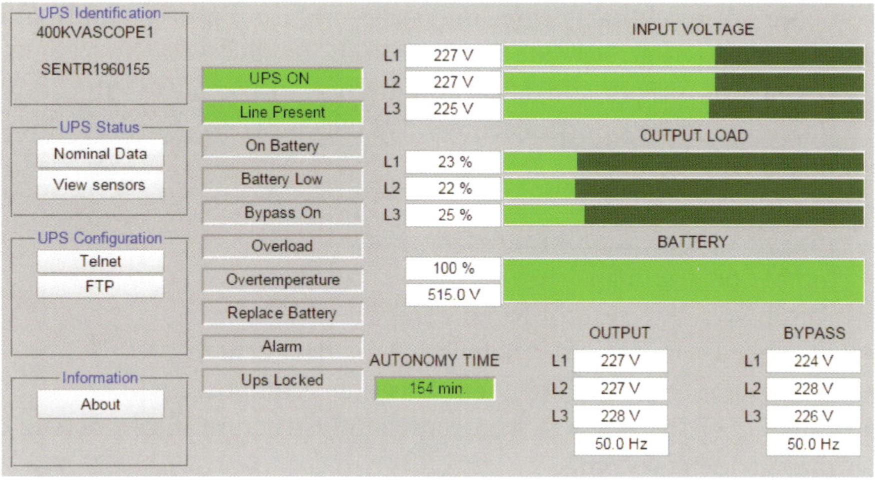

Fig. 1. The screen showing the state of the UPS.

Fig. 2. The CMC on the top of the rack.

System via SNMP. The CMC control units provide inputs and outputs, for measurement/control of power or speed control/monitoring of fan systems, while the CMS sensors monitor various parameters in the enclosures or the room where the racks are installed (see Fig. 2).

The software used to manage the CMC is RiZone. RiZone is the Data Center Infrastructure Management (DCIM) platform for all components in our data center infrastructure, from power supply and power distribution, to cooling, through to the security system. Thanks to its simple and fast configuration, RiZone was the ideal DCIM solution for our data centers. We chose RiZone for the following reasons:

- Energy optimization in the data center,
- Simple project management,
- Automatic detection of Rittal components,
- Increased security and reliability of the data center,
- Incorporation of building control system components,
- Linking of the physical data center infrastructure into a network management system,
- Linking and monitoring of all IT devices via SNMP protocol.

In the ReCaS data centers, we monitor the following parameters through RITTALs CMC devices: door opening; rack temperature; water loss in the rack; power consumption (watt, amperes) for each plug on the internal PDUs; chiller status (see Fig. 3).

Concerning the hardware inside the racks, our management network connects each node through a dedicated management network interface controller (NIC), not controlled by the server operating system, and capable of gathering data on the internal temperatures (CPUs, disks), on the fan speed, and other similar parameters which can describe the status of the server.

Managing distributed servers from a remote location is mandatory in our case, as ReCaS administrators must easily and effectively manage servers in locations that have no administrative IT staff. Such

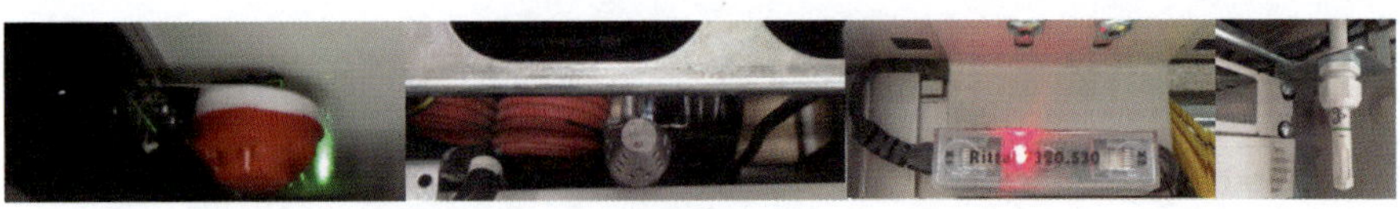

Fig. 3. (a) Fire sensor, (b) flood sensor, (c) intrusion sensor, (d) humidity sensor.

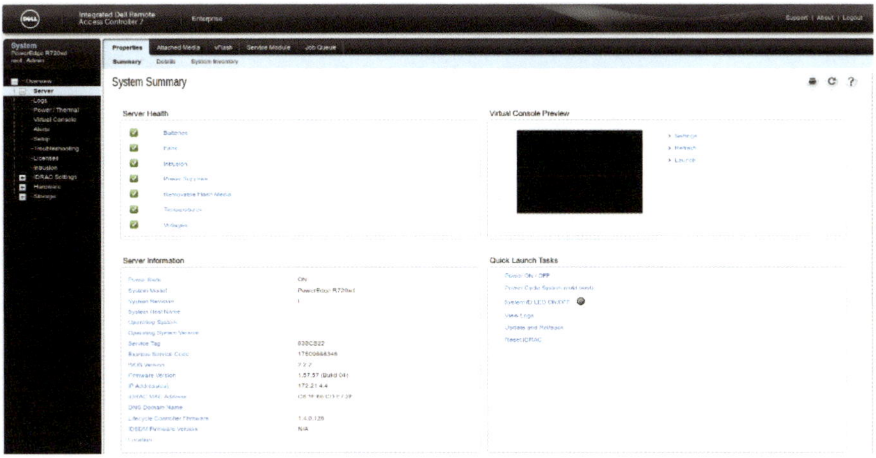

Fig. 4. The Integrated Dell Remote Access Control system (iDRAC).

scenario require remotely performing all server management operations and responding to server-down situations.

For all the DELL hardware, we used the integrated Dell Remote Access (iDRAC) proprietary protocol. Remote-access capabilities help to improve system administrator productivity and overall server availability by reducing administrator visits to the system and by allowing some operations on groups of systems instead of individual devices. The Dell Remote Assistant Card II provide our administrators with continuous access to servers. Administrators also achieve full control of the server hardware and operating system from any client system running a Web browser (Fig. 4), even if the server is down or hung.

Similar products are provided for our other servers provided by other Vendors (SuperMicro and HP): in Fig. 5, HP Integrated Lights-Out (iLO) system is shown and in Fig. 6, the Baseboard Management Controller (BMC) system is shown. In the ReCaS DataCenters, we monitor the following parameters through RITTALs CMC devices: door opening; rack temperature; water loss in the rack; power consumption (watt, amperes) for each plug on the internal PDUs, chiller status. The Dell remote access architecture consists of

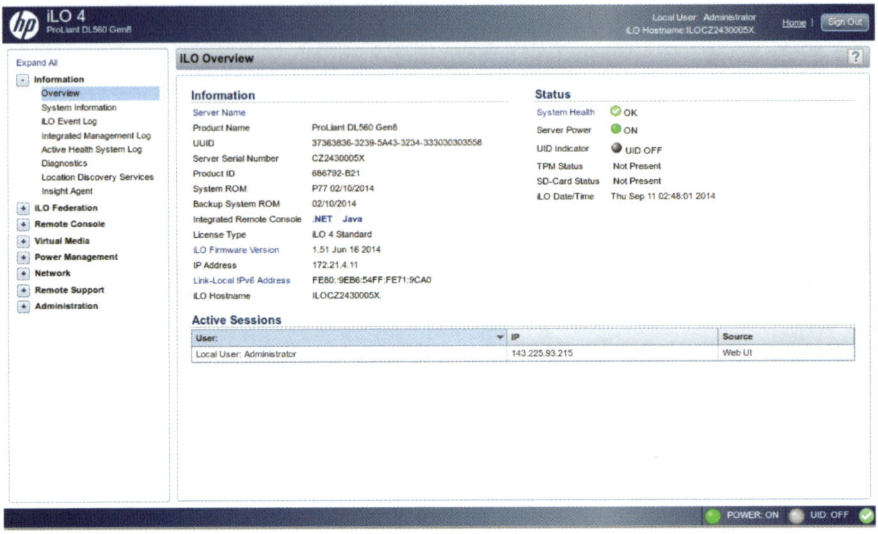

Fig. 5. The HP Integrated Lights-Out (iLO) system.

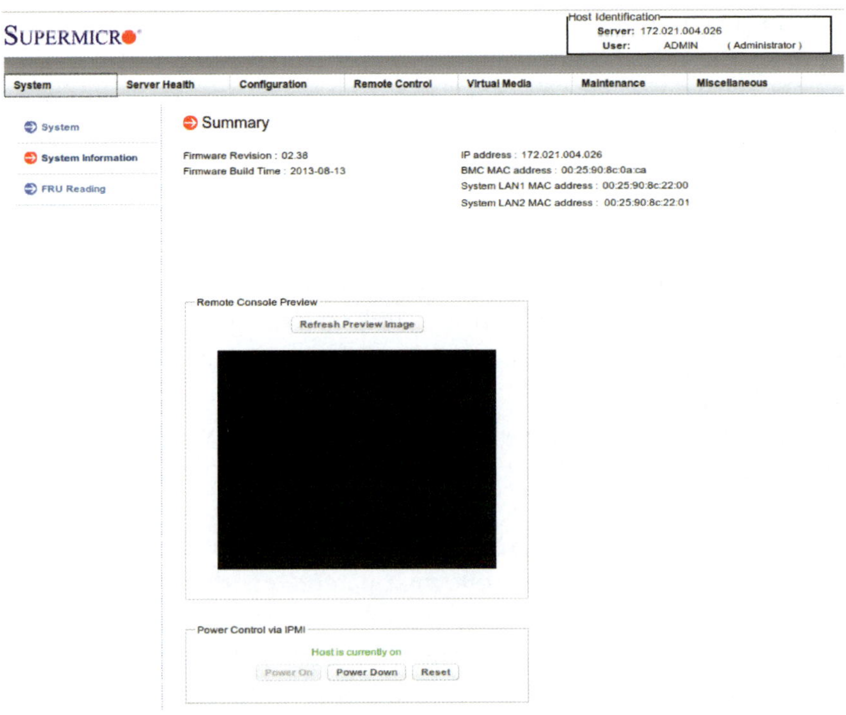

Fig. 6. The Baseboard Management Controller (BMC) system.

hardware and software components that allow administrators to do the following:

- Access a server after a server failure, power outage, or loss of a network connection (using a NIC card or modem);
- Remotely view a server's internal event logs and power-on self test (POST) codes for diagnostic purposes;
- Manage servers in multiple locations from a remote console;
- Manage servers by redirecting the console output to a remote console (graphic and text);
- Perform an orderly shutdown of a server for maintenance tasks;
- Diagnose a server failure and restart the server;
- Alert the administrator using alphanumeric page, numeric page, e-mail, or Simple Network Management Protocol (SNMP) trap when a server detects an error.

As part of the infrastructure monitoring, we are also concerned about the networking. With several kilometers of optical fibers put around the campus, there is the possibility to have some broken trunk, and hence to lose node reachability. For this purpose, we use the management interfaces of all the data switches, again connecting them to the dedicated out-of-band network, and monitoring the status (up/down) of the inter-switch links.

3. The Architecture of the Monitoring Software

The architecture of the monitoring system is based on a set of virtual machines, running on a pair of dedicated servers. The software, described in the next section, periodically queries all devices described above, to collect about 4,000 parameters every 5 minutes; other parameters come from traps originated by malfunctions and/or abnormal events. Each site has two or more virtual machines totally dedicated to monitoring and the management, configured in a failover and load balancing mode.

Naples is the centralized site that collects and manages all monitoring information (starting from the state of health of the monitoring framework itself), coming from all the other Italian sites and from its present form, each monitoring node connects with the very local nodes (e.g., worker node, storage node, etc.) than with the entire monitoring geographic network. The system has been designed also to interface with other European and rest of world sites.

One of the important innovations of this central monitoring system concerns the portal: the application container. The structure used reflects a particular model of web container, much more precisely is a portlet container. Each portlet is a container for a single application and that System has to offer different services and different authentication. The product used is Liferay. It is a "Service Oriented Architecture" (SOA), a software architecture that supports the use of web services; that ensures, in this way, the interoperability between different systems thus ensuring the use of individual applications as components of a single system. This allows to meet the demands of users in an integrated and transparent manner. Using this implementation strategy allows each application to have its own life, as the individual components of the portal do.

4. The Tools Used

The monitoring tools used are open source, they have been customized and extended according to the needs of our requirements. The monitoring framework used are the Ganglia subsystems (Figs. 7, 8 and 9), fully automated Nagios as: Nagios, NagVis, Centreon, and NaReTo; Nagios is the basic system of monitoring that is used to make queries at regular intervals on the monitored host, then publish the results via a web interface.

Nagios is very extensible because it can be accessed via plugins, is scalable. Centreon, however, is a very powerful tool that we have used for the global monitoring of services and for the Nagios configuration; NagVis is a plugin with which we associate a map of all entities and services monitored by Nagios, so we provide the user an interactive navigation; NaReTo is a subsystem that generates complete reports of their information from the data collected by monitoring of Nagios.

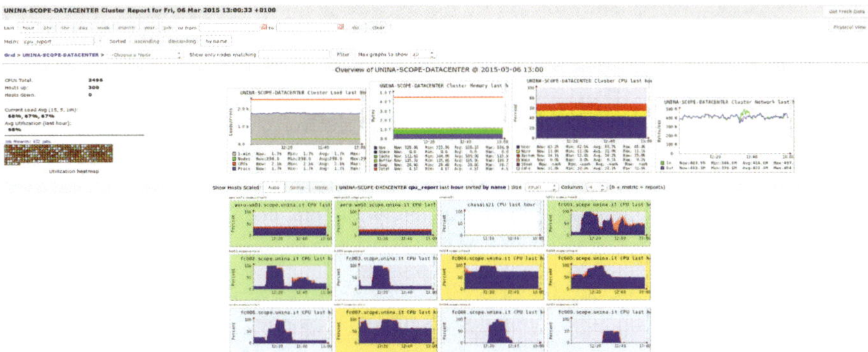

Fig. 7. The Ganglia user interface.

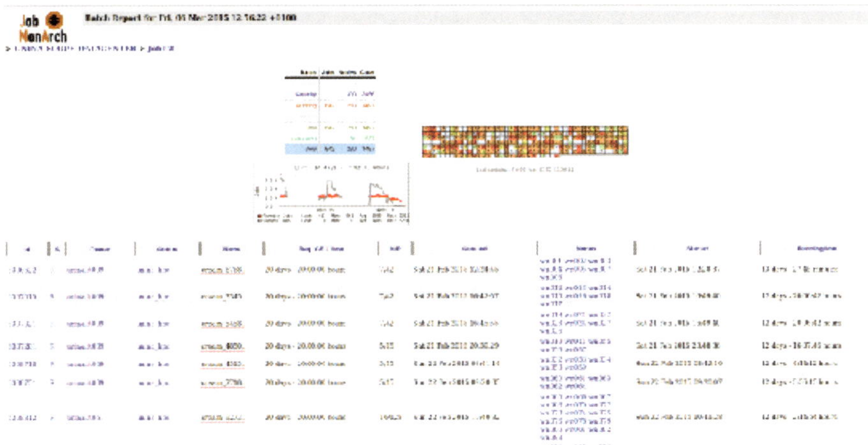

Fig. 8. Job monitoring report.

Besides the services of computational grids, our system enables monitoring the entire network and all the services of the network. Through these systems, we obtain information to monitor all apparatuses and the services associated with it. The status of a service is indicated by either OK, yellow = WARNING, red = CRITICAL. For each group of devices, a threshold of criticality is configured.

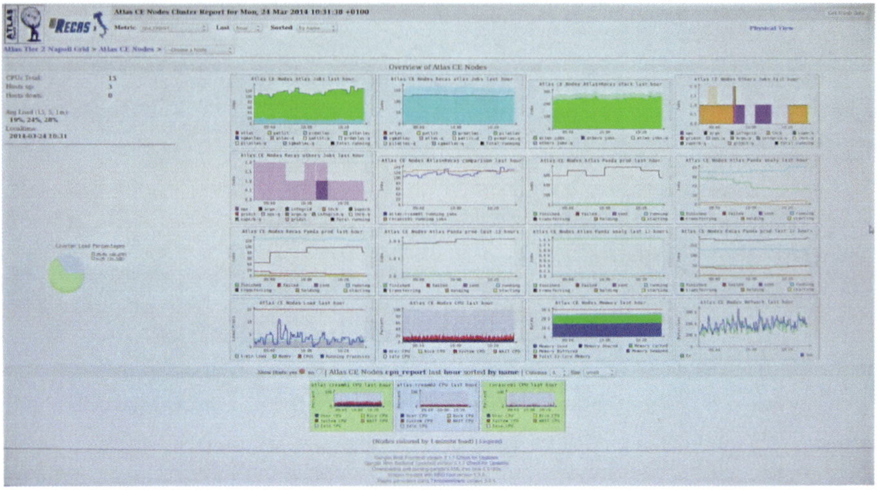

Fig. 9. The general monitoring dashboard.

Among the entities we monitored were storage systems, servers, switches, routers, environmental sensors, power systems, UPS, cards management of each server, application services of the highest level. Our system collects the parameters that allow the analysis of the efficiency, reliability, and degradation of the network being able to improve, in this way, the quality of service of the different applications. In this area, we have used different subsystems of monitoring aimed at satisfying every requirement described above.

5. The Database

Nagios, Centreon and Nagvis, to store the information read from the different remote devices using various databases. The goal is to retain the information permanently and make them easily available for any processing to be carried out by them such as parsing, statistical calculations, export and visualization by web protocols and web dynamic interface. By default, Nagios and Centreon uses several MySQL databases to store the information. The interactions between the hosts, modules installed on the monitoring server and the databases involved in the data persistence are described in Fig. 10. When the host sends

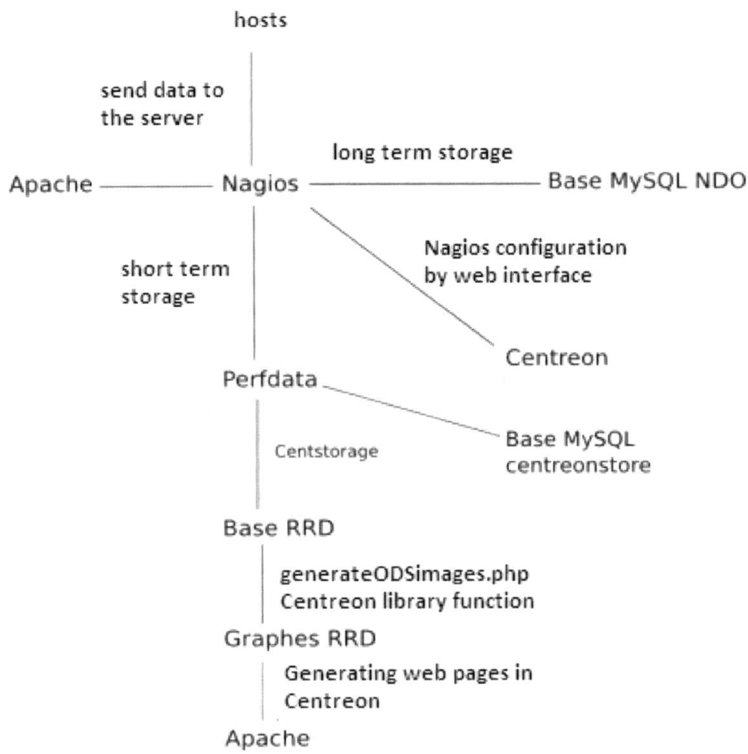

Fig. 10. Basic diagram of the software components of the monitoring architecture.

its own information to the server, in response to a SNMP request, to a Ldap query, to a NRPE command or to another path to request information, Nagios stores the results using two different strategies. The first strategy, also called long term storage, uses a MySQL database, called NDO, to store data from the hosts; the second one, also called short term storage, uses another MySQL database, called Perfdata. This database will also be used by Centreon as the storage from which information will be taken to be shown in the Centreon user interface.

The Perfdata will also be saved in the Round Robin Database (RRD), used by RRD Tool to generate the images which will be shown as dynamic charts in the web interface of Centreon. The fact that the same data are stored in different databases implies that the developer needs to know which of them should be used in the development of plugins and applications to be integrated into the monitoring server. As an example, to show information in a dynamic web page it is reasonable to think of taking the information from the RRD database and to insert them into the web page using the suitable PHP library function.

On the other hand, if the developer must implement any service to analyze the data and show usage statistics, the database to consider will be the Centreon or Nagios MySQL database where, as described before, are stored the long-term information. As an example, in our monitoring system, the following function implements a strategy to store information in the Round Robin Database storage using SNMP protocol and the guidelines to the scripting from the Nagios support:

```
#!/bin/bash
#usage: ./check_active_cores.sh <ip address> <warning threshold> <critical threshold>
#retrieving the number of inactive cores (if necessary, the snmp community must be set by changing the following snmp call)
inactive=`snmpwalk -Os -c n0npublic -v 2c $1 hrProcessorLoad |grep ": 0" | wc -l`
#retrieving the total number of cores
total_number=`snmpwalk -Os -c n0npublic -v 2c $1 hrProcessorLoad | wc -l`
#calculating the number of active cores
number=$(($total_number - $inactive))
#calculating the percentage of active cores
perc=`echo "($number / $total_number) * 100" | bc -l`
perc=${perc/.*}
#preparing the output
if [ $perc -ge $2 -a $perc -le $3 ]; then
echo "CHECK WARNING - $number of $total_number Active Cores | active_cores=$number"
exit 1
fi
if [ $perc -gt $3 ]; then
echo "CHECK OK - $number of $total_number Active Cores | active_cores=$number"
```

```
   exit 0
   fi
   if [ $perc -lt $2 ]; then
   echo "CHECK CRITICAL - $number of $total_number Active
Cores | active_cores=$number"
   exit 2
   else
   echo "CHECK FAILURE - no active Cores detected"
exit
```

This function uses the information present in the MIB of the device, in this case a node of the GRID, to obtain the number of active cores. His output will be stored in the RRD database and using the Perfdata format it will be used to generate the graphs with RRD tool. Another example shows how, in the monitoring system, we Ldap protocol to get the information about free slots of execution in the Grid:

```
#!/bin/bash
#CE list
ce=$(ldapsearch -LLL -x -H ldap://recasna-sitebdii.unina.
it:2170 -b "o=grid"
|grep GlueClusterUniqueID|grep GlueForeignKey|sort -u |
cut -d '=' -f 2)
sum=0
i=0
for x in $ce
do
echo $x
jobs[$i]=$(ldapsearch -LLL -x -H ldap://$x:2170 -b
"o=grid" GlueCEInfoTotalCPUs
| grep GlueCEInfoTotalCPUs | cut -d: -f2 | sed -e 's/^[
\t]*//' | cut -d ' ' -f1)
#echo ${jobs[$i]}
sum=`expr $sum + $(echo ${jobs[$i]} | cut -d ' ' -f1)`
(( i += 1 ))
done
echo $sum > /tmp/slots_na.dat
echo "OK - $sum slot disponibili"
   exit 0
```

Another aspect of the data collection and storage is how to collect data from external sites. The monitoring system expects a centralized portal, which will detail later, and that will collect data from all sites involved in the experiment. This implies the necessity to agree on a standard format for the transmission for the data from such sites to the central monitoring portal; the need to send this information via the HTTP protocol requires us to choose an interchange format common to all sites. We chose to use two equivalent ways: JSON and XML. Information are stored in XML file as the following example:

```
<batchsystem>
<availableslots>3594</availableslots>
<pendingjobs>590</pendingjobs>
<runningjobs>2734</runningjobs>
</batchsystem>
<timestamp>1394645404</timestamp>
<storage>
<available>100432351280</available>
<total>1560455344576</total>
<used>1460022993296</used>
</storage>
</bari>
```

The PHP function used to parse the information contained in the file just shown is:

```
<?php
$curl = curl_init('http://webmon.cs.infn.it/recas-prisma/
data.xml');
curl_setopt($curl,CURLOPT_TIMEOUT,10);
if(!curl_exec($curl)){
echo "HostUP=0 | host_up=0;";
}else{
$result = curl_exec($curl);
$xml_cs = simplexml_load_file($result);
$value = $xml_cs->batchsystem->availableslots;
echo "HostUP=".$value."| host_up=".$value.";";
}
?>
```

In the case of the data from external sites, it is important to point out that the data is not stored in the local database of the central server; they are only shown in the web interface (see next section for more details) for users and they are stored in the local databases on the monitoring server of the source site.

6. The User Interface: The Management Station

The work described leads to the creation of a powerful communication tool, with which any level of user is able to interact. So the user, both generic profile that expert can withdraw monitoring information of different types and with different levels of detail. Specifically the administrators of hardware and software systems, which, thanks to this system are able to get huge benefits in the management and control of all services deemed critical. Furthermore, through the implementation of the fundamental rules of software engineering such as extensibility, reliability and supportability, the system is subjected to minimal maintenance, and therefore, is mainly directed to "serve" and not to be "served": basic requirement of any automatic system that works closely with its users.

7. The Control Rooms

In the operations of data center management, several people are employed. Although the monitoring system is accessible remotely, for any processing to be carried to respond quickly in case of serious problems reported by the monitoring system.

For this reason, the data centers are equipped with two control rooms near it. These rooms have been installed and configured with everything that are needed to access and remotely configure the machines of the two data centers and to observe the status of the hosts using the monitoring system. In each of them, there is a monitoring station with a multi-screen display; these will show in real time all the information about the state of the parameters monitored in the two data centers and their variations, visually reporting any critical

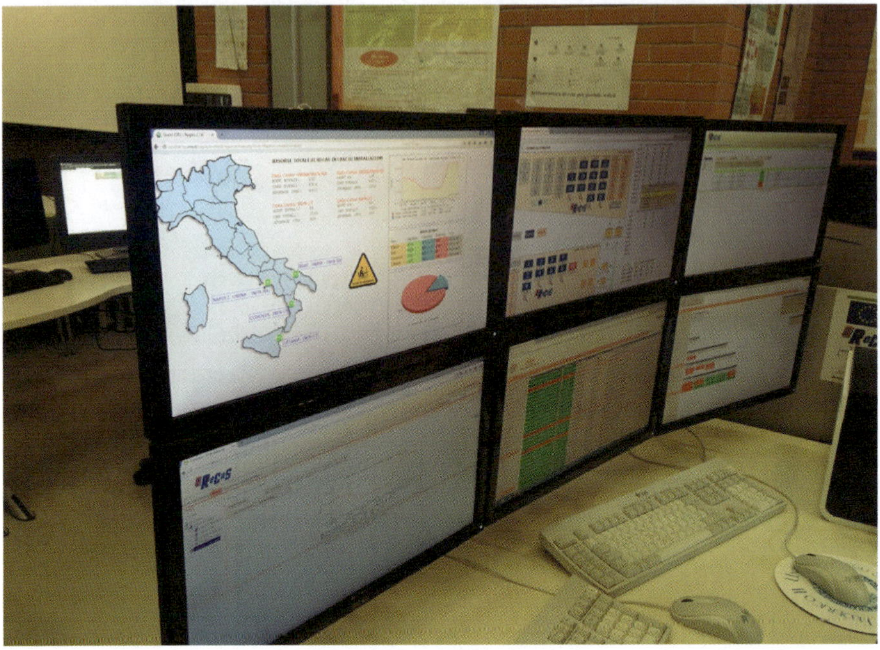

Fig. 11. The monitoring station in the ReCaS control room.

states and warnings that may occur during the lifetime of the data centers.

The two rooms (Fig. 11) will also have several workstations used for research and development activities related to the project. From them system administrators can access the monitoring server, change the configuration and add or remove hosts and services. Other usages may be related to the activities performed by students and Ph.D students as part of research activities connected to the ReCaS project.

8. Conclusions

Most of the hardware infrastructure, from power and cooling to network to rack, can be monitored with the tools developed by our group. These tools will be further exploited to extend their functionalities, as per requirements.

Acknowledgments

This work has been funded by PON 2007–2013, project code PONa3_00052.

References

1. C. Jarlskog, *CP Violation,* World Scientific, Singapore (1988).
2. B. W. Bestbury, *R*-matrices and the magic square, *J. Phys. A.* **36**(7), pp. 1947–1959 (2003).
3. P. X. Deligne and B. H. Gross, On the exceptional series, and its descendants, *C. R. Math. Acad. Sci. Paris.* **335**(11), pp. 877–881 (2002).

Chapter 8

Monitoring the ReCaS Project Resources

G. Andronico*, V. Boccia[†], D. Del Prete[†], E. Giorgi*, G. Maggi[‡,§], S. Naddeo[†,¶], S. Pardi[†,||], V. Spinoso[‡], B. Spisso[†] and A. Tarasio[†]

**Istituto Nazionale Fisica Nucleare, Catania Unit, Italy*
[†]*Istituto Nazionale Fisica Nucleare, Napoli Unit, Italy*
[‡]*Istituto Nazionale Fisica Nucleare, Bari Unit, Italy*
[§]*Dipartimento Interateneo di Fisica, Bari, Italy*
[¶]*University of Naples Federico II, Italy*
[||]*spardi@na.infn.it*

The *Rete di Calcolo per SuperB e altre applicazioni* (ReCaS) project, extensively presented in this volume, requires an accurate measurement of resources and their availability. This paper presents the work done by the ReCaS technical group in Naples, to implement a system entirely based on open source software, capable of monitoring the whole ReCaS distributed Infrastructure.

1. Introduction

In a previous paper in this volume, the *Rete di Calcolo per SuperB e altre applicazioni* (Computing network for SuperB and other applications,

ReCaS) Napoli group has presented the monitoring system for the infrastructure: the "dirty work" that has to be done, but with limited, if any, visibility to the user. In the present paper, we present a work which is just the opposite: a monitoring system to control the resources available in ReCaS sites at any time. The questions we have to give an answer for are: how many job slots are available in total in ReCaS, and at ReCaS, and at the four sites? How much of this computing power is used at the moment? How much storage space is available? We will present first of all the user interface, which in this kind of monitoring is the most important one: users want to see the information they need, not the way these information are collected. But immediately after, we will describe, for the skilled reader, the architecture of the software and the organization of the database.

2. Monitoring Needs

Our approach uses the open source tools and at the same time the integration and correlation of data. These operations do not always agree, especially in a network in constant evolution and where there is the need to maintain over the years the history of the measures collected without losing data. These requests impose the fundamental onerous task of the design and implementation of an information system that would act as the basis for the correlation and the temporal evolution of the data.

The ReCaS monitoring system includes tools for diagnosing and tracking problems of network infrastructure and network services, systems acquisition and visualization of traffic statistics and reporting. The interoperability of tools and integration provide the user with a fast problem diagnosis system. Many of the features of the ReCaS monitoring are publicly accessible to anyone via a web interface, for e.g., the visualization of traffic statistics, the services of the basic resources, the weather map backbone network, jobs running at each site, and much more. The weather maps are dynamic and interactive maps that allow a graphical view of the status of all services. On the map can be given a lot of information, from the level of the machine to the application layer.

Completeness in showing the data allows site administrators to have an important tool; so they have access to all the information in a centralized manner. The monitoring is understood as control, but also as a collection of statistics and data analysis: the suite makes it possible to collect and organize information about individual locations as available in monthly and annual reports.

3. The User Interface

In designing the user interface, we had in mind the idea of a simple, intuitive home page, reachable with one click from the ReCaS site (www.pon-recas.it), showing (on the left) the geographical locations of the four ReCaS sites, and their main dimensional parameters: hosts, cores, TB of storage. A green bullet shows that the site is up and running (and of course the bullet can become yellow or, hopefully in rare cases, red). The upper right part of the screen show the number of jobs running on the whole ReCaS infrastructure, in a graphical format and on an hourly basis, while beneath it a simple table shows the number of available batch slots, and the number of jobs queued and running. This is particularly important for monitoring the resources in cases where, for example, a site is participating in a Monte Carlo campaign, which needs thousands of job slots per day, or when an experiment is in the data acquisition phase, thus having to maintain a sustained data reconstruction rate. The same data are shown in a pie chart representation (Fig. 1).

The clickable map on the upper right corner gives access to a more detailed representation, in which the user can change the parameters of the graph, in particular the start and end time of the collected data.

4. The Architecture of the System

The architecture of the system allows the user to access all monitoring features in a direct and simplified way. It is a point of reference both for the control and the management. Its structure allows a natural extensibility suited to follow the evolution of the data center, both

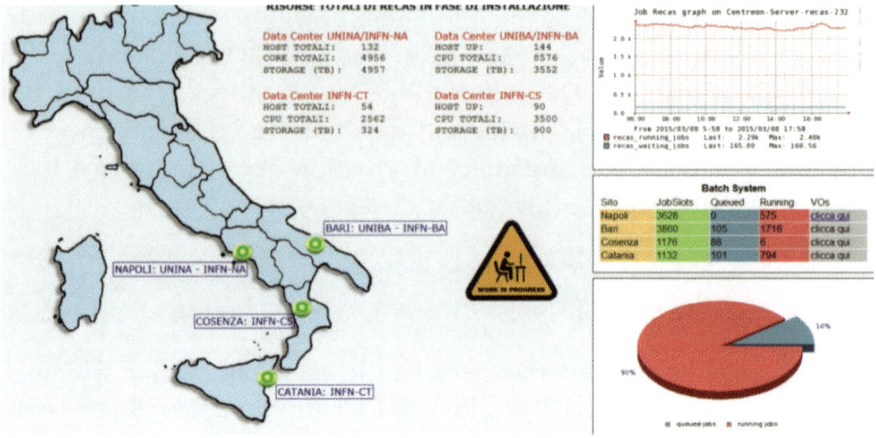

Fig. 1. The ReCaS monitoring dashboard.

locally and at a distributed level. The system we have created is based on a portlet container, thus allowing the integration of different functionalities and different subsystems of monitoring. The system provides the user with a global vision. The architecture is composed of macro components through a hierarchical navigation; this means you get to see all the services associated with the individual component up to the management of embedded devices.

The basic principle of the system is the modularity. Each site uses its own local system projects uses its own local system for monitoring; in more detail, Naples and Cosenza use Centreon, Bari and Catania use Zabbix.

Both systems use the same approach, based on the use of scripts to query hosts via standard protocols and interactive maps to represent information. We can access each of them through its entry point based on the HTTP protocol, but in order to maintain the concept of monitoring global view of the infrastructure, we created a single portal for monitoring. From this, the user will have access to the general map from which he can navigate to the details of the sites.

This portal (Fig. 2) is available at the following web link: http://www.pon-recas.it/web/guest/monitoraggio-infrastruttura.

Fig. 2. The ReCaS resources and services monitoring.

5. The Tools Used

The monitoring system consists of the following major subsystems:

- Nagios used for monitoring of all services,
- NagVis for interactive navigation,
- NaReTo used for the reporting of the monitoring data collected,
- Centreon used for advanced configuration of services,
- Weathermap configured for navigation of topological network and for the interactive data network monitoring.

The model chosen for the queries of services is based on subsystems NRPE (Nagios Remote Plugin Executor); it is delegated to each machine the responsibility of carrying out the plugin and not overloading the whole monitoring system.

6. The Interoperability Among the Various Sites

In the previous paragraph, we have explained the general architecture of the system. Now we will explain how the monitoring systems of the various sites interact with each other. In Fig. 3 we have shown the connection diagram between the sites; we can assume the monitoring system in Napoli as the central server and Bari as the focal point of the distributed system and Bari, Cosenza and Catania as external sites that sends information back to Napoli.

Two servers, physically located in the data center of Napoli, hosts the monitoring central server and the web server used to keep online the web portal of the ReCaS Project (http://www.pon-recas.it/).

This will be the access point from which users can view the general interface of the monitoring system (Fig. 4). The data shown by this interface are collected through two equivalent ways. The first one provides the use of JSON as the data interchange format, as we can see in this example:


```
{
"bari": {
"batchsystem": {
"availableslots": 3594,
"pendingjobs": 890,
"runningjobs": 2813
},
"storage": {
"available": 100432351280,
"total": 1560455344576,
"used": 1460022993296
},
"timestamp": 1394645404
}
}
```

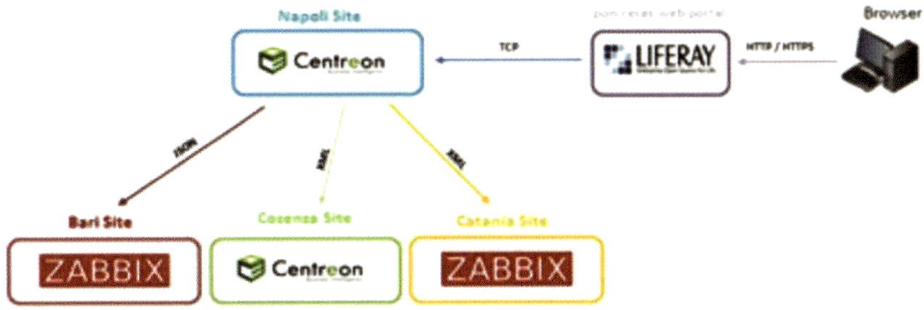

Fig. 3. The modularity and scalability of the system.

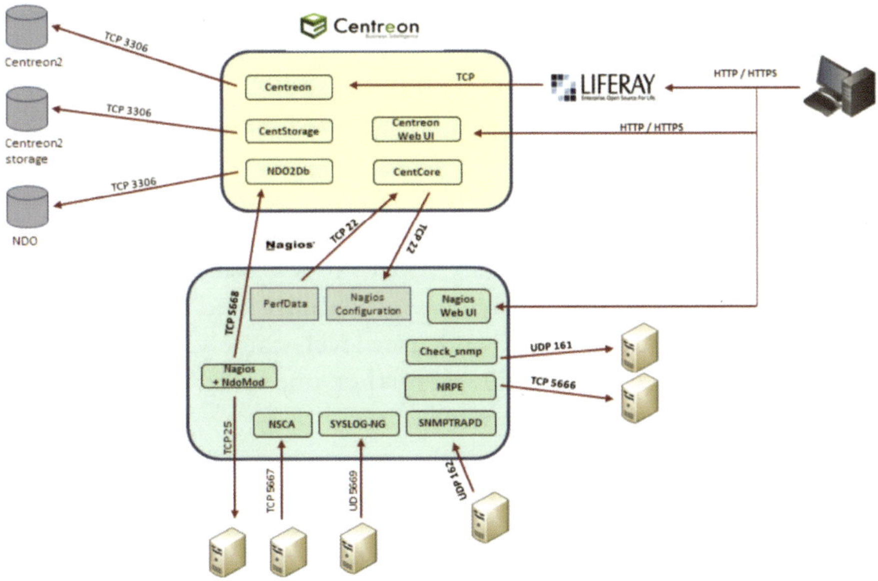

Fig. 4. The architecture and the interconnection between the system components.

The second one provides the use of XML, as we can see in the following example:

```
<bari>
<batchsystem>
<availableslots>3594</availableslots>
<pendingjobs>590</pendingjobs>
<runningjobs>2734</runningjobs>
</batchsystem>
```

```
<timestamp>1394645404</timestamp>
<storage>
<available>100432351280</available>
<total>1560455344576</total>
<used>1460022993296</used>
</storage>
</bari>
```


These files are stored on the local monitoring system of the external sites. The central system will implement the appropriate parser function which will include these information on webpages that will be displayed within the portal. Moreover the central monitoring system will store data in its databases as described before.

The interface between system and users is the Liferay Portal, that retrieves the various pages.

7. Conclusions

Using the tools that we have developed, it is possible to keep track and to monitor continuously the availability and the usage of the hardware resources at the end-user level, namely: jobs, storage, network traffic. This tool will be further improved to extend the functionalities, as the requirement arises.

Acknowledgment

This work has been funded by PON 2007–2013, project code PONa3_00052.

Chapter 9

IaaS Cloud Infrastructure@Bari: Implementation, Services and Use-Cases in the ReCaS DataCenter

M. Antonacci*, G. Donvito*,§, F. Giannuzzi†, A. Italiano*, S. Nicotri*, E. Tinelli†, R. Valentini*, F. Ventola† and G. Maggi*,‡

**INFN Sezione di Bari, Via Edoardo Orabona 4, 70126 Bari, Italy*

†Dipartimento Interateneo di Fisica, Università degli Studi di Bari, Via Edoardo Orabona 4, 70126 Bari, Italy

‡Politecnico di Bari, Dipartimento Interateneo di Fisica, Via Edoardo Orabona 4, 70126 Bari, Italy

§giacinto.donvito@ba.infn.it

In this paper, a general description of the cloud infrastructure Cloud@ReCaS, deployed at Istituto Nazionale di Fisica Nucleare (INFN) — Sezione di Bari and Physics Department, University of Bari, within the *Rete di Calcolo per SuperB e altre applicazioni*

(ReCaS) and *PiattafoRme Interoperabili per lo SMArt government* (PRISMA) national projects is presented.

The services are provided following the Infrastructure as a Service (IaaS), Platform as a Service (PaaS), and Software as a Service (SaaS) models, and their general features are shown.

Covered use-cases and communities involved in Cloud@ReCaS are described.

1. IaaS Platform Implementation

Part of the hardware resources of the *Rete di Calcolo per SuperB e altre applicazioni* (Computing network for SuperB and other application, ReCaS)[1] Data Center is offered as virtualized services through a modern *Infrastructure as a Service* (IaaS)[2] cloud platform: its main elements, hardware and software, have been selected and supplied in the framework of the two relevant PON projects, ReCaS and PRISMA[3] respectively, and jointly set-up by Istituto Nazionale di Fisica Nucleare (INFN) Sezione di Bari[4] and University of Bari.[5]

The software solutions aimed at the realization of a complete and robust IaaS platform have been selected from the open-source marketplace as a result of a deep study of the state-of-the-art of cloud and virtualization technologies. A preliminary comparison analysis of similar solutions has been conducted in order to assess their benefits and limits through intensive tests. As a result, the whole Cloud@ReCaS platform has been built around open-source components that have been integrated in order to provide highly-available, resilient and flexible solutions.

The IaaS platform is based on a highly customized deployment of OpenStack[6] as shown in Fig. 1.

The IaaS platform is composed of:

— Three "management" nodes, with highly available API, scheduler, storage and network services distributed over them. All the daemons are configured as resources in a cluster managed by the software Pacemaker/Corosync,[7,8]

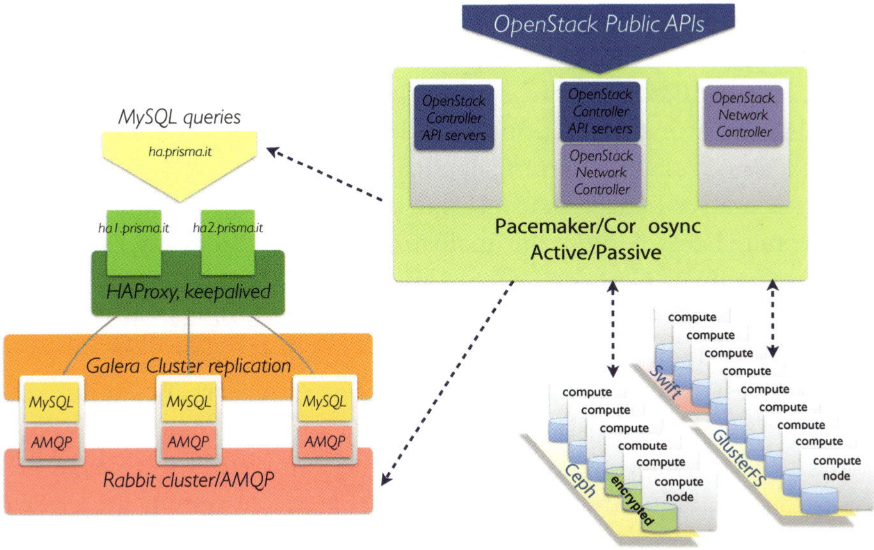

Fig. 1. IaaS Cloud Platform — Deployment. High availability has been implemented for both the backend services (the message broker and the database) and all the OpenStack core services. In case of failure, the stand-by node is automatically elected as the new active node. The storage is distributed over all the compute nodes and replicated with a factor 3 in order to provide high availability and scalability.

— Twenty KVM[9] hypervisors, on which the virtual machines run,
— Three nodes providing a highly available message broker RabbitMQ,[10] with queues mirrored on all the three nodes, and the Galera[11] cluster for MySQL. The client requests are served through two load-balancers based on HAProxy [12] and KeepAlived[13] exposing a virtual IP.

Each management node has 80 Gb of RAM, 24 Intel(R) Xeon(R) CPU E5-2630 0 @ 2.30 GHz, 6 2 TB SATA hard drives (some of which are used for distributed storage).

Each Hypervisor has 256 Gb of RAM, 32 Intel(R) Xeon(R) CPU E5-2650 0 @ 2.00 GHz, a principal 2 TB SATA hard drive and seven additional 4 TB SATA hard drives (some of which are used for distributed storage).

Each one of the three nodes providing the AMQP and MySQL services has 16 Gb of RAM, 8 Intel(R) Xeon(R) CPU E5410 @ 2.33 GHz, and 2 TB SATA hard drive.

Also the storage solutions have been chosen in order to ensure high levels of reliability and scalability.

As a matter of fact many cloud operators use separate compute and storage hosts. Unlike this mainstream trend, the storage solutions implemented at Cloud@ReCaS aim at the best cost performance scenario: this is achieved running compute and storage on the same machines, being able to dedicate as many hosts as possible to running instances. Each compute node is configured with a significant amount of disk space and a distributed file system (GlusterFS[14] and/or Ceph[15]) ties the disks from each compute node into a single file-system. In this case, the reliability and stability of the shared file-system are critical and define the effort to maintain the compute hosts. Tests have been performed to assess the stability of the shared file-systems changing the replica factor.

Severe performance and stability issues have been experienced using GlusterFS for Libvirt[16] and block devices backend in the production environment: many users complained about problems with their VMs frequently experiencing I/O errors and filesystem going in read-only mode.

These problems have been mitigated using a Ceph cluster for:

— storage of the running VMs enabling the live migration,
— storage of the virtual images (as primary Glance image store),
— implementation of several Cinder backends for block devices.

Currently, tests are ongoing on RBD and Ceph-FS protocols to run virtual machines and to implement the object storage.

2. Basic Cloud Services (IaaS)

The Cloud@ReCaS platform provides several basic services in order to support use-cases coming from both the local users and the EGI Federated Cloud[17] users.

Multiple authentication mechanisms based on the Identity Service implemented by OpenStack (Keystone) are provided: username/password credentials, X509 certificates, SAML2 Idp/LDAP external authentication. In particular, the module that allows the VOMS authentication[18] has been integrated and configured in order to extend the Keystone capabilities exploiting the external authentication method. The installation of this plugin is required inside the European Grid Infrastructure (EGI) Federated Cloud in order to grant access to the site cloud resources to the users of the supported Virtual Organizations (VO).

Cloud@ReCaS also provides different types of storage with related features and Quality of Service (QoS) that can fit specific use cases:

- The ***Block Storage*** service (based on OpenStack Cinder) allows the on-demand provisioning of volumes (block devices) that can be directly attached to the virtual machines; block storage tiering has been implemented in order to deliver different levels of service in terms of: performance, replica, I/O throughput and/or advanced features like data encryption. Users can choose among different volume types selecting the one that best fits the application requirements,
- The ***Object Storage*** service (based on OpenStack Swift) provides disk space where users can store unstructured data (virtual images, backup, video files, web content, etc.).

Users can access their data stored on the Object Storage using multiple protocols and application programming interfaces (APIs). Along with the native REST API implemented by Swift, Amazon S3[19] compliant API and Cloud Data Management Interface (CDMI)[20] interfaces have also been enabled on top of the Swift cluster in order to support application portability and cloud interoperability.

Cloud@ReCaS implements multiple network configurations, allowing the users to create private networks inside their tenants (ensuring isolation), allocate and associate floating internet protocols (IPs) and/or use directly public IPs.

3. Advanced IaaS Services

Data are important for businesses of all sizes. Therefore, one of the most common user requirements is the possibility to backup data, in order to minimize their loss, stay compliant and preserve data integrity. Implementing this feature is particularly challenging when the users come from the scientific communities that produce huge quantities of heterogeneous data and/or can have strict constraints. Currently OpenStack does not provide a complete solution for backup and disaster recovery. To address this limitation, simple homemade tools have been implemented at Cloud@ReCaS that take care of:

— performing automatic incremental backups of the Ceph pool used by Cinder (using the export-diff/import-diff feature),
— performing automatic backups of the Virtual Machines and attached block devices according to a flexible schedule that foresees also the rolling of the older items.

Moreover, a tool for the automatic full backup of the data stored by DBMS like MySQL and MongoDB[21] has been developed: the files containing the backup are compressed, encrypted if requested, and stored on the Object Storage.

Another frequent important requirement on data management concerns data security: some use-cases (for example the e-health project IPPOCRATE Living Lab[22]) involve the usage and manipulation of sensitive data that must be encrypted in order to be secured. To this purpose, users at Cloud@ReCaS are able to choose the volume type, encrypted or unencrypted, when they request the creation of a block device to be attached to a virtual machine. When encryption is requested, data are written to devices stored on a pool that supports it. Different kinds of encryptions are available, in order to support a wide range of use-cases. It is possible either to leave to the platform administrator the possibility of decrypting the data, or to make the encryption completely not accessible to others. In the first case, a stronger user support is possible, still keeping data secure, while the second case is more suitable for situations in which sensitive data are stored, whose access must be strongly restricted, e.g., as in medical applications.

Advanced virtual network services are implemented, as Load Balancing as a Service (LBaaS) and Virtual Private Network as a Service (VPNaaS).

LBaaS relies on the correspondent Neutron plugin and on HAproxy, which provides the end user with the capability of transparent load balancing the incoming traffic to different virtual machines (VMs) using all the most common protocols (TCP, HTTP, HTTPS, etc.).

Three balancing methods are currently available to distribute the 10 traffic among instances: (1) round robin, in which requests are simply rotated, (2) source IP, in which traffic is routed depending on the source address, and (3) connections, in which traffic is redirected on the instance with lowest load.

The service supports connection limits, to control the overall load, session persistence, and monitoring of the health of pool members.

VPNaaS allows to securely connect geographically distributed sites in a user-friendly way.

The service makes use of the Internet Protocol Security (IPSec) suite and Internet Key Exchange (IKE) Policy with advanced cryptography algorithms (3des, aes-128, aes-256, and aes-192), sha1 authentication and ESP, AH o AH-ESP transform protocols.

4. Platform as a Service

Cloud@ReCaS provides also high-level orchestration services based on the OpenStack Heat component that ease the interaction with the IaaS services through an orchestration engine able to deploy even complex applications (composed of many servers, load-balancers, volumes, etc.) instantiating templates written in a declarative language. Templates have been developed for some relevant use-cases and made available to the users:

— Database as a Service (DBaaS): users can easily deploy a MySQL or PostgreSQL[23] DBMS together with the management tools *phpMyAdmin* or *phpPgAdmin* respectively,

— Business Intelligence as a Service (BIaaS): users can easily deploy the Pentaho[24] Data Analytics suite including plugins like Saiku and client tools like Kettle Data Integration and Report Design,
— Content Management System as a Service (CMSaaS): users can easily install and configure a Content Management System (CMS) like WordPress[25] to develop their web sites.

5. Software as a Service

In order to serve scientific communities, as well as enterprises, the cloud IaaS infrastructure deployed at the ReCaS DataCenter located at INFN — Sezione di Bari, and in the physics department of the University of Bari, also provides a number of higher-level services, most of which are implemented within a Software as a Service (SaaS) paradigm.

A complete *Remote Desktop solution* has been realized, based on Linux and X2Go,[26] an open source tool for remote desktop management, which allows to create multiple desktop (or just per-application) sessions on the same instance, as well as actual desktop connections. In this way, the end user can access her/his workstation from any thin client connected to Internet.

A second implemented service is a *personal storage* solution *à la* Dropbox, based on the open source software OwnCloud.[27]

The service is built upon an OpenStack Cinder 1TB block storage backend, has a user friendly web interface and a desktop client, can be accessed through the WebDAV28 protocol, and is well integrated with Swift, allowing to link object storage and personal storage accounts.

The flexible scalability of the block storage backend is transferred to the upper level, the personal storage service, which can then host a very high number of accounts, serving the scientific community of a large sized university and beyond.

Multiple instances of *RStudio Server*[29] are available in production, with 32 CPU cores and 64 GB of RAM, which provide web access to a personal account of the R programming language interface RStudio.

In the same way, instances of the web Python framework *IPython Notebook*[30] are provided on demand.

Currently, Heat templates are under development for the management of such services, in order to increase scalability and elasticity of those services, taking into account auto scaling and resistance to failures.

In order to provide the software development community with a version control tool, a web-based *Git repository hosting service* has been deployed, based on the open source software Gitlab,[31] which provides web access to public and private repositories, with wiki and issue tracking features.

Last, a collaborative platform for LaTeX project has been realized, based on the open source software ShareLaTeX.[32]

6. Use-Cases Overview

IaaS, PaaS, and SaaS frameworks are currently used by a consistent number of heterogeneous users, coming from different realities, such as universities, research institutions, and enterprises.

The Desktop as a Service framework deployed in the Cloud@ ReCaS infrastructure and the SaaS version control tool Gitlab, are currently used by Postdocs, Researchers and Professors of the University of Bari, particularly in the Physics, Maths, and Computer Science Departments. One important goal consists in offering, in the close future, a complete enterprise-level virtual computing environment to every Ph.D. student, researcher or professor affiliated to the University of Bari.

Personal Storage as a Service and Object Storage are currently used by local academic users, as well as by public/private partnerships of universities and enterprises working on e-Learning, to host personal and public data.

Cloud@ReCaS offers advanced data analysis and programming environments, like Rstudio Server and IPython Notebook Server, to many scientific communities covering education (scientific computation graduate courses of the physics department of the University of Bari), and research in many fields, such as bioinformatics, genetics, financial data analysis, medical imaging and signal analysis. Particular effort is being put in supporting the high energy physics community of the LHC experiments ALICE and CMS, also within the EGI community, offering elastic data processing and Monte Carlo production tools, collaborative scientific writing environments (ShareLaTeX), as well as collaborating in the deployment of Virtual Analysis Facilities.[33]

Scientific communities can access the cloud resources of the ReCaS DataCenter also through the EGI Federated Cloud, which is a seamless grid of academic private clouds and virtualized resources, built around open standards and focusing on the requirements of the scientific community.

From a technical point of view, EGI Federated Cloud users can authenticate against the identity service exposed by the Cloud@ReCas using their X.509 certificate (the same used for Grid services) thanks to the Keystone VOMS extension and access the compute and storage resources using the standardized interfaces, namely Open Cloud Computing Interface (OCCI)[34] and CDMI.

The cloud site deployed at the ReCaS DataCenter is fully integrated with the EGI Accounting System based on Apel[35] and with the EGI Information System: the service endpoints of the cloud site are registered on the Grid Configuration Database (GOCDB) and site information (including the list of supported virtual images, the list of VM flavors, etc.) is published by the site BDII (Berkeley Database Information Index) in order to be availed by users who want to start VMs and/or allocate storage space on the cloud site.

Currently, the supported EGI use-cases involve many fields: high energy physics experiments, mainly from CMS VO; biodiversity research like the BioVel project[36]; hydrometeorology, like the DRIHM project[37] that provides services for predicting risks related to extreme flooding events; execution of scientific code (R and Octave jobs) through the Chain-Reds Science Gateway.[38] A particular effort has been spent to support the CSC team from the Finnish Elixir node in order to enable the Chipster[39] suite in the Federated Cloud environment. Chipster contains over 340 analysis tools for next generation sequencing (NGS), microarray, proteomics and sequence data. Users can save and share automatic analysis workflows, and visualize data interactively using a built-in genome browser and many other visualization tools.

7. Conclusions

In this paper, details have been given about the implementation, deployment and use of the medium/large scale cloud infrastructure Cloud@ReCaS, realized at INFN — Sezione di Bari and Physics Department, University of Bari, within the ReCaS and PRISMA national projects.

Features and services have been shown, provided along the IaaS, PaaS, and SaaS paradigms, aimed at supporting an heterogeneous set of communities such as enterprises, universities, research centers, and national and international organizations, as the European Grid Initiative.

The principal use-cases have been described, including data analysis and processing in the medical field, high energy physics experiments and bioinformatics, providing resources and services for research in computer science, educational purposes and eLearning.

References

1. Available at http://www.pon-recas.it/.
2. P. M. Mell and T. Grance. 2011. *SP 800-145. the NIST Definition of Cloud Computing,* Technical Report, (NIST, Gaithersburg, MD, United States).
3. Available at http://www.ponsmartcities-prisma.it/.
4. Available at http://www.ba.infn.it/.
5. Available at http://www.uniba.it/english-version.
6. Available at https://www.openstack.org/.
7. Available at http://clusterlabs.org/.
8. Available at http://corosync.github.io/corosync/.
9. www.linux-kvm.org/.
10. Available at https://www.rabbitmq.com/.
11. Available at http://galeracluster.com/products/.
12. Available at http://www.haproxy.org/.
13. Available at http://www.keepalived.org/.
14. Available at http://www.gluster.org/.
15. Available at http://ceph.com/.
16. Available at http://libvirt.org/.
17. Available at https://www.egi.eu/infrastructure/cloud/.

18. Available at https://keystone-voms.readthedocs.org. aws.amazon.com/s3/.
20. CDMI, "*Cloud Data Management Interface*", Version 1.0.2, June 2012. Specification, available at <http://www.snia.org/cdmi>.
21. Available at https://www.mongodb.org/.
22. Available at http://livinglabs.regione.puglia.it/web/guest;jsessionid=1962A4C98A7E634E960C97E3A8AEC4EE.
23. Available at http://www.postgresql.org/.
24. Available at http://community.pentaho.com/.
25. Available at https://wordpress.org/.
26. Available at http://wiki.x2go.org/doku.php.
27. Available at https://owncloud.org/.
28. Available at http://www.webdav.org/.
29. Available at http://www.rstudio.com/
30. Available at http://ipython.org/notebook.html.
31. Available at https://about.gitlab.com/.
32. Available at https://sharelatex.com/.
33. F. Colamaria *et al.*, The Italian PON ReCas Project, *VAF: A Virtual Analysis Facility exploiting PRISMA OpenStack Infrastructure at Bari*, World Scientific Publishing, Singapore.
34. OCCI, "Open Cloud Computing Interface," Version 1.1, June 2011. Specification, available at http://occiwg.org/about/specification/.
35. Available at https://wiki.egi.eu/wiki/APEL.
36. Available at https://www.biovel.eu/.
37. Available at http://www.drihm.eu/.
38. Available at https://www.chain-project.eu/.
39. Available at http://chipster.csc.fi/.

Chapter 10

A Cloud Environment for Catania ReCaS Site

G. Andronico*, S. Monforte, G. Platania and M. Fargetta

Istituto Nazionale di Fisica Nucleare,
Sezione di Catania, Italy
**giuseppe.andronico@ct.infn.it*

The *Rete di Calcolo per SuperB e altre* (ReCaS) project made available new and powerful resources to Catania data center, increasing the needed maintenance level. Some of the new problems are fronted, developing a new, powerful monitoring system, able to get information from almost every equipment part of the global data center. The monitoring system can support system administrator making evident problems, equipment involvement and, if the CAE, dispatching appropriate warnings.

A different set of management issues are related to managing resources itself. Nowadays a current approach to handle such a sort of management issues is based on cloud computing. In this paper, we report about Catania data center peculiarities and how we are planning to make use of them to develop a robust, affordable and resilient cloud environment.

1. Introduction

The *Rete di Calcolo per SuperB e altre applicazioni* (Computing network for SuperB and other applications, ReCaS) ReCaS project, described in Chapter 1, made possible to establish the Catania Data Center both in the basic infrastructure (see Chapter 4) and in the resources added. The new data center is able to satisfy more requests from the world of research, but it is more demanding in terms of management.

As far as the infrastructure is concerned, every equipment in the data center is part of a complex game and every problem in whatever small part of the equipment can have a remarkably dangerous effect for the data center business continuity. Reliability and resilience require continuous monitoring of every apparatus, usually by means of software systems able to obtain status information from all the equipment, to show summaries and to issue warnings if it is the case.

For the resources, namely computing power, storage, and network, monitoring is not enough. Every time a user needs resources to confront a problem usually these come with a specific setting and system administrators have to deploy and set up what is required. The more resources available the bigger the amount of work to manage them answering to user requests. Examples, in our case, come from system administrator requirements to specific requests from researchers.

For system administrators usual requirements are connected to services provided, and came from simple network services, as DNS, to complex platforms, as some e-mail solutions that requires not only computational power and storage but also specific network resources and settings. Moreover, services before being deployed in production should to be tested in controlled and isolated environment.

On the researcher's side, usual requirements are computer power, storage and simple ways to execute applications. This request can be usually satisfied with clusters, but sometimes applications can either introduce software conflicts, for example version of given libraries, or require hardware or software special environment, as parallel programming. Moreover, some large collaboration have specific requirements on minimum values for computing power, storage and network bandwidth.

To answer in a reasonable amount of time with the minimum of costs, the data centers are always using cloud computing. This model is largely based on resource virtualization and on demand provisioning. Key constituents are virtualization systems as KVM, XEN or VMWARE and other software layers, implementing automatic provisioning. Recently another part, Software Defined Network, was inserted to handle network virtualization and provisioning.

We developed a plan to deploy such a solution in the new Catania Data Center.

1.1. *The Catania Data Center and the Network*

The data center we have at the *Istituto Nazionale Fisica Nucleare* (INFN) division of Catania is made from two computing rooms. The bigger computing room, containing almost all the new ReCaS resources is called *Sala Grid* and is located in the basement of the building. This room is served by all the needed services, such as UPS, chillers and diesel generator and hosts around 31 racks where storage and computers are accommodated. Racks are organized in a linear set of 10 racks and 2 isolated aisles. The other computing room is placed on the third floor of the building and mainly contains resources dedicated to local services. It requires smaller chilling power (provided by standard conditioners) and share UPS and diesel generator with *Sala Grid*. This room hosts six racks to accommodate storage and computers and three racks for network and phone equipment.

The two rooms are independently connected to the Italian Research and Education Network that the Consortium GARR supplies the INFN division of Catania with. Moreover, more than 12 optical fibers certified to work on 100 Gbps links are used to connect the rooms with each other.

In Fig. 1, the connections scheme is as follows.

The presence of two independent links gives us some opportunities. Firstly, it allows the configuring of two border routers to work in fault tolerance. To such an aim, we are working in collaboration with Consortium GARR to implement BGP between our apparatus and the ones in the local Point of Presence (POP). In this way, if one of the routers suffers a fault for technical or management reasons, all the

Fig. 1. Scheme of network connections between the two computing rooms and the external network.

traffic can be rerouted through the remaining one. In our plans, once this BGP is up and working, we will study how to rearrange links and have the two devices working in load balancing. At the same time, we are working to set up an internal BGP using the direct links between the two routers. This internal link is present to ensure direct connections between local LAN in *Sala Grid* and the other local LAN, and with the internal BGP we can afford a better level of network virtualization.

1.2. *The Cloud Infrastructure*

Our choice for the cloud infrastructure was OpenStack, given the enormous momentum it gained in the last few years, and the enormous active community users who access its support and from which we could confront new problems.

OpenStack is made from several services implementing different services that work synergistically. The basic services are:

- Keystone, the authentication and authorization service,
- Nova, the virtual machines (VMs) scheduler towards the OpenStack Computes,
- Glance, the cloud storage system where VM images and templates are saved,
- Cinder, the cloud storage system providing on demand storage to users for VMs,
- Swift, object storage for users,
- Neutron, the network element handling network provisioning.

It is known how to install and setup these services in High Availability (HA), and we want to take advantage from the two computing rooms to try to ensure some sort of fault tolerance. The idea is to have two sets of physical machines, each of them equipped with Citrix XenServer to implement a safe virtualization environment where to deploy VMs with OpenStack basic services. The counterparts in the two systems are configured in HA Active–Active. This

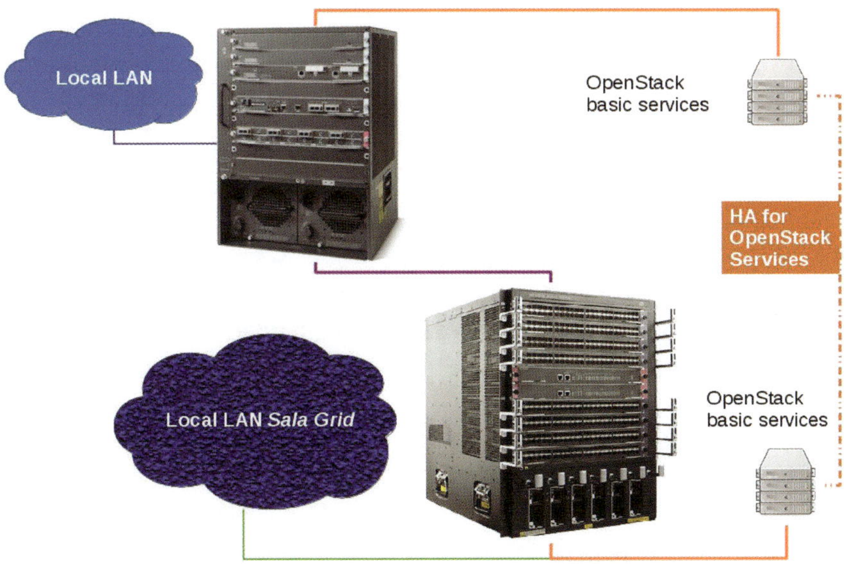

Fig. 2. Network schema and OpenStack basic services.

means that, for example, there are two instances of Keystone, one for every virtualization system, working in HA Active–Active. The same is for all the other services. At the same time, the two systems are located in the two computing rooms, so that if something happens to one of the computing room, OpenStack basic services are yet available.

This set-up is not so easy to deploy, and needs some work on the network side. In detail, to better exploit the network, we are using the Distributed Virtual Router service with Neutron, enabling us to better usage of network bandwidth.

This redundant system controls all the computers, that is the physical machines hosting VMs dispatched from OpenStack and the storage. Network requests will be answered by means of VLAN so the traffic can be separated. In this way, when an existing or new user asks for new resources, they can be provided and the network arranged to fulfill specifications. To automatize this part of the work we are using Horizon, the web interface of OpenStack, and are planning to use the Catania Science Gateway Framework to implement a Science Gateway to make simpler the cloud usage.

2. Conclusions

The ReCaS project gives us the opportunity to potentiate our data center. To simplify the management of the data center, we are deploying a cloud infrastructure in order to automatize most of the provisioning and obtain a better fault tolerance level that would be able to answer to greater kinds of requests in simpler ways.

This already required a big amount of work to understand, plan and start the deployment. More work will probably be needed to operate this solution and handle updates, but certain other activities were simplified and we can concentrate on adding functionalities and developing new features for users.

Chapter 11

User Recruitment and Support: A Viral Marketing Approach Based on the Word-of-Mouth Influence

G. B. Barone[*,§], V. Boccia[†], D. Bottalico[*], R. Campagna[*,†] and L. Carracciuolo[‡,¶]

[*] *University of Naples Federico II, Center of Information Services, Naples, Italy*
[†] *National Institute of Nuclear Physics, INFN Seat of Naples, Italy*
[‡] *National Research Council, Naples, Italy*
[§] *gbbarone@unina.it*
[¶] *luisa.carracciuolo@cnr.it*

The deployment, management and total cost of ownership of large computing environments always involve huge investments. These systems, once in production, have to meet the needs of users belonging to large and heterogeneous communities whose demands are often conflicting. Only an efficient and effective use of these systems can pay back the investment carried out. The infrastructure needs to be sustained also by means of the building of an "appropriate" community of users (the "catchment area"), to be, firstly, recruited and then driven to a conscious use of resources. Based on the experience we gained at the University of Naples Federico II, in the activities related to "support users and applications" on the SCoPE system, we "enjoyed" to write our "recipe". Since 2008, also in the framework of the ReCaS project, that recipe represents our way to recruit new users and to consolidate the "catchment area" in terms of both numbers and awareness in the use of the infrastructure.

1. Introduction

The management of complex infrastructures for computing and storage requires to carry out many activities. Among these, are certainly considered all those operations that are intended to ensure the efficiency and effectiveness of the entire system. These operations do not include only actions related to the *mere* functioning of computing and storage services (to cite somewhat), but all the activities aimed to "optimize" the use of the infrastructure (from the point of view of both the system manager and the final user) whose users belong to large and heterogeneous communities whose demands are often conflicting.

All said, the support of users on complex infrastructures is critical and not simple:

- there is no single yardstick of goodness and management of such a service,
- there is no formula "equal for all",
- all the users have to be satisfied,
- the infrastructure has to work efficiently to be sustainable.

Who manages and supports large and heterogeneous users communities on complex infrastructures provides one-of-a-kind operational effort and needs heterogeneous areas of expertize to provide a highly flexible service. Similar experiences, in international contexts, seem to validate as success stories approaches that are, such as ours, multi-level, and multi-disciplinary.[2]

The paper describes the strategies and results that emerged after the experience we gained at the University of Naples Federico II, in the activities of "support to users and applications" on the SCoPE system, during the homonymous project.

Section 2 refers to how it has managed the users support during the SCoPE project and how the events have imposed that something had to change in user support after the end of the project; Section 3 describes the "recipe" that we use to recruit and support users and the best practice to be followed in this sense. Section 4 refers to results and conclusions.

2. The Origins: The SCoPE Project and the Support to the Involved Scientific Communities

The SCoPE project started, at University of Naples Federico II, at the beginning of the year 2006 with the aim to create a supercomputing general purpose infrastructure, based on the paradigm of Grid Computing and the latest technologies of distributed computing, to support basic research and small and medium enterprises activities (SME).[1] The project was completed within three years, with the construction and the deployment of a data center which is now completely integrated in national and international GRID infrastructures (IGI, EGI[5,6]).

The projects identified as "strategic" the following four scientific research areas:

- Particle Physics,
- Astrophysics,
- Materials and Environmental Science,
- Life Sciences.

Two SCoPE Work Packages (WP3 and WP4) were respectively dedicated to:

(1) the basic level user support and management (account creation, authentication/authorization management, etc.),
(2) the advanced level user support to develop innovative codes for the SCoPE computing environment and to integrate them into applications codes of the above cited scientific users.

The WP3 implemented some strategies to support and to train the users of the SCoPE infrastructure through tutorials with a demonstration of first access. Moreover, generalist workshops were periodically organized to provide technical information about the infrastructure usage (e.g., how to access, execute application, and manage data).

In the WP4, a different and more advanced support was "experimented" in order to:

- understand the characteristics of applications of interest for each community,
- identify the needs of various applications in terms of hardware and software resources.

In this context, an "application form" was prepared and submitted to users to collect all the above information in an organized way. This activity was designed to guide and support the involved scientific communities towards using the computing platform in an effective and efficient way.

The support activity was delivered by a group of people, with different areas of expertize: from the system to the middleware management, from the design to the implementation of algorithms for HPC and distributed environments. The "supporters" were able to provide organic answers to the different needs of each community also in terms of scientific software to make available on the computing resources.

We called "SCoPE Toolkit" the overall set of such software that includes libraries, software packages and other tools necessary to execute users applications. The preliminary release of the SCoPE Toolkit was realized thanks to the results of a survey, carried out under the PON SCoPE WP4 with the aim to identify the initial set of libraries, packages and other software tools.

2.1. *Toward the "Recipe" for User Recruitment and Support*

During the SCoPE project, periodic meetings with users were scheduled accordingly to the project rhythms but, at the end of the project, the interest in the massive and constant use of the infrastructure, mainly induced by the timing of the project, decreased physiologically.

SCoPE didn't expect any recruitment activity for users to guarantee the sustainability of the infrastructure (the users communities were already identified and involved by the project objectives).

Really *the continued support for the user facing services will depend much more on the commitment the users of those individual services are willing to make to their support.*[3] In other terms, for large infrastructures, services and resources availability can be sustainable only if a large number of users require and use them.

So, an infrastructure, especially if born with a public financial support (that usually does not fund the post-project infrastructure maintenance), needs to be sustained by means of an "appropriate" catchment area (i.e., with a suitable number of users with respect to the available resources) also in prevision of participation to new funding calls.

Moreover, both to justify the public investment made and to ensure the usefulness of the infrastructure for 5 years or more after the end of the project, a "maintenance strategy" has to be realized.

So, at the end of the SCoPE project, the real challenge was to consolidate the already existing communities of users, while trying to widen the users' audience.

In the experience gained within the SCoPE project, self-referential and generalist workshops revealed themselves unsuccessful, since they are too focused on the description of the infrastructure (potentiality, services, mode of use, ...) and little or not at all *eye-catching* with respect to communities traditionally "far" from this type of computing environments. Such events are, in fact:

- too "technical" and they may scare potential new users because they could emphasize more the difficulties of using than the benefits that they may derive,
- unsuitable to truly know the needs of potential community ("tell me what you do and like") and then unsuitable to show them, case by case, the benefits that they could obtain from the use of distributed computing infrastructure ("what we can do to their").

Hence the need to *reinvent* the user support by developing a real "recipe" for the recruitment and management of the user communities.

With this aim, it was necessary to modify/integrate/consolidate the actual *modus operandi* to make available a system that could be considered *of production* and no more *project-aimed*.

So the key elements of this approach were and are:

- the creation of a multi-disciplinary group (the so called Gruppo Tecnico Trasversale — GTT), which, starting from the core support staff born within the SCoPE project, is able to evolve itself sharing different skills, both technological and scientific, to better meet the needs of different scientific communities; this group benefits also from people by Istituto Nazionale di Fisica Nucleare (INFN), Consiglio Nazionale delle Ricerche (CNR), Departments and Centers of the University Federico II. In an economic metaphor, the GTT represents the panel where the producers and consumers discuss the problems, and related solutions, by placing the heterogeneous experiences of everyone at the disposal of all to contribute to a global improvement for everyone,
- the editing of the Service Levels Agreements (SLA) document,[4] reporting the types, methods and timing of support offered,
- the consolidation of the applicative middleware through the continuous enrichments of the SCoPE Toolkit (the current release is the number 3) on the basis of new requirements expressed by users community,
- the enabling of several communication channels among the supporters and the final users (user guide, software guide, web site, skype contact, social networks, administrative email, etc.).

3. The "Recipe" of Unina to Enlarge the Catchment Area

All said, in this section we finally describe the strategy used to enrich the catchment area. We use *the viral marketing approach*, widely spread in Economics, essentially based on the presence of a "contact person" and on the following steps:

(1) ***Let's identify the contact person inside the community (the infection starts)*:** this action is useful to increment each existent users group by means a "contact person" who already is a SCoPE loyal user. These users are aware of the activities delivered by their scientific colleagues and speak "the same language". For this reason, they are the most effective "evangelists" because they act as spokesmen of their own positive experience with the infrastructure towards friends and/or colleagues.

(2) ***Let's organize a preliminary meeting*:** after the first contact with a new potential user community, the SCoPE staff organizes a preliminary meeting with the aim to know "what the users do and how they do it." All the obtained information will be used to identify all the community's application characteristics and needs.

(3) ***Let's build the support team*:** the SCoPE staff identifies, within the GTT, the set of people having skills to be able to give the most suitable answer to the needs expressed by the community.

(4) ***Let's draw up a feasibility study*:** the support team identifies and performs all the operative steps to support in the "best way" the new community on the infrastructure, e.g.,:

    1. By integrating new libraries and software in the actual applicative middleware.
    2. Eventually, by proposing alternative software solutions or configurations to exploit the infrastructure at the best.
    3. By making the community conscious about the chance to face new scientific challenges having more powerful computing resources and different computing paradigms.

(5) ***Let's be operative to deploy the user case study*:** the support team:

    1. Requests the community for a "meaningful" case study and deploys it on the infrastructure.

2. Informs the community about "what it needs to do to become an user" e.g., the invitation to — please carefully — read the start guide to become familiar with the infrastructure.

(6) ***Let's collect feedback*** from the new community in order to describe best practices to be shared with all users

1. By writing a GTT internal report, describing all the activities made and related the first results from the community.
2. Eventually, by updating all the user guides.

(7) ***The "infection" goes on...***: all the users belonging to the new community are new potential "evangelists" ... *the word-of-mouth is the basis for the "infection"!*

The activity described has been revealed essential to recruit new communities and to realize a "place" where communities can confront and also collaborate in interdisciplinary contexts.

In fact, once the new community has become loyal, it is involved in GTT's activities and this allows SCoPE staff to keep in touch with all the users ensuring the support activity continuation at all the support levels defined in SLA document.[4]

And if the contact person is not available? In this case, the SCoPE staff plays the role of the "infector" by participating at least some of the events organized e.g.,:

- by potential new communities,
- by vendors of application software.

Also in this case, *the word-of-mouth is the basis for the "infection"!*

4. Results and Conclusions

The strength of the used formula was, and remains, the fact that the provider of the support is a group of people with different skills, able to provide organic answers to the needs of each community.

The result of the described work has been a meaningful enrichment and diversification of the catchment area are shown in Fig. 1 in respect to the few research areas involved into the SCoPE project.

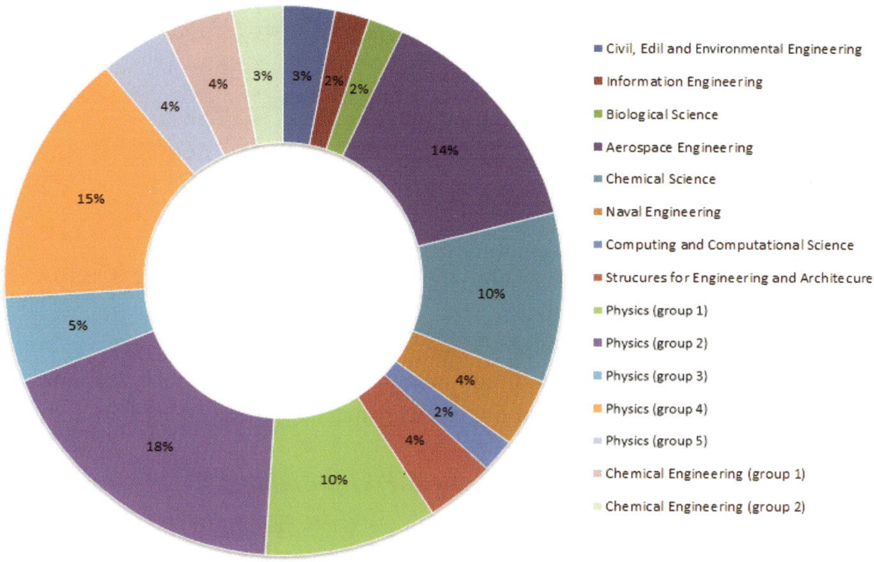

Fig. 1. Major scientific communities using SCoPE/ReCaS resources.

Anyway some of those *historical scientific communities* are yet SCoPE loyal users (e.g., some high energy physicians, some chemists and chemical engineers and computational scientists). Among them, we found the contact person who help us with the loyal new users. If we lost some of the *historical scientific communities*, other users from different scientific fields came up, thanks to the *word-of-mouth* within the University: naval and aerospace engineers, structural and civil engineers, information engineers and computing scientists, to cite the "major".

The wide catchment area of the SCoPE users can be considered one of the key points to support the participation of the University of Naples, with the ReCaS project, to the past "Research and Competitiveness 2007–2013, MIUR call 254/Ric." aimed to improve existing computing infrastructures:

- to capitalize on the results of the programming PON 2000–2006,
- to accommodate the growing need, from many areas of scientific research, to dispose of large computing resources and storage in distributed computing environments.

The ReCaS project obtained that funding, so confirming the efforts and the results achieved with SCoPE.

References

1. L. Merola on behalf of S.Co.P.E. Project, *The S.Co.P.E. Project*, in Proc. of the Final Workshop of the Grid Projects of the Italian National Operational Programme 2000–2006 Call 1575. Edited by Consorzio COMETA, 2008, pp. 18–35.
2. *The Caltech Advanced Computing Center mission and user support*, available at http://www.cacr.caltech.edu/main/?page_id=24, http://www.cacr.caltech.edu/main/?page_id=67, http://www.cacr.caltech.edu/main/?page_id=76.
3. *EGI Workshop on User and General EGI Sustainability* 24–26 January 2012, Amsterdam. Available at https://indico.egi.eu/indico/conferenceDisplay.py?confId=709
4. G. B. Barone, V. Boccia, D. Bottalico *et al., Services provided by the Data Center and service levels of user support*. Available at: http://143.225.20.120:8080/documents/10181/15460/SIS_sla_ev.pdf/38dcd0ac-0d50-4d57-bf56-61aa8d2ae0cd.
5. Italian Grid Infrastructure. Available at http://www.italiangrid.it/.
6. European Grid Infrastructure. Available at http://www.egi.eu/.

Applications

Chapter 12

Applications on the ReCaS Project Infrastructure

V. Boccia*,¶, M. Lapegna†, A. Monaco‡, S. Monforte§, S. Pardi* and A. Tarasio*

**National Institute of Nuclear Physics, INFN seat of Naples, Italy*
†Department of Mathematics and Applications University of Naples Federico II, Italy
‡National Institute of Nuclear Physics, INFN, Bari, Italy
§National Institute of Nuclear Physics, INFN, Catania, Italy
¶vania.boccia@na.infn.it

The cost to realize large distributed computing infrastructures is very high, but the role of these infrastructures for science is very important: indeed, each year a very big part of applied research is made by using computational simulations.

The ReCaS infrastructure is realized by means of the local infrastructures of four seats in Southern Italy, each of them already with its own user community, well supported locally.

Here, we describe the effort made to make the census of all users communities with the final aim to support them well not only locally but also on the entire distributed infrastructure.

1. Introduction

Here we report about the state-of-the-art overall applications using the *Rete di Calcolo per SuperB e altre applicazioni* (Computing network for SuperB and other applications, ReCaS) infrastructure.

This document provides a characterization of the users applications on the basis of the paradigm used to access the resources (direct, grid, cloud), the type of parallelism exploited by the applications (multi-core or multi-node) and related technologies (CPU versus GPU) and finally the required software layer. In Sec. 2 scientific applications for each of the four sites of the ReCaS infrastructure are reported, in Sec. 3 the tool used for the census is described while in Sec. 4 the census results are reported.

Finally, in Sec. 5 both the software layer, built to support well all the users on the whole infrastructure, and the user support strategies are described.

2. The Majors Users, Seat for Seat

2.1. *Naples*

Several applications run on the Naples infrastructure. Users that most intensively use systems to the following scientific fields:

- biology: study of genome of plants, animals, drug discovery,
- physics: seismology, quantum mechanics, statistical mechanics for complex systems, astrophysics,
- high energy physics: ATLAS, Belle, CMS, LHC experiments,
- civil and structural engineering: simulations to verify structural strength to earthquakes, study of the consequences of earthquakes,
- numerical analysis: models, algorithms and software for scientific computing applications,
- informatics: data mining, e-information retrieval, log analysis, information classification for Big Data,
- naval and aerospace engineering: design of ships and airplanes,
- fluid dynamics simulations,

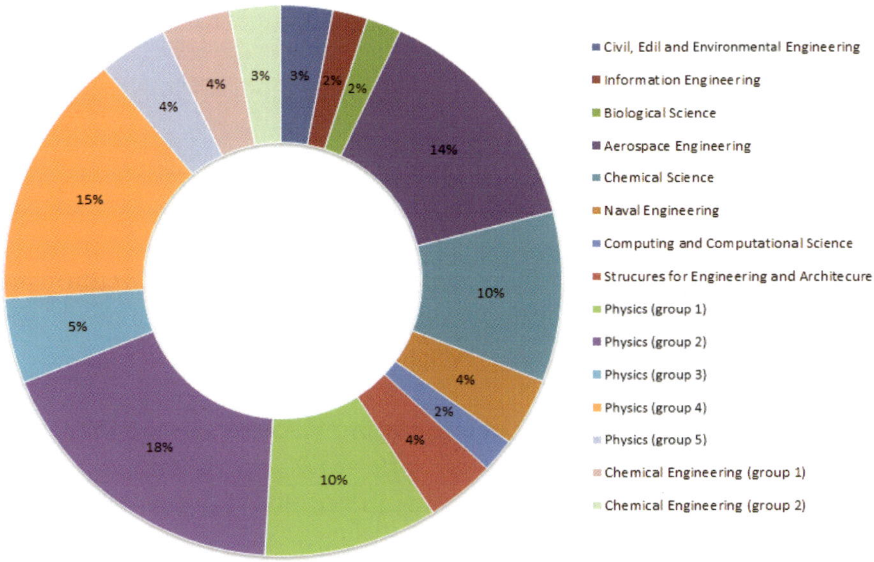

Fig. 1. Major users using resources of the Naples site: scientific fields.

- chemistry and chemical engineering: fluid dynamics studies, molecular dynamic, new materials discovery and study, *Ab-initio* modelling, inorganic computational chemistry.

Figure 1 represents the percentage of computing resources usage, detected during the last year.

2.2. *Bari*

Several applications run on the Bari infrastructure. Among them, those Users who most intensively use systems to the following scientific fields are:

- high energy physics experiments: five experiments ALICE, T2K, FERMI/GLAST, CMS, PAMELA,
- bioscience experiments: study of biodiversity, Next Generation Sequencing (NGS), phylogenetic analysis, simulations of biochemical

systems. Support of international experiments (Elixir, LifeWatch, Biovel),

- medical imaging analysis: hippocampal segmentation of 3D MRI brain scan, reconstruction and image recognition,
- theoretical physics applications: study of fluid dynamics and QCD, study of elastic and mechanical properties of disordered systems,
- other experiments: SAR images study, study and prevention of earthquakes, Weather forecasting, Numerical analysis, mathematical modeling of economic systems, Km3Net.

In Fig. 2 the percentage of computing resources usage, detected during the last year is represented. The HEP experiments use more than 80% of the CPU (3000) and about 90% (about 1500 TB) of storage.

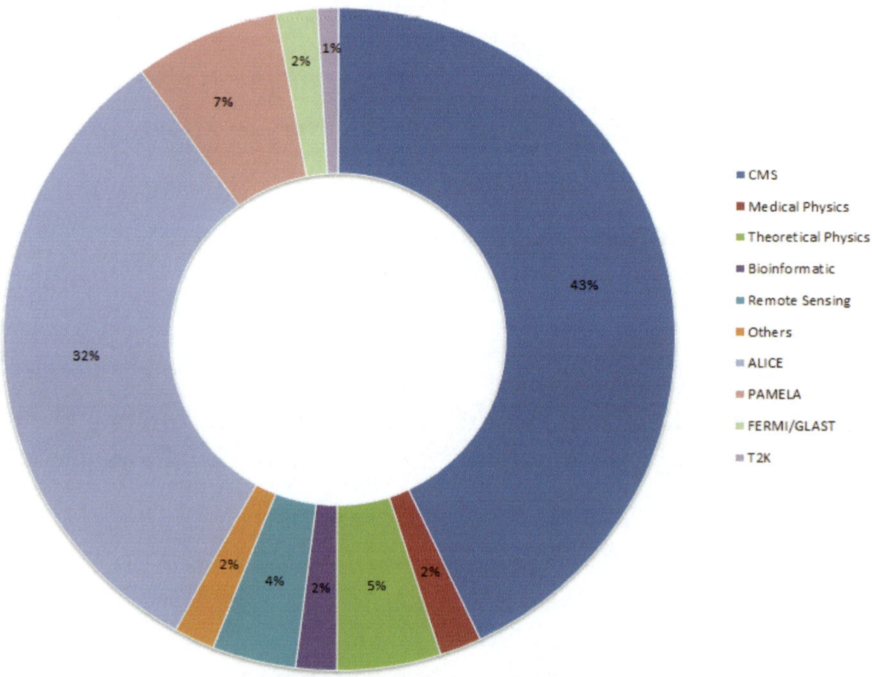

Fig. 2. Major users of the Bari site resources: scientific fields.

2.3. *Cosenza*

Several applications run on the Cosenza infrastructure. Among them, users that most intensively use systems to the field of high energy physics and include both the ATLAS and the Belle experiments.

2.4. *Catania*

Several applications run on the Catania infrastructure. Among them, users that most intensively use systems to the following scientific fields:

- high energy physics: ALICE, LHCB, and CMS are supported experiments,
- life and earth Sciences: most intensively used application among all the supported includes GROMACS to perform molecular dynamics simulations, GridSPM providing statistical parametric mapping analysis specifically designed for SPECT and PET neurological clinical images, WRF next-generation mesoscale numerical weather prediction system, FRATT for Hadronic models validation, ClustalW for multiple sequence alignment,
- agriculture: integration within the Catania Science Gateway of the WebGIS-enabled Italian Soil Information System (ISIS), developed by the Agrobiology and Pedology Research Centre of the Italian Agricultural Research Council.

3. The Tool Used for the Census: Limesurvey

LimeSurvey is a free and open source on-line survey application written in PHP distributed under the GNU General Public License that enables users to develop and publish on-line surveys, and collect responses, without doing any programming. LimeSurvey also provides basic statistical and graphical analysis of survey results. In addition, it is possible to export survey results to third party products (i.e., Microsoft Excel, IBM SPSS) for more in-depth analysis and reporting.

The survey, realized a stable service, is available online at: http://grid.ct.infn.it/surveys/index.php/291536/lang-it.

4. The Census Results

The survey showed that there are over 40 users groups. A first classification shows that half of the users belonging to at least one of the Virtual Organization, the other half is made from local users as shown in Fig. 3.

Among the users, 41% requires direct access directly to computing resources by means of a direct interaction with the Local Resources Management System (LRMS), 51% comes from GRID contexts and 8% need to use cloud resources (see Fig. 4).

Data show that 85% of the users' applications are based on code developed on their own using open-source software, while about 15% of users use commercial codes.

VO	Bari	Catania	Cosenza	Napoli
atlas	–	–	x	x
alice	x	x	–	x
lhcb	–	x	–	x
bessill	–	–	–	x
biomed	x	x	–	–
compchem	x	x	–	x
cms	x	x	–	x
belle	–	–	x	x
gridit	x	x	x	x
argo	x	–	–	x
virgo	x	–	–	x
pamela	x	–	–	x
biovel	x	–	–	–
glast.org	–	–	–	–
t2k.org	x	–		
aginfra	–	x	–	–
Km3Net.org	x	–	–	x
eumed	–	x	–	–

Fig. 3. The VO supported in each site of the ReCaS infrastructure.

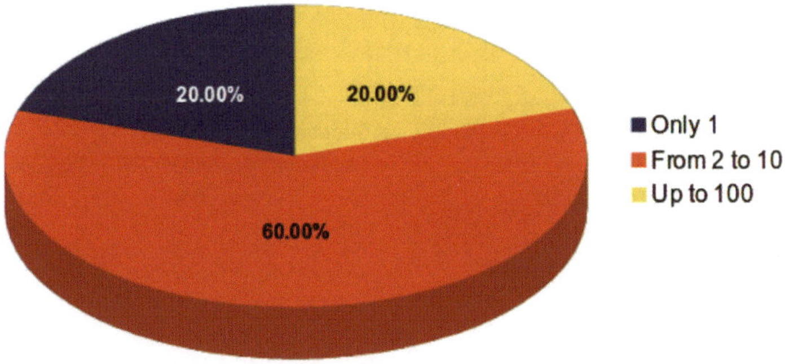

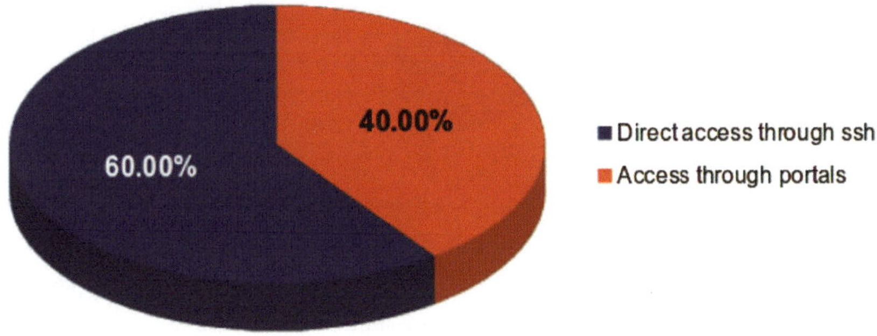

Fig. 4. From top to bottom: percentage of users requiring only 1, 2–10 and up to 100 VMs; percentage of users who prefer to access VMs through SSH or portals.

The local disk space required by the applications varies from a few GB to a dozen TB, while the local disk space request (for which is generally required POSIX access protocol or via SRM) can reach up to 1 PB.

The RAM requirement is 2 GB/core, while the number of all concurrent processors required is at the same instance is highly variable and depends on the strong heterogeneity of the detected applications.

Data show that while 65% of the applications are sequential, about 35% of users use parallel codes, implemented using classical and more modern paradigms (see Fig. 5).

Moreover, in Fig. 4 are shown Cloud Computing users' needs both in terms of number of required virtual machines (VMs) and in terms of preferences in the access to their VMs.

The complete set of scientific fields that, actually run their own applications on one or more ReCaS sites is shown in Fig. 6.

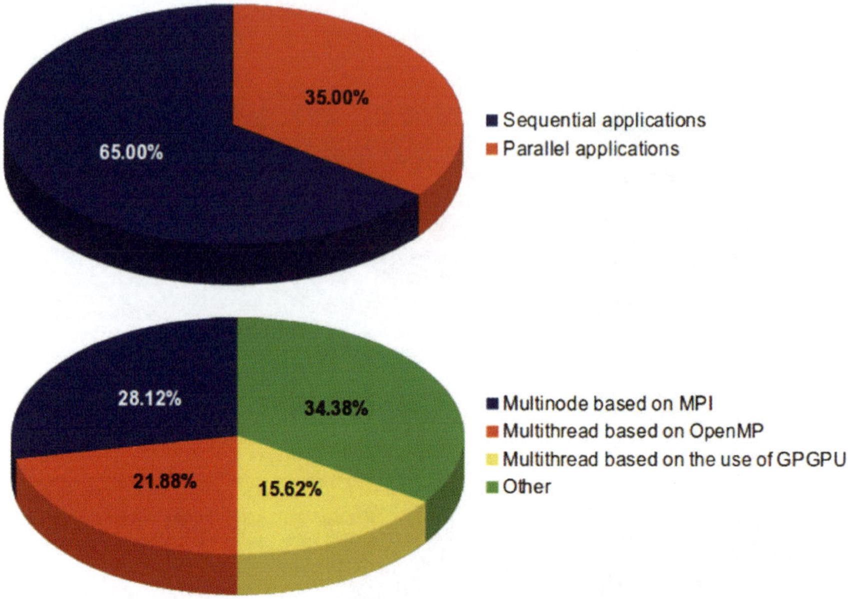

Fig. 5. Percentage of parallel applications and information about used paradigms (other means are a combination of different parallel paradigms).

4.1. *Cloud Users*

A Cloud Testbed is being set-up in Bari, Naples and Catania as part of the PON-PRISMA project.[2] This project, in fact, aims to develop an innovative cloud computing platform, starting from OpenSource component already available in order to achieve the realization of a complete IaaS+PaaS open platform, of large capacity, secure, robust, elastic and federated, capable of providing performance and features comparable to that of proprietary systems.

The Cloud Testbed offers:

(a) IaaS Services: the possibility to require specific services on VMs,
(b) "Cloud Storage": services based on WebDav and ownCloud;
(c) PaaS Services: Database, Application Server, Web Server;
(d) A system of project and ticket management based on REDMINE.

Scientific field	Institution					
High Energy Physics (HEP) and Astrophysics	ATLAS (INFN)	CMS (INFN)	ALICE (INFN)	LHCb (INFN)	BELLE 2 (INFN)	KM3NeT (INFN)
	FERMI/GLAST (INFN)	VIRGO (INFN)	T2K (INFN)	PAMELA (INFN)	ARGO (INFN)	
Analysis of biodiversity, phylogenetic analysis, NGS, molecular dynamics simulation	CENGE-Biotechnology (Univ. of Naples Federico II)	Institute of biomembranes and Bioenergy (CNR)	Institute for Biomedical Technologies (CNR)	Institute of Virology of Plants (CNR)	Dept. of Health Sciences (UMG)	Dept. of Computer Science (Univ. Bicocca of Milan)
	Department of Pharmacy (Univ. of Bari A. Moro)	Institute for Biomedical Technologies (CNR)	Department of Chemistry and Chemical Engineering (Univ. of Naples Federico II)	Department of Botany (Univ. of Naples Federico II)		
Simulations of seismic, meteorological, fluid dynamics	Institute of Volcanology (CNR)	Civil and Structural engineering (Univ. of Naples Federico II)	ARPA Puglia	Dept. of Physics (Univ. of Bari A. Moro)	Group IV (INFN)	Aerospace and Naval engineering (Univ. of Naples Federico II)
Analysis of Neuroimaging, classification of digital images	Dept. of Physics (Univ. of Bari A. Moro)	Dept. of Physics and Chemistry (Univ. of Palermo)	Dept. of Computer Science (Univ. of Naples Federico II)	nextNMR (INFN)		
Estimates of Bayesian models	Department of Economics (Università Roma Tre)					
Study and propagation of waves	Telecommunications Engineering (Univ. of Naples Federico II)					

Fig. 6. ReCaS applications.

The cloud computing services are used by different communities (about 30 projects) for a lot of activities:

(1) Mathematics and numerical analysis,
(2) Biology and bioinformatics,
(3) Medical applications,
(4) Development of services "general purpose" with the use of web server and database,
(5) Activities of Business Intelligence for Healthcare through provision of an environment based on Pentaho.

5. The Solutions Adopted to Support all the Recas Users

As said, the survey was realized with the aim to support in the best way all the ReCaS users on the whole infrastructure. An important aspect to be considered, in this aim, is the chance to provide users with all the software needed by their applications at each site where their application can run. On the basis of the experiences during past PON projects (e.g., SCoPE[1]), we built the so-called ReCaS Toolkit, which is described in Sec. 5.1.

5.1. *The ReCaS Toolkit*

As ReCaS Toolkit, we refer to the applicative middleware installed on the ReCaS infrastructure. It includes a set of libraries, software packages and other tools that are for the users the necessary instruments to execute their own applications. Starting from the survey, submitted to the users of each seat, the complete set of libraries and software useful to the users has been revealed and installed on thc wholc infrastructure. The set of software resources included now in the ReCaS Toolkit release 1.0 are shown in Fig. 7.

Scientific libraries and software for applications	
Supported languages and compilers	C/C++, R, Python, Java, Fortran, Perl, bash, Supporto per OpenMP, CUDA
Workflow managers	Taverna work flow, Loni Pipeline, Galaxy
Commercial software	Mathematica, Matlab, IDL, StarCCM+
Libraries and software opensource	Biopython, R, ITK, FSL, Root Cern, GEANT, Fluka, WPS, WRVF, HMMER, emboss, MrBayes, VMD, NAMD, OpenCV, FANN, Beast, NAMD, FFTW, GROMACS, Schrodinger Drug Discovery, StochKit2, BLAST, HNMER, MIRA, FASTA, CLUSTAL, Model Test, Visual FM, Lib SVM, OpenFoam, Numpy, WPS, WRVF, proprietary code for HEP experiments) PETSc, SLEPC, Blas, Lapack, GSL, FFTW, Octave

Fig. 7. Compilers, tools, libraries, packages and PSE of the ReCaS Toolkit release 1.0.

		Napoli	Bari	Cosenza	Catania
Web site		X	X	X	X
Facebook page		X	X		X
Twitter contact		X	X		
Administrators unique email contact		X	X		X
Users mailing list		X	X		
Skype contact		X	X		
Ticketing system		X	X		

Fig. 8. Tools to support users at each ReCaS site.

5.2. *The User Support*

The user support is a key-point in the medium and long-term sustainability of large and distributed computing infrastructures, as the ReCaS is. However, this kind of activity is anything but simple and needs a lot of people and the enabling of several communication channels among the supporters and the final users. Currently, on the ReCaS infrastructure, all the communication tools listed in Fig. 8. Tools to support users at each ReCaS site are available.

Acknowledgments

We want to thank the collaborators for having shared their experience, in particular the administration team of the SCoPE datacenter at experience, in particular the SCoPE administration team at University of Naples Federico II.

References

1. G. B. Barone, D. Bottalico, V. Boccia *et al., Middleware applicativo: lo SCoPE Toolkit*, Italian National Conference e-Science2008, 2008.
2. Available at http://www.ponsmartcities-prisma.it/.

Chapter 13

Science Applications in Bari ReCaS Farm

A. Monaco* and P. Notarangelo

Istituto Nazionale Fisica Nucleare, Sezione di Bari via E. Orabona 4, 70126 Bari, Italy

**alfonso.monaco@ba.infn.it*

Nowadays we are witnessing a striking increase in the volume of data generated in life science applications and, consequently, a growing demand for computing resources for their analysis and interpretation. Grid/Cloud technologies appear to be able to meet these demands. For example, those technologies are able to provide easily and seamlessly the needed computational power as well as the storage resources to record the data produced. The main focus has to be in the direction of reducing the complexity of their usage: in this sense, the exploitation of innovative cloud technologies can yield a fundamental advance. In this paper, we present a dedicated job submission tool (JST) which allows to manage the submission and monitoring of life science applications on different infrastructures such as: dedicated servers, local batch systems, IaaS resources and EGI grid infrastructure. JST has recently undergone a major rewrite, in particular for what concerns the interaction with the user;

2 WEB services interfaces (*Simple Object Access Protocol* — SOAP and *Representational State Transfer* — REST based) have been added. In this way, it is now possible to exploit the JST capabilities also within a workflow manager like Taverna, LONI and Galaxy and Liferay scientific gateway.

1. Bari ReCaS Farm

The PON ReCaS[1] has the objective of strengthening the scientific computing infrastructure in Italy through the construction of four data centers in South Italy: Bari, Catania, Italy, Naples. The computing power of the previous Bari Computer Center for Science (Bc^{2}S) has been increased by almost a factor of three. Currently Bari ReCaS Farm has a total of more than 12,000 CPU core, four PB of disk space shared by all compute nodes and 10 Gbts of bandwidth.

The ReCaS Farm support several life science communities and projects (medical physics, Elixir project,[2] LifeWatch[3]), etc. and it can be accessed by European Grid Infrastructure (EGI)[4] grid-enabled users. In particular, the bioinformatics community uses 2.8% of the farm resources (to which one must add the local use), while the computational chemistry community uses up 3.4% of the farm resources.

The center offers:

(a) the possibility to require specific services on virtual machines,
(b) the possibility to execute parallel jobs based on MPI library,
(c) facilities to manage the execution of a bunch of independent jobs (automatic grid scheduling) also by means of WebServices,
(d) facilities to perform distributed data processing using HADOOP,
(e) "Cloud Storage" services based on WebDav and ownCloud,
(f) a system of project and ticket management based on REDMINE,
(g) several Web and MySQL servers.

In this paper, we present a brief description of the life science applications that are available in Bari ReCaS Farm and the technical solutions that we have implemented to support them. The core of this solution is the job submission tool (JST). The JST is a tool, developed years ago, to allow the submission on grid and local computing infrastructure of large numbers of jobs and to keep track of all of them in an unattended way.

Recently JST has been improved to submit jobs on cloud infrastructure.

Thanks to the JST capabilities it is now possible to exploit a lot of web and client features to support, in term of computing, the life science researchers in their scientific analysis.

The main user interface applications available on ReCaS Farm are Galaxy and Taverna workflow manager system and ReCaS science gateway.

2. Services Provided by Bari ReCaS Farm

On the Bari ReCaS farm, several services have been implemented to address the needs of the various life science communities:

- service for robust alignment,
- phylogenetic inference,
- support to phylogenetic inference,
- evolutionary model selection,
- metagenomic analysis,
- neuro-imaging analysis,
- environmental analysis (ARPA Puglia).

JST is integrable and interoperable in several tools such as applications (Scientific Gateway, etc.), clients, web browsers and workflow manager systems (Galaxy, LONI, Taverana, etc.). By the standard technology of the web services, supported by any programming languages, it is possible to use the JST services. This is possible because the JST frontend has a RESTful and SOAP web service implementation.

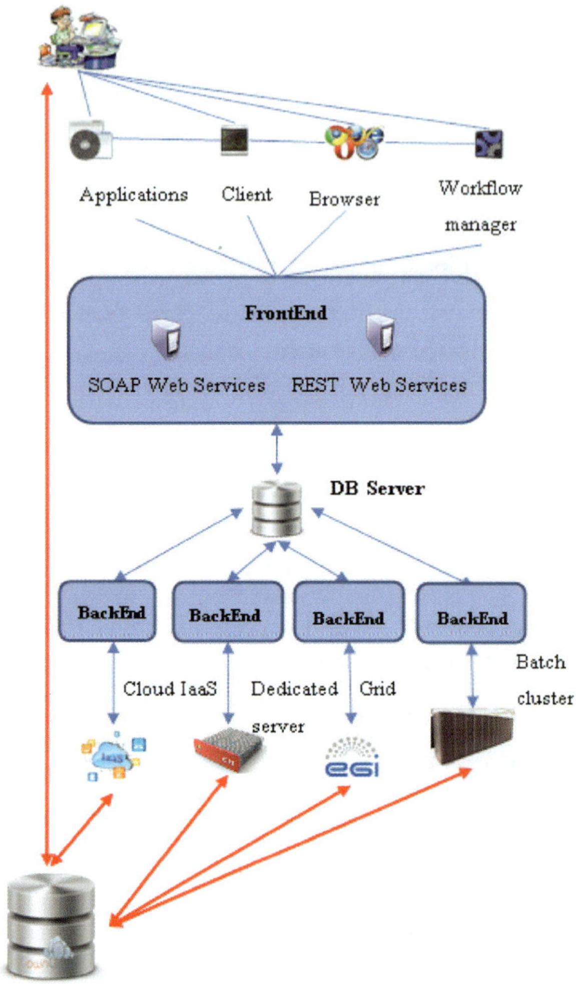

Fig. 1. Job submission tool scheme.

3. Job Submission Tool (JST)

The JST[5] was born to simplify the submission, management and book-keeping of large number of jobs required by particular bioinformatics use case. JST offers two web services interfaces — SOAP and REST, this way it is possible to exploit the JST capabilities also within a work-flows manager like Taverna,[6] LONI[7] and Galaxy[8] and ReCaS Liferay scientific gateway.[9]

The JST architecture is shown in Fig. 1. The system is made of three macroelements:

(1) Database contains the list of tasks (job) to be executed and the configuration parameters.
(2) FrontEnd used to fill the TaskList via Web Services (RestFul/Soap).
(3) BackEnd is an application daemon, that runs in the background and submits jobs, empting the TaskList.

***JST FrontEnd*:** The FrontEnd is a web service layer that provides several access methods. This layer has two implementations: SOAP web service technology and RESTful technology. The user can interact with the FrontEnd in several ways (client applications, a Web browser, workflow engine or a combination of the above).

***JST BackEnd*:** The application called "BackEnd" is based on a MultiThread strategy and is constantly running. It regularly queries the TaskList to understand if there are jobs to be executed and, if there are, submits them to the computing resources for which it is configured. The behavior of the BackEnd instance and of the jobs depends on the type of submission:

- interactive instance or job submission on a dedicated server,
- in all other cases is the job itself, when it start executing on a Worker Node of the batch cluster, or on a worker node of the grid infrastructure or on a Cloud computing resource, that updates the TaskList information.

However, the BackEnd checks regularly that all jobs have been successfully completed; in this case it sends to the user a notification mail. Also several instances of the BackEnd can be executed simultaneously, each configured in order to take care of specific job submission (Interactive or Dedicated server, Batch cluster, EGI GRID infrastructure, CLOUD IaaS). The BackEnd takes care of the job scheduling and user notification.

4. JST Integration

A generic instance of JST is used in order to exploit all the available resources needed to schedule and execute the different steps needed by the supported applications. Both the JST itself and the computational resources are provided exploiting modern IaaS cloud computing technology, in order to be able to guarantee the scalability and the reliability needed by a service that is publicly exposed to the end-users.

5. Galaxy Workflow Manager

Galaxy is a bioinformatics Workflow Management System (WFMS), written in Python by which it is possible to execute and to share biomedical analysis. It is a web-based application in order to avoid the installation of local client software by the user, but he/she can easily use Galaxy workflow manager by means of a common web browser.

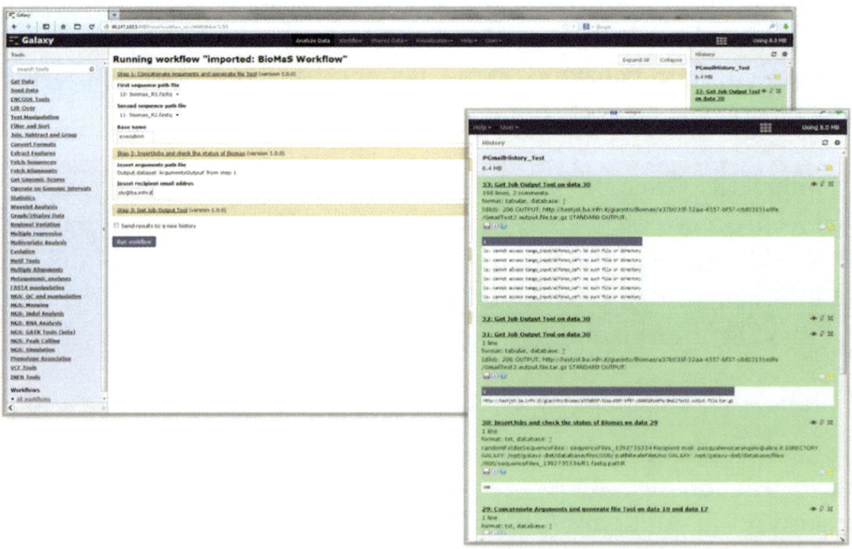

Fig. 2. Galaxy workflow manager.

On the INFN computational resources, a modified version of Galaxy[10] is installed in order to support metagenomic pipeline called BioMaS. In particular, in this Galaxy instance, an *ad hoc* tool has been developed and customized. This tool is characterized by three macro module: upload input files, submit analysis, recover result.

The Istituto Nazionale Fisica Nucleare (INFN) group has developed, within Galaxy workflow manager, the submit analysis module. This new component uses the REST interface capabilities provided by JST. Using this simple building block, the very end-users should be able to build new workflows exploiting the basic application provided as SaaS (Software as a Service) by mean of JST interface, and composing them to build more complex and high-level analysis.

In the INFN Galaxy instance, the Galaxy BioMaS workflow package has been created and shared with the BioMaS users. These users can import and run the workflow in your private Galaxy environment through the web interface.

In this instance, a BioMaS workflow specially created to perform the analysis is already available. In Fig. 2 an example of a BioMaS workflow submission in the Galaxy instance is shown.

6. Java Portlet in ReCaS Science Gateway

We have also realized web interfaces for a lot of life science applications.

The ReCaS Science Gateway portal is developed on Liferay framework and based on Catania Science Gateway.[11]

In this framework, the developers can build a simple java portlet component with the JST facilities. This way the complexity of dealing with the different computational platform is hidden by JST and the developers could concentrate the effort on providing simple and powerful graphical interfaces.

The following life science applications actually are available:

(a) BioMaS (Bioinformatic analysis of Metagenomic AmpliconS): a metagenomic pipeline developed to equip the biomolecular researcher involved in taxonomic studies of environmental microbial communities with a comprehensive and user-friendly workflow including all the fundamental steps for the NGS amplicon-based metagenomic analysis, by guiding his path from raw sequences to final taxonomic identification.

(b) MSA–PAD[12]: is a multiple DNA sequence alignment framework designed to align conserved protein coding DNA sequences. The application accounts for either single or multiple protein domains coding sequences and uses this information in assembling its output. It is mainly useful for comparative genomics by the possibility to align genomes having different genic organization (i.e., bacterial genomes). It also takes into account and/or aligning genes' exons, including those undergone intron loss or gain, respecting genomic organization.

(c) Medical Physics[13]: is a series of fully automated algorithms for the analysis of structural T1 weighted brain magnetic resonance scans. In particular, the proposed workflows perform brain extraction and hippocampal segmentation.

(d) Complex network analysis: the goal of this tool is to individuate a complex network structure in multi-variate data, thus allowing the user to investigate or unveil the presence of communities or similarity clusters.

The ReCaS Science Gateway users, opportunely pre-authorized, can choose the kind of the applications to use. This way the user is able to submit a new analysis and easily check the status of own jobs.

When the job is finished, the user receives the mail notification sent by the JST backend with the final status of the job and the link where the user can download the output results.

References

1. Available at http://www.pon-recas.it/.
2. Available at http://www.elixir-europe.org.
3. Available at http://www.lifewatch.eu/web/guest/home.
4. Available at http://www.egi.eu.

5. G. Donvito, S. Vicario, P. Notarangelo *et al.*, *The BioVeL Project: Robust phylogenetic workflows running on the GRID*, EGICF2012-EMTITC2-029.
6. Available at http://www.taverna.org.uk.
7. Available at pipeline.loni.ucla.edu.
8. Available at http://galaxyproject.org/.
9. Available at https://recasgateway.ba.infn.it.
10. Available at http://galaxy.cloud.ba.infn.it:8080.
11. Available at http://www.catania-science-gateways.it/.
12. B. Balech, S. Vicario, G. Donvito *et al.*, MSA-PAD: DNA multiple sequence alignment framework based on PFAM accessed domain information, *Bioinformatics*, pp. 1–3, (2015).
13. S. Tangaro, N. Amoroso, M. Boccardi *et al.*, Automated voxel-by-voxel tissue classification for hippocampal segmentation: Methods and Validation, *Phys Med.*, **30**, pp. 878–887, (2014).

Chapter 14

Users Applications in the PRISMA–Napoli Cloud

M. Alfano*, D. Del Prete, D. Michelino, S. Pardi and R. Vela

Istituto Nazionale Fisica Nucleare, Napoli Unit, Italy
**m.alfano@na.infn.it*

The PRISMA–Napoli Cloud is one of the nodes of the distributed IaaS infrastructure in Southern Italy, created to study the adoption of cloud technologies by local Public Administrations.

In this work, we present the general architecture of the Cloud, then we focus on the application components deployed at IaaS and PaaS level, characterizing their current use cases and possible evolutions.

More specifically, we will show a set of pilot services and setup created in collaboration with the City of Piano di Sorrento, partner of the PRISMA project for the testbed.

The results obtained show the concrete possibility of the Public Administration to use external clusters through the hybrid-cloud paradigm to manage local services and to integrate new functionalities.

1. Introduction

The PRISMA project[1] aims at implementing a multi-regional cloud infrastructure and at creating a testbed to support local public administrations in their daily activities. It is realized through few clusters located in a set of sites in the regions of Campania, Puglia, and Sicilia. These are connected through widely supported protocols and an advanced geographical network.

The project involves private companies, research institutes, and public administrations as pilot user communities.

The installation in Campania is managed by the *Istituto Nazionale Fisica Nucleare* (INFN)–Napoli and takes advantage of the ReCaS infrastructure that provides all the services needed to host the PRISMA resources, including: rack, cooling system, uninterruptible power supply (UPS), and access to the wide area network (WAN).

The supported user community is primarily the city of "Piano di Sorrento" that enacts a set of hybrid cloud use cases. This cloud integrate the existing local infrastructure with a new range of services.

More specifically, these services are: a monitoring system over a private VPN as a Service component; an e-learning platform, and a set of images created to be instantaneously deployable on the IaaS or through a PaaS/SaaS approach.

The rest of the paper is structured as follows: in Sec. 2, we describe the PRISMA–Napoli OpenStack based cluster by characterizing all the components and the front-end interface. In Sec. 3, we present the general requirements of the applications we tested and a set of images provided to the final user through the OpenStack Glance image service; in Sec. 4, we analyze three used cases: a monitoring system, an e-learning platform and WebTV.

Finally, in Sec. 5 we draw some conclusions and explain the ongoing activities and future applications.

2. The Cloud Architecture

The cloud architecture is totally based on the OpenStack[2] cloud computing software platform for IaaS and it is implemented on a dedicated cluster. The production farm is composed of eight servers configured as follows:

- two manager nodes that constitute a mini cluster to host the front-end and management services of the cloud infrastructure in a basic

virtualized environment. The cluster, internally, uses a shared network file system based on Ceph.[3]
- two storage nodes that implement a core network file system and an object storage area Ceph, configured with the replica 2 option, and shared with all the nodes of the cluster for a total of 44 TB.
- four compute nodes, mainly used as hypervisors that host the cloud infrastructure virtual machines. Each server provides 64 cores and 256 GB of RAM.

In addition, two 64 core servers, connected with the production farm, are used to test the latest OpenStack release and as a testbed to experiment new components and technologies.

With regard to the network infrastructure,[4] each node of the cluster implements a two network fabric: a 10 Gbps network for traffic among the hypervisor and storage units and as a main network for the virtual machines. In addition, a 1 Gbps link is used for management and monitoring purposes.

The basic operative system for the production site is Scientific Linux version 6.5, while the pre-production environment uses Scientific Linux version 7.

The hardware architecture of the production cluster is shown in Fig. 1.

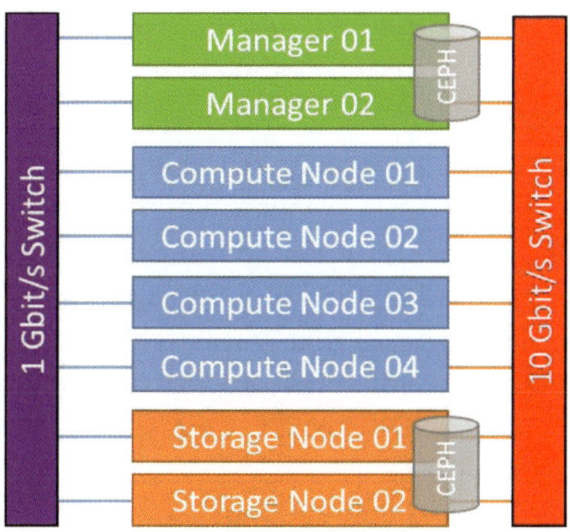

Fig. 1. The schema of the main hardware component of the Cloud testbed.

The Cloud Configuration

Our IaaS infrastructure is based on the OpenStack Cloud Management System (CMS). We are currently using the Icehouse release in production and the Juno release for the pre-production environment.

OpenStack core services are implemented on four Virtual Machines on the basic cluster, composed of the two manager nodes. The Ceph network file system shared between the two servers allows migrating the VMs in live mode. This functionality guarantees almost zero downtime in case of unexpected failure of one of the nodes or during upgrade activities and scheduled down-time of a specific server.

The four OpenStack infrastructure VMs are:

- *Cloud Controller:* a central machine that runs the Glance Image Service, the Keystone Identity Service, Nova-api, Nova-scheduler and Nova-conductor for instances management and Neutron for Network Services,
- *OpenStack web interface*: a dedicated VM hosting the standard dashboard, Horizon, customized with new functionalities,
- *Volume Manager*: virtual server that hosts the Cinder block storage service. Cinder uses the Ceph storage system as a backend,
- *Web frontend*: the main public portal of the project, based on the Content Management System Liferay.

In our production cloud, we take advantage of the KVM virtualization platform configured on each compute node. These nodes host the OpenStack Nova compute component as well, which interacts directly with the cloud controller. In this way, all the computing power can be managed from a single entry point by means of the OpenStack REST API or using the web dashboard.

The two storage nodes provide a large storage area based on a separate Ceph implementation that works as a back-end for the Glance image service, and to provide block storage and object storage.

3. Products and Services

The Cloud offers a set of basic images already configured to instantiate complete high level services ready to be used with minimal post

Table 1. List of the images available on the cloud of PRISMA–Napoli.

Image name	Service description
Debian 7	Operative System — Basic Debian installation version 7
SL7	Operative System — Basic Scientific Linux machine version
Arch_Linux_Cloud	Operative System — Basic Arch Linux Machine
CentOS6.5	Operative System — Basic CentOS machine version 6.5
Observium_SL7	Automatic SNMP Monitoring System
FAN 3.0	Monitoring System based on Nagios powered with Nagvis Map tool and Centreon control panel
pfsense	Open source router and firewall
SL7–MariaDB	Basic implementation of Maria DataBase
Joomla 3.3.6	CMS based on Joomla to create web content
LAMP	An already configured Linux Apache MySQL Php environment to host web content
Wordpress	CMS based on Wordpress to create web content
Alfresco	Collaborative tool to share documents
Moodle	A well-known e-learning system written in PHP
WebTV_Cloud	A Live streaming platform
Owncloud 8–SL7	A Dropbox-like cloud storage system

configuration activities. The images are provided directly on the user page in a sort of free-market place.

In Table 1, we show the list of the images available and a description of the basic characteristics.

3.1. *Network-as-a-Service, VPN-as-a-Service and Port-Forwarding*

In order to provide services to external entities, the INFN–Napoli cloud implements a VPN service in order to establish a direct connection to the LANs of remote infrastructures.

To achieve this goal, a Virtual Private Network as a service (VPNaaS) facility has been implemented. This component connects the cloud to the City of Piano di Sorrento, a pilot institution participating in the experimentation.

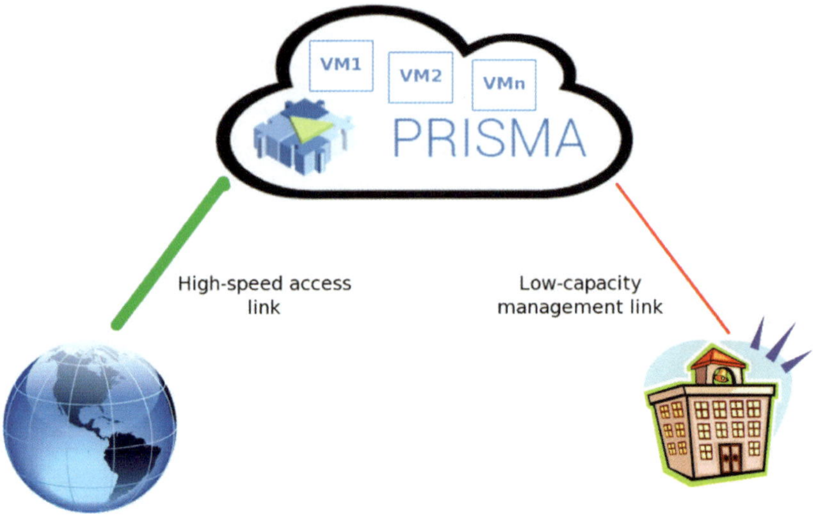

Fig. 2. Representation of a use case of the experimentation with the city of Piano di Sorrento.

More specifically, the VPNaaS service creates an IPSec tunnel between the Neutron router and a frontend machine, based on CentOS 6.5, located in the city network.

Moreover, a Neutron extension was implemented. It enables you to use the public IP of the virtual gateway to access cloud VMs from the internet. This component performs a port-forwarding, avoiding the need to reserve a dedicated public floating IP to each VM.

This extension is successfully working in our production environment and is going to be reviewed to become part of the OpenStack code upstream.

This plugin was specifically developed to enable the city of Piano di Sorrento to expose public services from the Cloud (Fig. 2), thus taking advantage of the high-speed internet connection available at INFN–Napoli.

3.2. *WebTV*

In the past, the monthly city council meeting has been live streamed on the web via the low-speed internet access of the city of Piano di

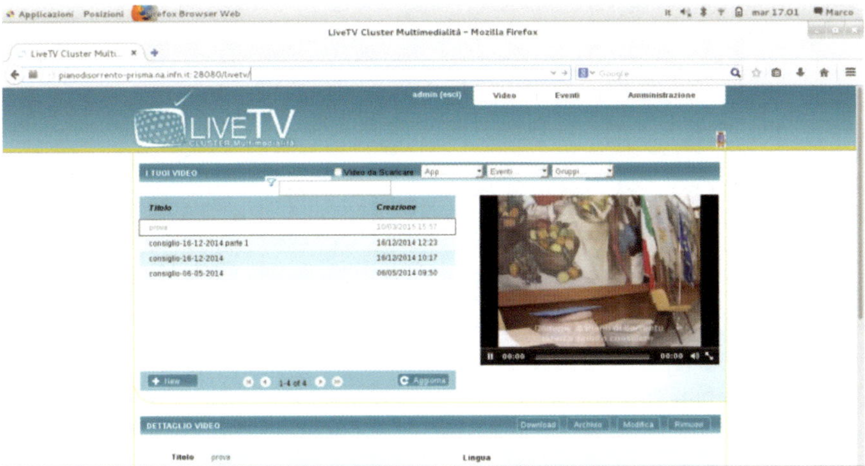

Fig. 3. The Web TV user interface during a video streaming session with the city of Piano di Sorrento.

Sorrento. Today, a VM uses the broadband access available on the cloud, allowing you to set up a high quality live streaming.

The VM available on the PRISMA Napoli Cloud is based on Ubuntu and the software Red5. The VM is directly connected to the City LAN, thanks to the service of VPNaaS developed, thus receiving the video flow from the meeting room.

Furthermore, the WebTV interface (Fig. 3) is publicly available thanks to the port-forwarding extension described above. Once again, taking advantage of the bandwidth at INFN Napoli, a notable amount of users will be able to watch the live streaming with good levels of quality and fluidity.

3.3. *Monitoring as a Service*

A monitoring system has been implemented on the PRISMA infrastructure. It is based on Fully Automated Nagios (FAN),[5] a suite of monitoring tools using Nagios as a core component. This service is implemented in a virtual machine with a network connection that uses the VPNaaS to communicate with the servers of the city of Piano

Fig. 4. NagVis representation of host and services status (on the left) and the geographical distribution of the monitored resources (on the right).

di Sorrento data center. The service allows a system administrator to monitor any hardware component at the remote site.

Moreover, it is used to check the status of the connections between the external data center and some peripheral sites (e.g., municipal police stations, schools, etc.).

Moreover, the system is provided with a geographical overview (Fig. 4) of the different sites that gives a summary of their status and virtual panel showing information about every single device and many details about monitored services. In addition to that, a set of tools allows to store and analyze historical data and plot significant trends. The whole system helps the administrator in his daily troubleshooting work.

Example of the monitored services include: average central processing unit (CPU) load, disk and memory usage and network traffic.

3.4. *e-Learning on Cloud*

The need to train the local community on the usage of our OpenStack based cloud infrastructure has led us to develop an e-learning platform. The open source Moodle Content Management System has been used as a main container of a set of video guides (Fig. 5) recorded for this purpose.

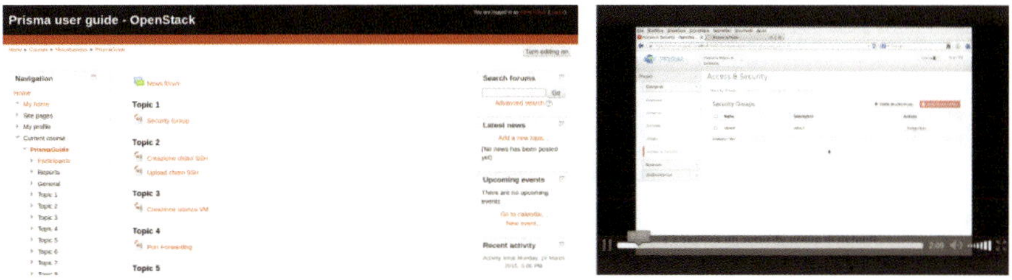

Fig. 5. The Moodle user interface and a video guide on the e-learning platform.

Thanks to the flexibility of the Moodle framework, we can easily provide any kind of learning material (such as videos, texts, audio lessons, downloadable software and much more). Moodle represents a useful tool to rapidly set up a course aimed at administration employees and citizens.

The requirements of the VM hosting the e-learning system are comparable to those of the WebTV use case. In fact, we took advantage of the port forwarding feature to use the large bandwidth provided by the public part of the cloud.

3.5. *Owncloud for Cloud Storage*

The PRISMA service suite has been enriched with a basic Owncloud-as-a-Service image.

Owncloud offers a Dropbox like cloud storage service that enables users to easily backup their files and replicate them on any personal device thanks to a simple client-side software.

Owncloud is deployable on the PRISMA Napoli infrastructure both as an ephemeral instance and as a persistent volume, using Ceph as a backend for data storage in a reliable and efficient way.

This service can be either used as an intranet private resource or as a public cloud storage thanks to the port-forwarding feature.

Owncloud is currently under test with the City of Piano di Sorrento in order to simplify data sharing among the administration employees.

4. Experimental Results

One of the parameters that can be used to evaluate the whole infrastructure is the service continuity (Fig. 6). It can be measured thanks to the monitoring system implemented to check the status of all services and devices that compose the cloud infrastructure.

More specifically, we can study thanks to the historical data, the amount of time that the core components (both hardware and software) have been up and properly working.

Our analysis began in November 2014, when the experimental activities with the city of Piano di Sorrento started, in order to validate the cloud implementation.

Since the Cloud was put in production, the core services availability met ranges from 99.0% to 99.6%. Moreover, in the last three months, its value has stabilized on 100% due to an intense tuning activity of all components.

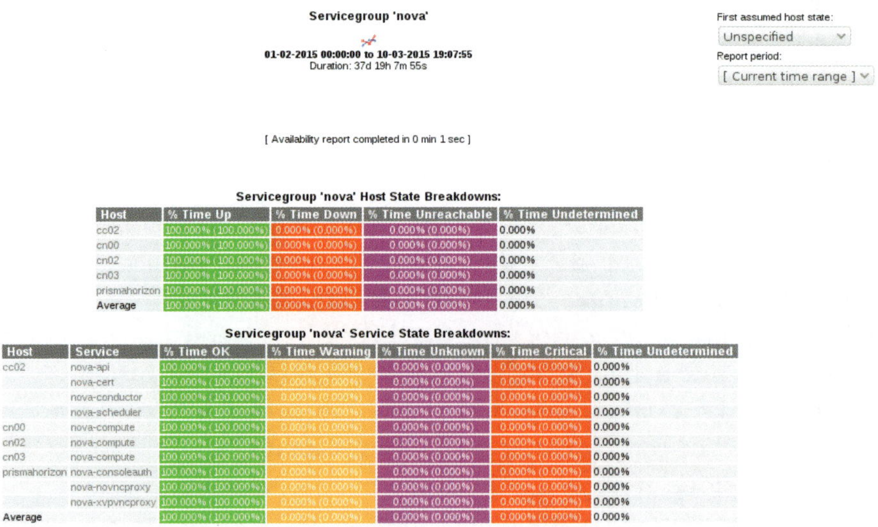

Servicegroup 'nova'

01-02-2015 00:00:00 to 10-03-2015 19:07:55
Duration: 37d 19h 7m 55s

First assumed host state: Unspecified
Report period: [Current time range]

[Availability report completed in 0 min 1 sec]

Servicegroup 'nova' Host State Breakdowns:

Host	% Time Up	% Time Down	% Time Unreachable	% Time Undetermined
cc02	100.000% (100.000%)	0.000% (0.000%)	0.000% (0.000%)	0.000%
cn00	100.000% (100.000%)	0.000% (0.000%)	0.000% (0.000%)	0.000%
cn02	100.000% (100.000%)	0.000% (0.000%)	0.000% (0.000%)	0.000%
cn03	100.000% (100.000%)	0.000% (0.000%)	0.000% (0.000%)	0.000%
prismahorizon	100.000% (100.000%)	0.000% (0.000%)	0.000% (0.000%)	0.000%
Average	100.000% (100.000%)	0.000% (0.000%)	0.000% (0.000%)	0.000%

Servicegroup 'nova' Service State Breakdowns:

Host	Service	% Time OK	% Time Warning	% Time Unknown	% Time Critical	% Time Undetermined
cc02	nova-api	100.000% (100.000%)	0.000% (0.000%)	0.000% (0.000%)	0.000% (0.000%)	0.000%
	nova-cert	100.000% (100.000%)	0.000% (0.000%)	0.000% (0.000%)	0.000% (0.000%)	0.000%
	nova-conductor	100.000% (100.000%)	0.000% (0.000%)	0.000% (0.000%)	0.000% (0.000%)	0.000%
	nova-scheduler	100.000% (100.000%)	0.000% (0.000%)	0.000% (0.000%)	0.000% (0.000%)	0.000%
cn00	nova-compute	100.000% (100.000%)	0.000% (0.000%)	0.000% (0.000%)	0.000% (0.000%)	0.000%
cn02	nova-compute	100.000% (100.000%)	0.000% (0.000%)	0.000% (0.000%)	0.000% (0.000%)	0.000%
cn03	nova-compute	100.000% (100.000%)	0.000% (0.000%)	0.000% (0.000%)	0.000% (0.000%)	0.000%
prismahorizon	nova-consoleauth	100.000% (100.000%)	0.000% (0.000%)	0.000% (0.000%)	0.000% (0.000%)	0.000%
	nova-novncproxy	100.000% (100.000%)	0.000% (0.000%)	0.000% (0.000%)	0.000% (0.000%)	0.000%
	nova-xvpvncproxy	100.000% (100.000%)	0.000% (0.000%)	0.000% (0.000%)	0.000% (0.000%)	0.000%
Average		100.000% (100.000%)	0.000% (0.000%)	0.000% (0.000%)	0.000% (0.000%)	0.000%

Fig. 6. The uptime of the compute hosts and services from February 2015 as seen from the Nagios user interface. Almost all of them score around 100%.

5. Conclusions

A Cloud infrastructure has been set up and put in production in November 2014. Several services have been deployed and tested with the help of the community of the city of Piano di Sorrento. The enacted use cases demonstrated the effectiveness of the developed components. In particular, the VPNaaS and the port-forwarding extension demonstrated a possible key component to simplify the setup and usage of a hybrid cloud by the public administration.

The cloud can be easily extended with additional functionalities and could be used in different contexts, host new applications and support new communities.

Moreover, the developed components can also be decoupled from the infrastructure and integrated in other environments and cloud management systems.

Acknowledgments

This work has been funded by PON 2007–2013, project code PON04a2_A; details on the project can be found at the URL accessible through the following QR code.

References

1. The Prisma Project, available at http://www.ponsmartcities-prisma.it.
2. The OpenStack, available at http://www.openstack.org/.
3. The Ceph, available at http://www.Ceph.com.

4. M. Alfano, D. Del Prete, D. Michelino *et al.*, *Network Management in Cloud computing for Public Administration: A Practical Use Case* at 2014 IEEE/ACM 7th International Conference on Utility and Cloud Computing — 978-1-4799-7881-6/14, pp. 768–773.
5. D. Del Prete, S. Pardi and G. Russo, A Grid monitoring model over Network-Aware IaaS Cloud Infrastructure, *Int. J. of High Performance Computing and Networking*, 7, pp. 195–204 (2013).

Physical Science

Chapter 15

Computing Activities in High Energy Physics: Atlas Tier 2 at INFN–Napoli

G. Carlino*,‡, P. Castellano*, R. Cevenini*, A. Doria*,§, R. Esposito*, L. Merola*,†, S. Pardi * and G. Russo*,†

**Istituto Nazionale Fisica Nucleare, sezione di Napoli, Italy*

†*Università di Napoli Federico II, Italy*

‡*g.carlino@na.infn.it*

§*doria@na.infn.it*

The computing infrastructure built for the RECAS Project in Naples, at the Istituto Nazionale Fisica Nucleare (INFN) and University Federico II, is used in an extremely profitable way for the computing of the ATLAS high energy physics experiment.

ATLAS is one of four experiments running at the Large Hadron Collider (LHC) at the European Organization for Nuclear Research (CERN). It studies the basic forces that gave birth to the Universe, through proton–proton collisions at high energy. Its distributed computing system includes numerous centers for processing and archiving data in Grid, geographically distributed and organized in a hierarchy of levels. The Level 0 ("Tier 0") is located at CERN and an important "Tier 2" center is active at the INFN of Naples.

1. Introduction

The ATLAS experiment[1] is one of the multi-purpose detectors at the Large Hadron Collider (LHC),[2] located at European Organization for Nuclear Research (CERN), Geneva (CH). LHC provides proton collisions at a center-of-mass energy of 14 TeV with a frequency of 40 MHz. The overall ATLAS detector consists of about 140 million electronic channels and the expected average event size is about 1.2 MB.

ATLAS carried out its first phase of data taking from 2010 to 2012 that led to the discovery of the Higgs boson mass. The second phase, Run-2, has started in March 2015 and will go on until 2018, allowing in-depth investigation of the boson's properties and thereby of the origin of mass.

Given the complexity of the experiment, the ATLAS detector produces huge amounts of data (several PB/year). In addition, an equivalent large sample of simulated data is necessary for each complete physics analysis. The organization of data management on such a large scale needed new computing paradigms, based on Distributed Grid Computing. The LHC data are managed by the World Wide LHC Computing Grid (WLCG) Collaboration. The data are distributed around the world: each country involved in LHC experiments hosts in its computing centers a fraction of the data and grants access to physicists from all over the world. The sites are organized in different level (Tier-0/1/2/3), each level having its own duties.

2. Atlas Computing Model

In the ATLAS computing model,[3] the sites providing computing facilities are organized in a hierarchical four-level tiered structure. The Tier 0 facility based at CERN is responsible of the first-pass processing and archiving of the primary raw data and their distribution to the Tier 1 centers, where data are reprocessed and made available to Tier 2s. Each Tier is connected to a set of Tier 2 and Tier 3 sites, grouped on regional basis in so called "clouds."

According to the ATLAS Computing Model, the Tier 1 receives and stores permanently on tape a 10% fraction of ATLAS data in several formats: real RAW data and simulated HITS data, Event Summary Data (ESD), Analysis Object Data (AOD), Derived Physics Data (DPD). An additional fraction of the AOD and DPD data dynamically placed by the ATLAS Production and Distributed Analysis (PanDA) System is stored on the Tier1 disk area.

The Tier1 is mainly used for central production activities, such as simulation and data processing. The Tier 2s are medium size computing centers designed for the Monte Carlo simulation and for the user analysis of the derived data samples dynamically placed in the sites.

Tier 3s are dedicated to the final steps of data analysis, via Grid or interactively, and code testing and development and in some cases they also contribute to the production of simulated data.

In the first period of the data taking activities, the data were pushed hierarchically from the CERN Tier 0 to the Tier 1s and eventually to the Tier 2s within the same regional cloud.

In order to optimize the resource utilization, the data flow was recently modified and the data are now dynamically placed on sites, pulled by the Tier-1s and Tier 2s according to data popularity; only the more accessed datasets are effectively replicated.

A further evolution of the computing model came from the high performance of the recent global network connections. A subset of Tier 2 centers showing a high level of reliability, continuity of services and good network connectivity are included in a mesh network structure together with all Tier 1s. These sites are known as directly attached Tier 2s. Inside this network, the data distribution does not need to follow the tier hierarchy but can flow between any two sites.

The ATLAS Italian cloud[4] consists of many computing centers (Fig. 1): the Tier 1, located at the INFN CNAF laboratory (Bologna), four Tier 2s (Frascati, Milano, Napoli, Roma-1) and eight Tier 3s. The Italian sites are located in the Universities and INFN departments and differ in capacities, setups and purposes. Recently the Universities of Witwatersrand and Johannesburg (ZA) and Thessaloniki (GR) joined the Italian cloud as new Tier 3s.

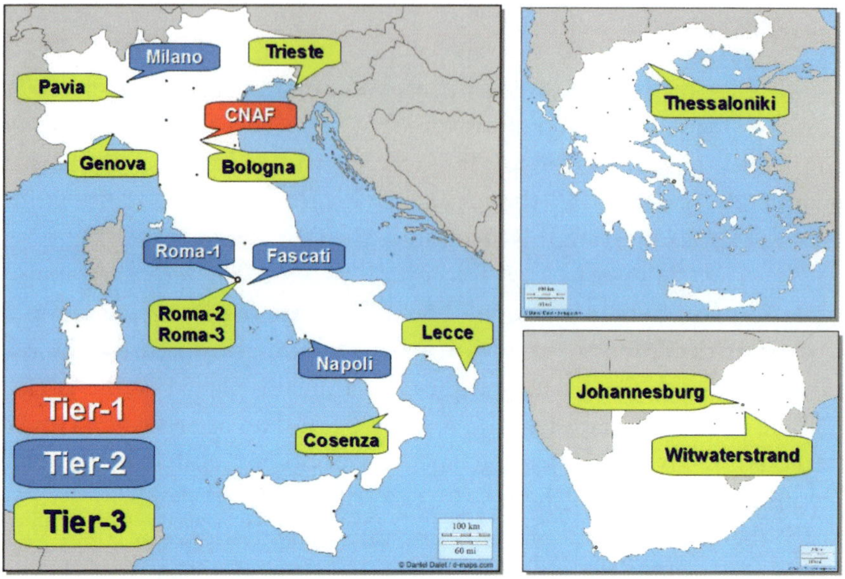

Fig. 1. Tiers distribution in the Italian Cloud.

3. ATLAS Activities at INFN–Napoli

The research activities of the ATLAS Naples group are focused on:

- Higgs data analysis and search for physics beyond the standard model,
- Development, implementation and analysis of the first level muon trigger detectors based on RPC detectors,
- Development and characterization of *Micromegas* detectors for the Phase 1 Upgrade of the Muon Spectrometer in the small eta region,
- Activities in the general organization of the ATLAS computing, experimentation in networking and distributed computing,
- Management of a Tier 2 center for the ATLAS distributed offline computing.

4. ATLAS Tier 2 at INFN–Napoli

The Tier 2 in Naples is part of the Grid of ATLAS distributed computing sites since 2005. Its activities include Monte Carlo simulations, data reprocessing, muon detector calibrations and user analysis of the physics data archived locally.

In order to commit its data and activities to a computing center, the experiment requires the site to meet high levels of availability and reliability. These requirements are fulfilled through a wide range of management tools, monitoring systems, and optimization practices, that will be described below.

4.1. *Resources*

In 2015, Tier 2 reached a total processing power of 210 multi-core CPUs, both Intel Xeon and AMD Opteron, corresponding to about 2,700 cores, and a storage area of about 2 PB. These resources have been funded every year by INFN since 2005 and increased in 2014 with resources funded by ReCaS PON Project.

Infrastructure and facilities of SCoPE[5] and *Rete di Calcolo per SuperB e altre applicazioni* (Computing network for SuperB and other applications, ReCaS) DataCenters are used to accommodate Tier 2 resources applications running simultaneously.

4.2. *Technologies*

4.2.1. *HW Technologies*

The computing resources are made of rack mountable multi-processor servers, equipped with Intel Xeon or AMD Opteron multi-core CPUs, x86_64 architecture. The installed operating system is Scientific Linux, a rebuild of Red Hat Enterprise by Fermilab and CERN, for primary use within the High Energy and High Intensity Physics community.

The storage systems (Fig. 2) are based on SAN (Storage Area Network) technology and are provided with Fiber Channel controllers and connections, with a minimum throughput of 4 Gbps. SATA II disks are used, in RAID6 configuration.

The local area network (LAN) relies on a 10 Gbps optical fiber infrastructure based on a star topology, whose center is an HP 10508 router with 280 Ethernet ports at 10 Gbps. The same router provides

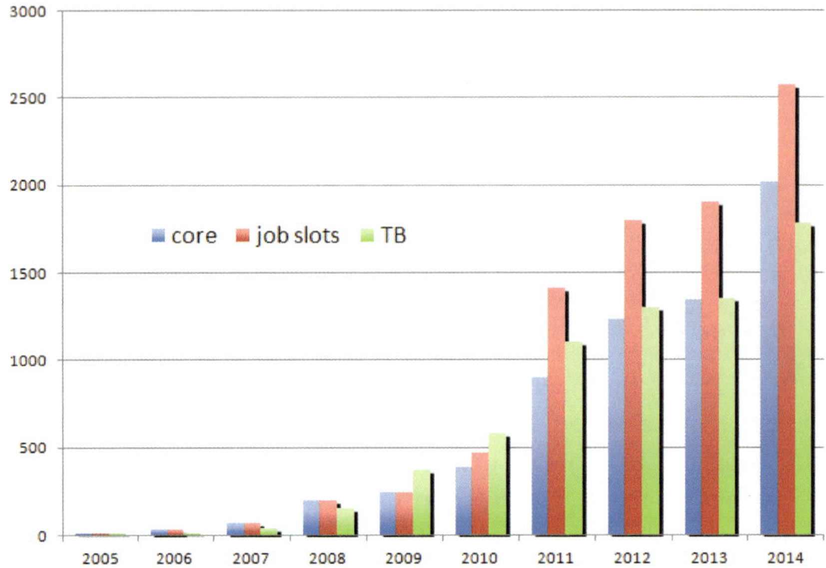

Fig. 2. Increase in the years of the resources at the ATLAS Tier 2 of Naples, in terms of disk space (TB), number of cores and total number of applications running simultaneously.

the 10 Gbps wireless area network (WAN) connection to the GARR backbone, that will be upgraded to 20 Gbps in the current year and to 40 Gbps starting from 2016.

4.2.2. *SW Technologies*

The computing technology chosen for the Tier 2 activity is based on the GRID paradigm, in the environment of Word Wide LHC Computing Grid (WLCG) and a wide range of Grid services is implemented.

The jobs from the Grid are received via Computing Resource Execution And Management (CREAM) Computing Elements and the computing resources are managed with the Torque Batch system and the MAUI Scheduler.

The storage elements are implemented with Disk Pool Manager (DPM),[6] a storage management system developed at CERN, that offers a simple way to create a disk-based Grid storage element and supports relevant protocols (SRM, gridFTP, RFIO) for file management and access.

The management of such a large amount of resources requires the usage of the most recent tools for automatic installation, configuration and monitoring. For the system management, we choose an infrastructure consisting of two tools, Puppet and Foreman: Puppet[7] is a tool designed to manage the configuration of systems declaratively. Foreman[8] is a complete lifecycle management tool that allows administrators to automate repetitive tasks and proactively manage servers.

The Puppet/Foreman infrastructure is used to implement automatic installation of the servers (mainly Scientific Linux and CentOS), to configure the services and to keep the systems updated. It consists of a main server where all needed configuration files (modules) are stored and shared to all the machines requesting them. The Puppet/Foreman server also acts as Preboot Execution Environment (PXE) server. Beside the standard modules, custom Puppet modules were added to install and configure a wider range of machines. Installation of utility packages, services configuration, automatic disk detection and Redundant Array of Inexpensive (Independent) Disks (RAID) configuration upon install, and informational emails are some of the features implemented with the new developed modules.

4.3. *Monitoring*

The Ganglia Monitoring System[9] is the main active monitoring tool in the Naples Tier 2, with over 200 machines monitored. A lot of scripting (PHP, SNMP queries and RRDtool) was developed to watch on critical statistics:

- Batch servers (Fig. 3): graphs and statistics on Grid queues and job slots used,

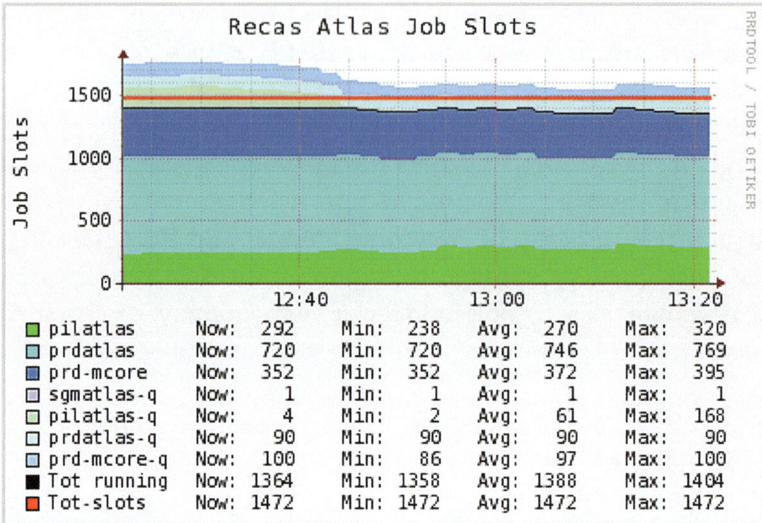

Fig. 3. Occupied JOB slots per group, evolution in time.

- Panda Monitor plug-in (Fig. 4): graphs and statistics imported from the CERN production monitoring website.

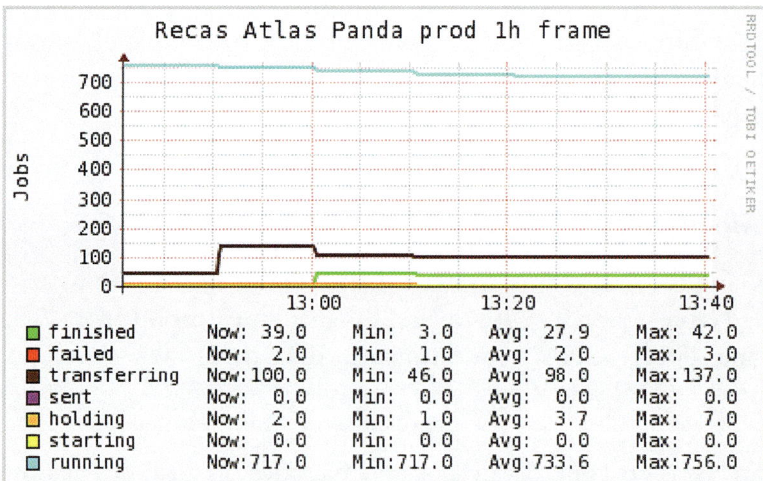

Fig. 4. Number of jobs in the different run states.

- Rack sensors (Figs. 5 and 6): graphs and statistics on temperatures and PDUs power usage.

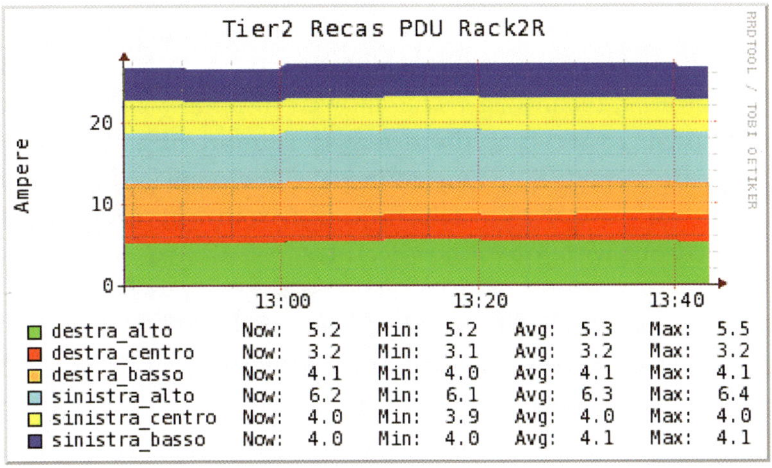

Fig. 5. Current drawn by each Power Distribution Unit (PDU) in a rack.

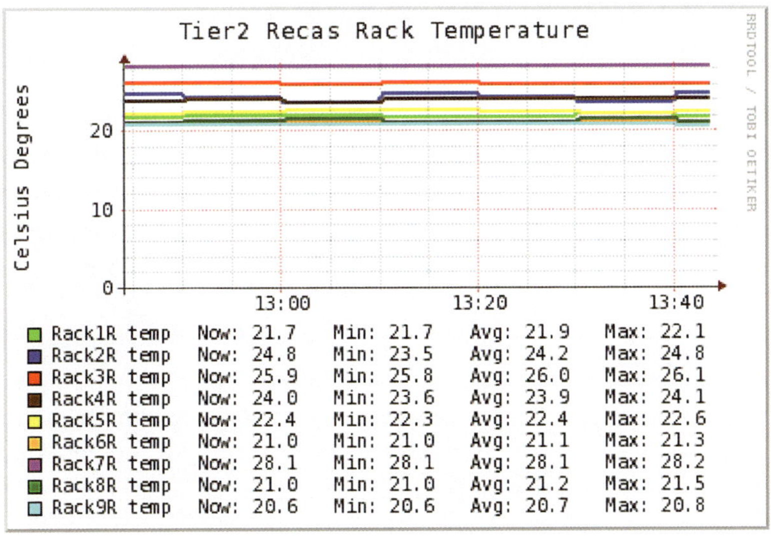

Fig. 6. Internal rack temperature for all racks.

4.4. *Virtualization*

In the last few years, the way of organizing the site services has been deeply revamped and restructured. A large number of old physical servers has been replaced by few powerful multi-processor servers, each running multiple virtual machines based on KVM (*Kernel-based* Virtual *Machine*) technology. This allowed to optimize the HW usage and to implement service redundancy, thus improving site reliability.

Lately the default KVM instances have been replaced by the more advanced PROXMOX Virtualization Environment (Fig. 7),[10] still KVM based but with lots of new features:

- Online migration, backup and restore capabilities,
- Scheduled backups with email notifications,
- Snapshots and rollback of the VMs,
- Cluster infrastructure, entirely manageable from a single web interface,
- New (better) console on the machines, no longer requiring java plug-ins,
- GlusterFS and NFS file-systems support,
- "On the fly" VLAN tagging and bridging,

The distributed File System GlusterFS[11] is used to store the VM images. Several GlusterFS replica volumes were deployed to provide HA (High Availability) features to the virtual machines. The replica volumes are file-systems shared between two or more machines over the network to provide redundancy of the data stored in it.

4.5. *Performance*

The Naples Tier 2 is deeply involved in all the experiment activities and it is one of the most performing sites in ATLAS. As an example, we show the CPU consumption of the site in the last year of operation: compared to all the ATLAS Tier 2s (Fig. 8) the Naples Tier 2 is the sixth after some very big centers in US and Germany, while at the Italian level (Fig. 9) it's not too far from the Tier 1 performance.

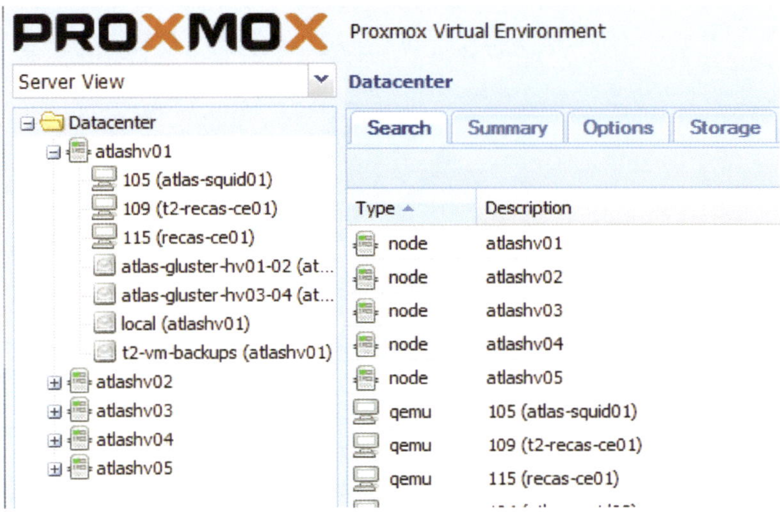

Fig. 7. PROXMOX graphical interface.

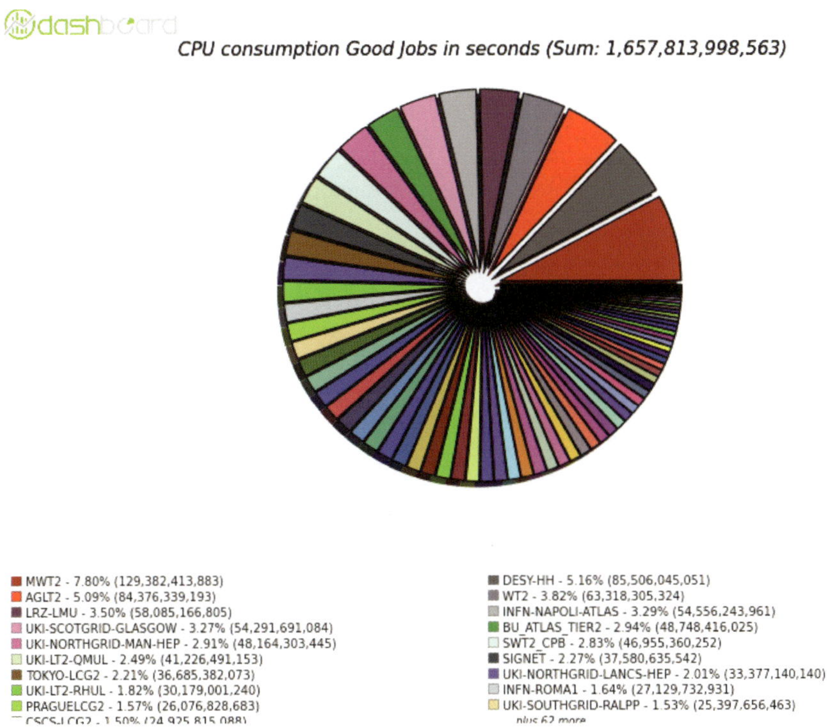

Fig. 8. ATLAS CPU consumption (all sites).

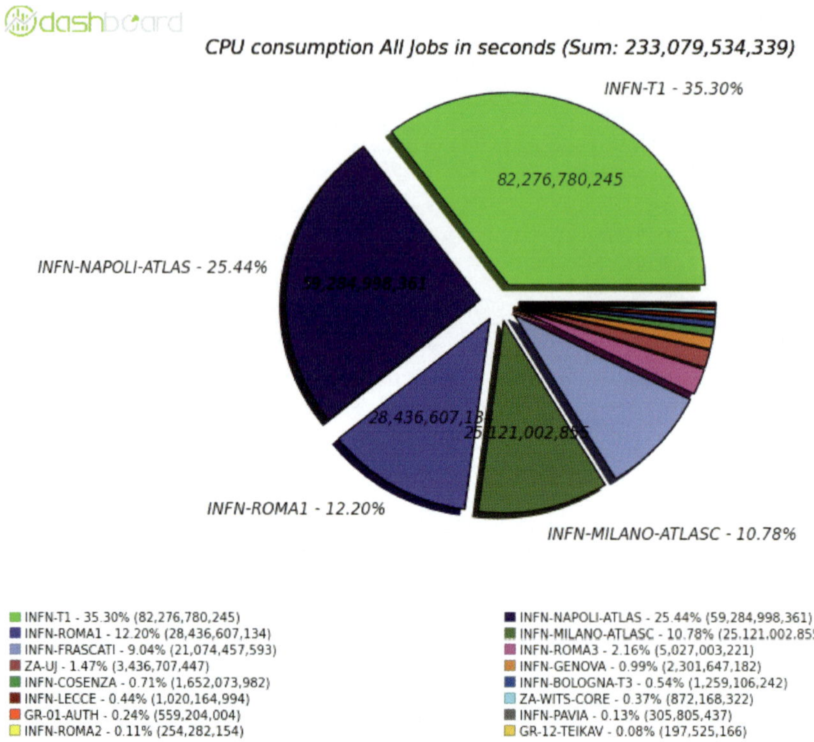

Fig. 9. ATLAS jobs CPU consumption (Italian sites).

5. Integration of the Tier 2 with the Recas Infrastructure

The computing room at INFN — Department of Physics (Fig. 10) has been restructured thanks to the RECAS project. Electrical and cooling systems have been updated and 14 new water-cooled racks, equipped with a sensor system for remote monitoring and alarm management, have been installed in the room. Each rack contains an optical patch panel wired with 10 Gbit/s fibers to the central HP 10508 router.

This infrastructure hosts computing and storage resources that are made available to the communities of ATLAS and Belle II

Fig. 10. INFN-RECAS computing room.

experiments, to other projects and to local users through Grid and Cloud protocols.

Four racks are dedicated to host ATLAS resources, in addition to the four racks of the same type already existing in the room and to the six racks used by Tier 2 in the SCoPE[5] Data Center, realized with the funding of a previous PON 2000–2006 program.

As already mentioned, the ATLAS Tier 2 shares its resources with the international community in the World Wide LHC Computing Grid, that leads the computing operations for all the LHC experiments. A new Computing Element, heading 23 worker nodes for a total of almost 1,500 cores, has been added to manage in Grid the RECAS resources dedicated to the Tier 2. Two new DELL storage systems offer disk space of about 400 TB, integrated in the DPM data management system that offers a unique endpoint to access the site storage from the Grid.

The Monitoring system based on Ganglia, the automatic installation system based on Puppet and Foreman and the virtualization technique with PROXMOX, described in Sec. 4, have been developed and implemented in synergy among the ATLAS Tier 2 and the other

RECAS infrastructure INFN users, in order to minimize the effort needed to operate system administration tasks.

References

1. The ATLAS Collaboration, The ATLAS Experiment at the CERN Large Hadron Collider, *JINST* **3**, p. S08003 (2008).
2. E. Lyndon , *The Large Hadron Collider*, *New J. Phys.* **9**, p. 335 (2007).
3. S. Campana (on behalf of the Atlas Collaboration), Evolution of the ATLAS distributed computing system during the LHC long shutdown, *J. Phys., Conf. Ser.* **513**, p. 032016 (2014).
4. L. Rinaldi, G. Carlino, A. Doria, L. Merola, G. Russo *et. al.*, *ATLAS computing activities and developments in the Italian Grid cloud*, *J. Phys., Conf. Ser.* **396**, p. 042052 (2012).
5. L. Merola, The S.Co.P.E. Project, *Proc. Final Workshop of the Grid Projects of the National Operational Programme*, 2000–2006 Ed, Consorzio Cometa (2010).
6. M. Hellmich, F. Furano, D. Smith, R. Brito da Rocha *et al.*, DPM — efficient storage in diverse environments, *J. Phys., Conf. Ser.* **513**, p. 042025 (2014).
7. Available at http://docs.puppetlabs.com/puppet/3.7/reference/.
8. Available at http://theforeman.org/
9. Available at http://ganglia.sourceforge.net/
10. Available at https://www.proxmox.com/
11. Available at http://www.gluster.org/

Chapter 16

The Belle II Simulation Campaign at ReCaS

V. Boccia*, G. De Nardo*,†, D. Del Prete*,
P. Guida*, S. Pardi*,‡ and G. Russo*

**Istituto Nazionale Fisica Nucleare, Napoli Unit, Italy*
†*University of Naples Federico II, Italy*
‡*spardi@na.infn.it*

The paper describes how an international collaboration uses the newly built *Rete di Calcolo per SuperB e altre applicazioni* (ReCaS) infrastructure. The Belle II experiment, carried out in Japan but with a strong support from Italian scientist and from *Istituto Nazionale Fisica Nucleare* (INFN), provides extensive resources that are used since October 2014 for the Monte Carlo Campaign; up to 4000 computing hours per day were used in the 4th campaign, for a continuous run of 50 days.

1. Introduction

The *Rete di Calcolo per SuperB e altre applicazioni* (Computing network for SuperB and other applications, ReCaS) project[1] was funded to support scientific research and international collaborations. One of the first scientific collaboration that gained benefit from the ReCaS infrastructure is the Belle II[2] Physics experiment at KEK laboratory in Tsukuba, Japan.

2. The Belle II Experiment

The new SuperKEKB[3] accelerator (Fig. 1) and the large detector Belle II constitute a milestone in the investigation of the matter excess (CP violation) in the Universe. In SuperKEKB, bunches of particles of matter (electrons) and their anti-particles (positrons), with energies up to 8 giga electron volts, are brought to collision at rates which are 40 times larger than in the previous KEKB accelerator. The particles being created in the collisions and the decay products formed are measured and analyzed in the Belle II experiment (Fig. 2). With the high statistics provided by SuperKEKB, scientists are hoping to finally find deviations from the predictions of the Standard Model. So called B mesons, composed of a "heavy" B quark and a "light" anti-quark play a special role here.

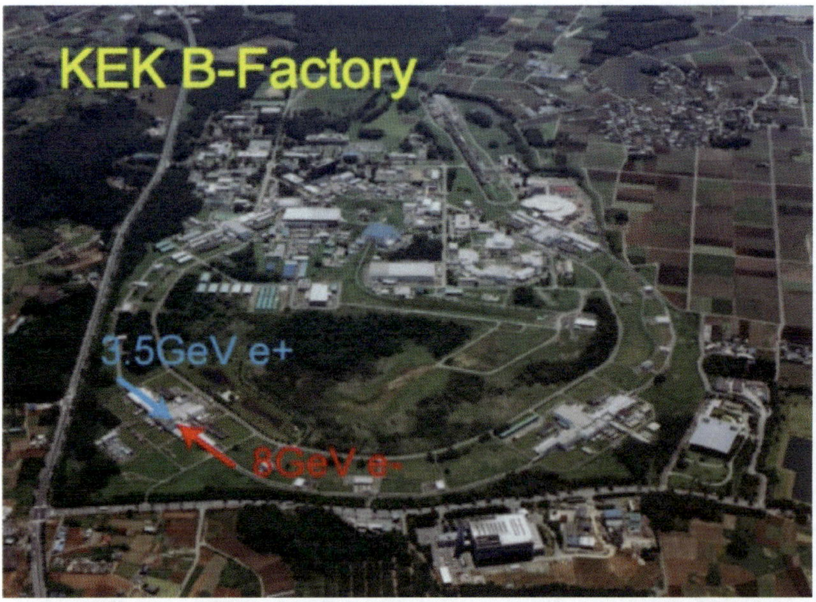

Fig. 1. An aerial view of the KEK B-factory site in Japan.

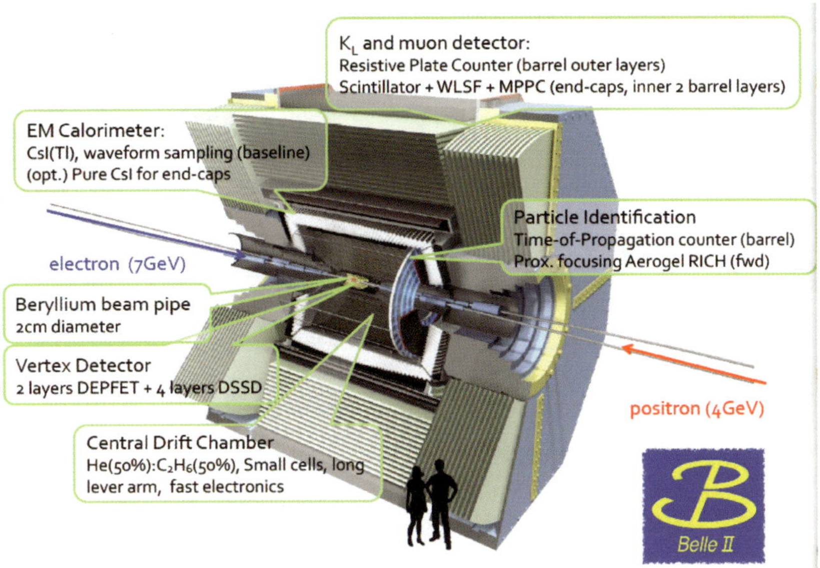

Fig. 2. A pictorial view of the Belle II detector.

In SuperKEKB, the energies of the electrons and positrons are chosen such that exactly one B meson and one anti-B meson are produced per collision.

Due to the high density of the colliding bunches, made possible by the extremely small beam cross sections of SuperKEKB, the B meson pairs are produced in unprecedented large numbers. For this reason, the SuperKEKB accelerator is also called a "Super B factory".

3. The Belle II Collaboration

The Belle II collaboration (Fig. 3) consists of about 600 people, belonging to about 100 different institutions in 23 countries. A very widespread community, that needs a strong interaction by means of collaborative tools and a strong coordination for the distributed computing activities.

The collaborative service tools, used extensively by all scientists, are maintained mainly at KEK (Japan) and Pacific **Northwest National**

Fig. 3. The Belle II international collaboration.

Laboratory (PNNL)[4] (USA), and are listed at https://belle2.cc.kek.jp/~twiki/bin/view/Computing/Services.

4. The Belle II Computing Model

The most important thing that was agreed upon in the collaboration is the computing model.[5] The Belle II computing model assumes a tiered structure, based on: (i) Raw Data Centers; (ii) Regional Data Centers; (iii) Monte Carlo Production Sites; (iv) local resources.

The Raw Data Center is the site where the raw data is recorded and processed; "Raw Data Center" also serves as "Regional Data Center" and "MC Production Site". There are two centers of this kind: the KEK Data Center, the host laboratory where the raw data is recorded from the experiment and processed, and the PNNL Data Center, another raw data center where the raw data is processed in parallel with KEK. When the raw data reprocess becomes necessary because of an update of the reconstruction software and/or the detector constants, it will be done at PNNL.

The Regional Data Centers are, at the moment, the following: GridKa (Germany), CNAF (Italy), KISTI (South Korea). These centers are large computing centers where a copy of the mDST data is stored; a "Regional Data Center" also serves as "MC Production Site".

The MC Production Site is the computing site where a proportional share of the MC production/reconstruction and physics analysis is performed. All the other computing sites belong to this. This is categorized into three types according to the used technology: (i) GRID site, a site that is operated with a standard GRID middleware (e.g., European Middleware Initiative (EMI), Open Science Grid (OSG)); the ReCaS site belongs to this category; (ii) Cloud site, a site that operates with a standard Cloud infrastructure; the ReCaS site will evolve toward this category; (iii) Computing cluster sites, sites with a standalone computer cluster which is accessible with the ssh protocol from the internet and available through a batch system such as LSF, TORQUE.

The Local resources consist of computing resources in institutes and universities, which are used for *n*tuple-level user analysis.

The model for data is:

- All raw data is stored at KEK.
- One copy of the raw data is transferred to PNNL.
- Raw data is partially processed at KEK to produce DST that is used to estimate the detector constants.
- Raw data is processed at KEK to produce *mDST* data. PNNL may process the raw data to lower the KEK workload.
- All *dst* and *mdst* files are stored at KEK.
- A full *mdst* dataset is stored at PNNL.
- A full *mdst* dataset is stored at each Regional Data Center.
- The raw data of one year should be processed within 11 months.
- At least one previous version of mdst files should be kept in case of reprocessing.

The model for generic MC is:

- MC is generated at MC Production Sites (including Regional Data Centers).

- A sample corresponding to six times the data integrated luminosity is generated.
- The MC files for one year of data should be generated within 5–11 months.
- The MC files are stored at GRID site where they are produced.
- The MC files produced at Cloud and Computing Cluster sites are moved to other GRID site.[6]
- One copy of all MC files is replicated at another site. (European group would like to have one copy of all MC in European sites.)
- At least one previous version of MC files should be kept in case of reprocessing.

The model for reprocessing is:

- The number of raw data reprocesses is assumed to decrease as a function of years.
- The raw data reprocess is done at PNNL.
- The free resources at KEK (if exists) may be used.
- The reproduction of MC samples is done at MC Production sites.

The model for analysis is:

- The production of official *n*tuples and signal MC files used by a working group are coordinated by selected people. The output is stored at the GRID sites where they are produced. Further copies may be distributed to other sites.
- Users can submit grid jobs themselves to access data and MC files at grid sites.
- The output files (*n*tuples) are copied to storage on Local Resources. Grid sites may provide temporary limited storage resources.

5. The Computing and Storage Requirements

Storage estimation is based on:

- RAW data.
- mDST after data taking.

- mDST during data reprocessing.
- mDST — Monte Carlo related the data.
- mDST — Monte Carlo related data reprocessing.

The current parameters for data estimation are:

- Event Size × RAW data: 300 Kb.
- Event Size × mDST: 40 Kb.

The foreseen rate is the following:

(× 10⁹)	2016	2017	2018	2019	2020	2021	2022
Events/year	1.2	3.1	29.6	70.7	87.8	89.3	93.5
Integrated	1.2	4.3	33.9	104.6	192.4	281.7	375.2

The above data are the basis for determining the CPU and storage needs, which are plotted below in Fig. 4:

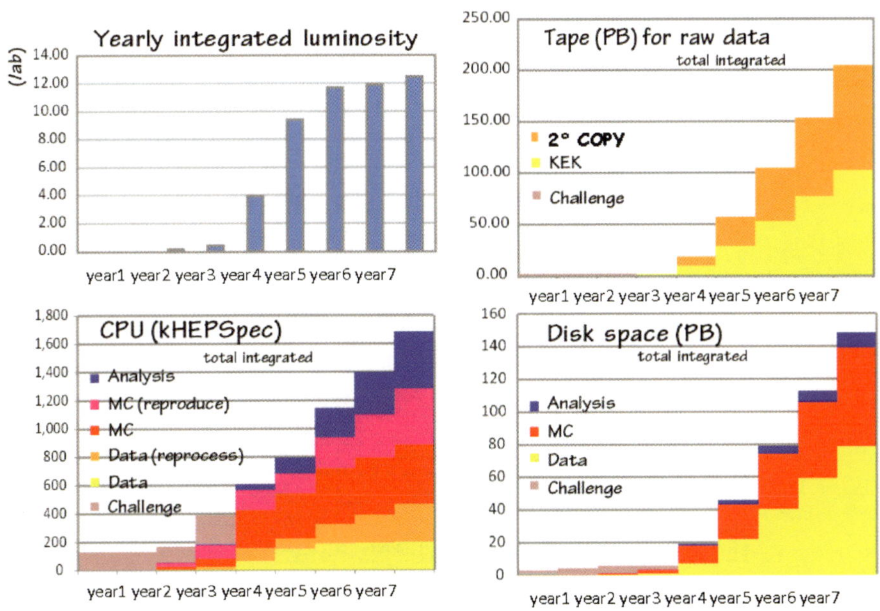

Fig. 4. The Computing needs of the Belle II experiment.

6. The Available Resources

Table 1 summarizes the resources available at Naples for the Belle II simulations; some of them are dedicated resources, others are available on a best-effort basis.

Moreover, all nodes have a 1 GbE port dedicated to management and monitoring, out-of-band, and 2 * 1 GbE ports for backup. A storage system with 400 TB of raw capacity is also dedicated to the Belle II activities. The data center is constantly monitored with the support of the software described in other chapters of this volume.

The data centers are connected to the external network with two connections at 2×10 Gbps. The data center is constantly monitored with the support of some software: Nagios, Centreon, Nagvis. That is

Table 1. The available resources in Naples.

ReCaS-A	*rack*	*type*	*cores*	*HepSpec*
	rack 3	non-pledged	120	1041
	rack 4	non-pledged	120	1041
	rack 5	non-pledged	120	1041
	rack 9	non-pledged	120	1041
	rack 10	non-pledged	120	1041
	rack 11	non-pledged	120	1041
	rack 12	non-pledged	50	37
	rack 18	non-pledged	320	5000
	rack 22	pledged	128	2144
	rack 23	pledged	416	4062
	rack 24	non-pledged	96	1080
	rack 25	non-pledged	228	2289

ReCaS-B	*rack*	*type*	*cores*	*HepSpec*
	rack 02	pledged	384	3366
	rack 06	pledged	384	3366
	rack 08	non-pledged	320	2805

ReCaS-A	total pledged cores, INFN	6732
ReCaS-B	total pledged cores, UNINA	6206
ReCaS-A	total non-pledged cores, INFN	6732
ReCaS-B	total non-pledged cores, UNINA	14652

Napoli pledged	**13 kHS**
Napoli non-pledged	**21 kHS**

several monitoring systems[7] that alert for any abnormalities, and by sending email to all supervisory staff.

The staff has supervisory task 24/24 hours and 365/365 days a year. The response time to provide solutions is within 12 hours.

7. The ReCas Contribution

In October 2014, when the fourth MC campaign started, Napoli contributed initially only with the old SCoPE resources (Fig. 5). This limited the number of jobs executed per day to roughly 2000, an appreciable number but not enough to guarantee that the Italian group could attain the anticipated percentage of 10% of all the computing needs.

Starting from October 20, after a solicitation form the Italian computing leader, F. Bianchi, we started switching on our servers, configuring them in the necessary batch queues. In 10 days, the number of jobs executed per day sped up to an astonishing 23,000 as shown in

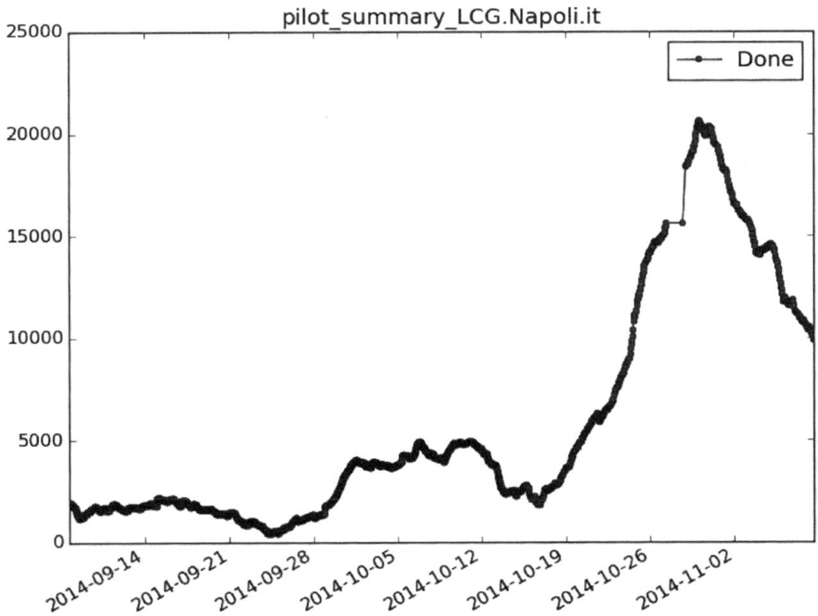

Fig. 5. The contribution to 4° MC by ReCaS Naples at the end of 2014.

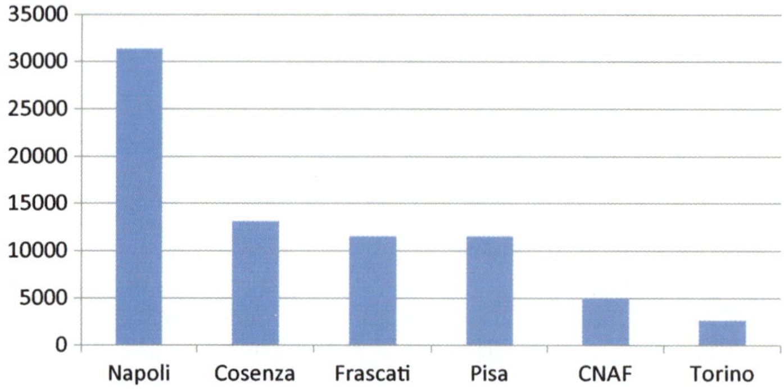

Fig. 6. The contribution to 5° MC by ReCaS Naples.

the above plot. Of course, these jobs summed up with the others in Italian Belle II sites, namely CNAF, the Tier 1 site, and the Tier 2 federation which includes Torino, Frascati — and Naples as well.

The results of the fifth MC campaign highlighted the strong contribution of the Naples site as shown in Fig. 6, which has established a contribution of 11.3% at the international level (Fig. 7).

8. The Networking

In order to support the intense activity of data-movement, network is one of the key resources to take in account. In fact, the computing model of the Belle II experiment requires an efficient and reliable network that interconnects all ReCaS sites.

Therefore, all the sites that support international experiments, including ReCaS, have migrated to the new GARR-X network. ReCaS is connected by a high-capacity optical network spread nationwide, managed by the GARR Consortium.

On this infrastructure, the Consortium GARR has realized backbone geographical connections on which dedicated end-to-end services can be configured. In this way, you can draw interconnections geographically guaranteed bandwidth and minimum delay, enabling the transparent transport of Ethernet traffic, particularly suitable for the realization of LAN–LAN connections between data centers that enables ReCaS efficiency of datasets transmissions for the Belle II experiment.

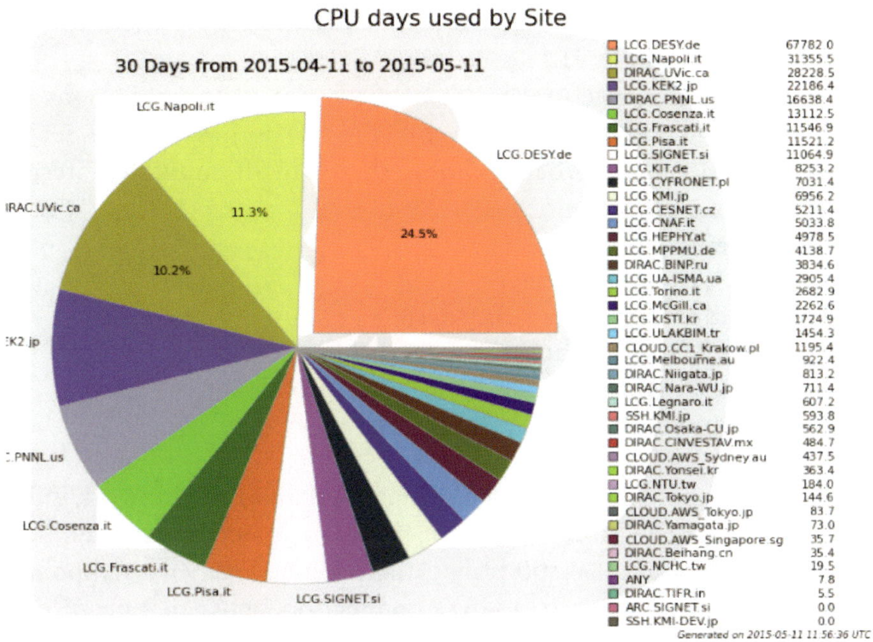

Fig. 7. Belle II 4th campaign MC — April–May 2015.

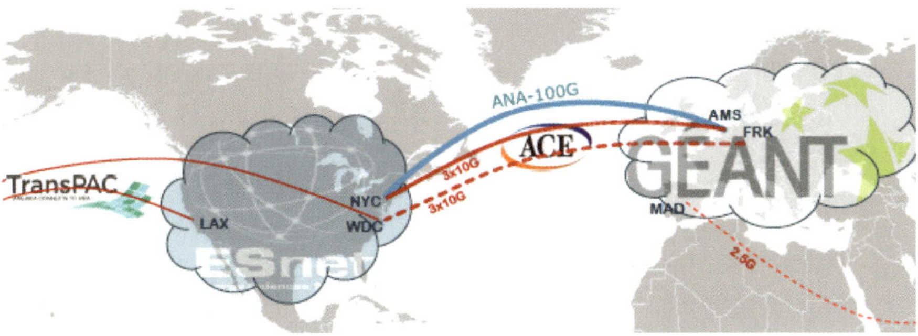

Fig. 8. Scheme of the international connections currently available.

In practice, ReCaS Napoli is directly connected to the GARR with a rate of 10 Gbps, with an imminent update at 40 Gbps. Moreover the site is also connected at LHCONE, an overlay network over the general network infrastructure for e-science, that allows to isolate and manage the specific traffic generated by high energy physics experiments, including Belle II.

From an infrastructural point of view, the geographic network supporting the Belle II experiment (Fig. 8) is based on a set of Trans-Asia and Trans-Atlantic links, shared with multiple international scientific communities.

The current speeds are:

- 2 × 10 Gb/s Tokyo–Los Angeles,
- 2.5 Gb/s Madrid–Mumbai,
- 3 × 10 Gb/s Amsterdan–NY,
- 3 × 10 Gb/s Frankfurt–Washington.

In order to promote an international update of the network infrastructure, the Belle II collaboration has participated as pilot application in testing the ANA-100 trans-Atlantic link (Fig. 9), a temporary network path that offer 100 Gb/s connection between Europe and America (Fig. 7).

The sites involved in the activity have been PNNL (USA), ReCaS-Napoli, CNAF (ITALY) and KIT (GERMANY). The main goals of the testing were:

- Validate the routing of the ANA-100 Trans-Atlantic link,
- Study and tune the performance of current Belle II data transfer tools.

The major part of the tests, has been performed between May and June 2014. Results have confirmed the effectiveness of the ANA-100

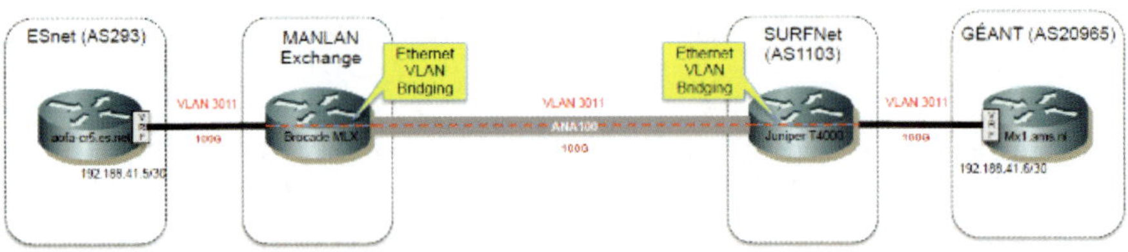

Fig. 9. A representation of the ANA-100 Trans-Atlantic link.

link, and the possibility to saturate multiple 10 Gbit/s flows on multiple storage systems, among European and USA sites.

The tests have been performed with the following tools:

- Traceroute was used to confirm the routing to each DTN,
- Iperf was used to do initial network transfer rate tests,
- gridftp and/or srm-copy was used to test sites,
- FTS3 server at GridKa was used to schedule data transfers.

The results have demonstrated that the 100 Gbps link can fulfil the requirements for the future data movement strategies.

9. Conclusions

Starting from 1st March 2015, the ReCaS project will support the Belle II Monte Carlo campaign with a guaranteed job execution rate of 50,000 per day, using resources not only at Naples, but also at Cosenza.

References

1. The ReCaS Project. Available at http://www.pon-recas.it. (2015).
2. The BELLE II Project — Collaboration. Available at http://belle2.kek.jp. (2013).
3. The SuperKEKB Project. Available at http://www-superkekb.kek.jp. (2011).
4. The Pacific Northwest National Laboratory. Available at http://www.pnnl.gov. (2015).
5. D. Del Prete, Computing at Superb, PoS (ICHEP2012) 488. *Proc. of Science, 36th International Conference on High Energy Physics* (ICHEP2012 Melbourne, Australia, July 4–11, 2012. CNUM C12-07-04.
6. D. Del Prete, S. Pardi and G. Russo, A grid monitoring model over network-aware IaaS cloud infrastructure, *International Journal of High Performance Computing and Networking* 7(3). DOI: 10.1504/IJHPCN.2013.056524.
7. D. Del Prete, S. Pardi and G. Russo, *A Centralized, Extensible, Multilayer Monitoring System for Distributed Infrastructures,* The Atlas Tier 2-Naples Experience, 21–24 June 2011.

Chapter 17

The ReCaS Infrastructure for the Neutrino Astronomy with KM3NeT

C. Bozza*,†,‡ and P. Migliozzi†,§

**Università degli Studi di Salerno, Salerno, Italy*
†Istituto Nazionale Fisica Nucleare, Napoli, Italy
‡kryss@sa.infn.it
§migliozzi@na.infn.it

The goals and principles of the KM3NeT project are presented. The large needs in terms of computing power are met thanks to the sizable contribution of the *Rete di Calcolo per SuperB e altre* (ReCaS) infrastructure hosted at the University of Naples. The KM3NeT computing model is reviewed and details are shown of how it is implemented and supported by means of ReCaS facilities and technologies.

1. The KM3NeT Neutrino Telescopes

KM3NeT[1] is a large research infrastructure that will consist of a network of deep sea neutrino telescopes in the Mediterranean. The main scientific field is neutrino astroparticle physics. Moreover, the KM3NeT location in the deep sea offers interdisciplinary opportunities for continuous, real-time measurements, e.g., for marine biology,

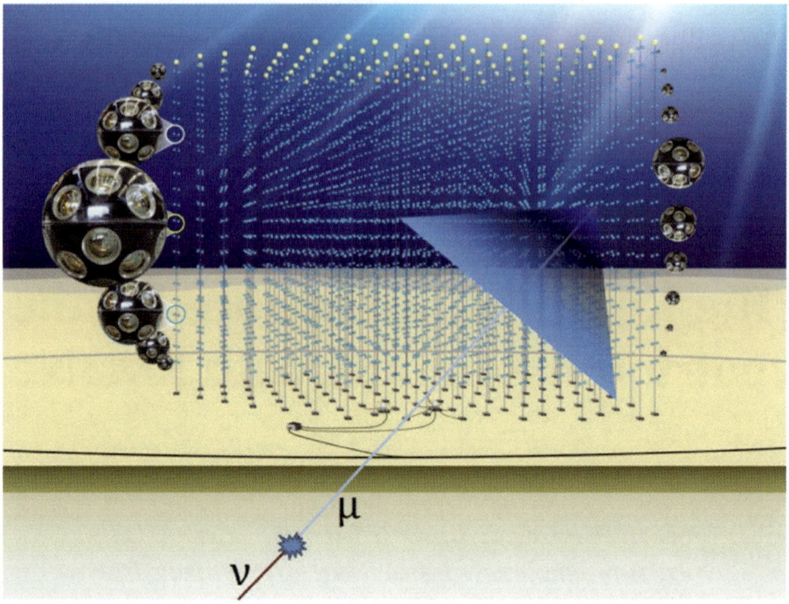

Fig. 1. Concept of the detection of a muon produced in an interaction of a neutrino with the rock beneath the detector. The muon continues in almost the same direction as the neutrino producing Cerenkov light sampled by the photo-sensors of the apparatus.

oceanography or environmental sciences. The scientific goals are the study of astrophysical objects by detecting their high-energy neutrino emission and the investigation of neutrino properties by measuring neutrinos produced through cosmic-ray interactions in the atmosphere.

The high-energy neutrino telescope of KM3NeT (Astroparticle Research with Cosmics in the Abyss, ARCA) is located offshore Capo Passero, Italy. A schematic view of the apparatus and the detection principle of the neutrinos is shown in Fig. 1.

The IceCube collaboration has recently reported a first signal of neutrinos with energies of the order of PeV.[2] This signal with energies up to 2 PeV strengthens the motivation for the construction of KM3NeT. Several types of astrophysical objects have been proposed as sites where hadrons are accelerated to extreme energies. The interaction of these particles with matter or radiation in or near the source produces pions and subsequently high-energy neutrinos. These neutrinos propagate with almost no interaction and may reach the Earth undisturbed. From the observed neutrino direction, the sources may be identified. The low-energy neutrino telescope of KM3NeT (Oscillation Research with Cosmics in the Abyss, ORCA) is located offshore Toulon, France. It aims at the measurement of the neutrino mass hierarchy. Indeed, it has been recently suggested[3] that the neutrino mass hierarchy can be experimentally determined from the oscillation pattern of atmospheric neutrinos passing through the Earth by measuring the two-dimensional arrival pattern of neutrinos in energy and zenith angle, in the energy regime of about 3–20 GeV.

Both arrays are made of thousands of Digital Optical Modules (DOMs). The optical modules are arranged along flexible strings kept vertical by a submerged buoy. ARCA, to be constructed near Sicily, consists of six building blocks of 115 strings each, with 18 optical modules per string, in its final configuration. Before its completion, 1/4 building block is foreseen for phase 1, 2 building blocks for phase 2; ORCA (near Toulon, France) is 1 block of 115 strings, but with much smaller horizontal and vertical spacing. The DOM will detect the Cherenkov light emitted in the sea from charged particles originating from collisions of the neutrinos and the interacting medium. Each DOM consists of 31 PMTs of 3″ diameter inside a 17″ diameter glass sphere.

The timing and the charge determination are the main parameters that determine the detector resolution. The measurement of the arrival times of the photons on the PMTs is a crucial parameter since it affects the accuracy of the event reconstruction. The charge estimation is based on the Time over Threshold (ToT) values of the PMT pulses and the number of hit PMTs on each DOM. The accuracy of this estimation affects the reconstruction energy resolution of the charged particles emitting the detected photons. Dark current rate and after pulses can also affect the telescope performance.

The first prototype of a DOM of the future KM3NeT neutrino telescope has been deployed in the deep waters of the Mediterranean

Sea. The data taking and the first results about rate measurements are summarized in Ref. 4.

2. The KM3NeT Computing Model

The computing model adopted by KM3NeT is derived from Large Hadron Collider (LHC) experiments. The choice is natural in the case of a multi-site research organization, with modular and scalable detectors deployed near the coasts of France, Italy and Greece. One building block will be made of 115 Detection Units (DUs) of 18 Digital Optical Modules (DOMs) each. For ORCA, one building block is foreseen; for ARCA, 1/4 building block is foreseen for phase 1, 2 building blocks for phase 2, and 6 building blocks for the final phase. Both ARCA and ORCA will share a similar organization of data and a similar data flow, differing only for the analysis and simulation steps to reflect the different research fields.

Data will be generated by the DOMs anchored offshore on the seabed. Most of them will be signals from photomultipliers; but since the structure holding the DOMs is not rigid, and is subject to the action of water flow, the instantaneous detector geometry is to be worked out by processing the data of hydrophones listening to arrival times of signals from acoustic beacons; in addition, instruments in each DOM monitor the temperature, current direction and strength, DOM orientation and acceleration. For physics and astrophysics goals, optical (PMT) data have the highest information content, and reliable estimates on overall data size can be obtained by considering only them, whereas datasets of other kinds mostly provide logging and monitoring information. The task of data and workload distribution is obtained through a three-tier scheme, as depicted in Fig. 2.

Tier 0 corresponds to the shore stations that collect data from the seabed, apply triggering algorithms to filter them, compute online calibrations and perform quasi-online reconstructions. Tier 1 is embodied by the computing center (such as ReCaS, see later in the text) that work out calibrations, reconstructions and simulations offline. Tier 2 includes local computing clusters where simulation and analysis tasks are performed. Calibrated data (including the raw data) are envisaged

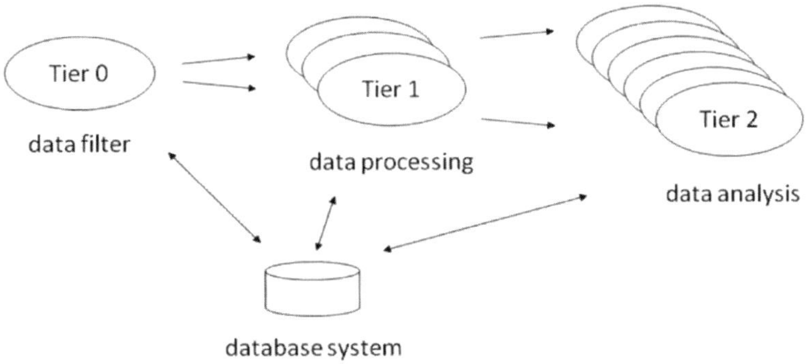

Fig. 2. Three-tier scheme for data reduction. The database system includes a relational database for monitoring and logging a storage system for raw data, calibrated data, physics and simulation data.

to take 750 TB/year/building block. Reconstruction is expected to reduce this dataset by a factor 5. Approximately the same size (150 TB/year/building block) should be taken by simulation data.

Data from the shore stations are sent to computing centers (CCIN2P3 — Lyon, CNAF — Bologna, ReCaS), which provide both rolling storage for data processing (e.g., ReCaS, CCIN2P3) and final storage of reconstructed events and simulation datasets (CCIN2P3, CNAF). CCIN2P3 and ReCaS host the relational database systems used for monitoring and logging, which contains also position reconstruction data.

Tier 2 is made of local computing resources mostly devoted to analysis and specific simulation tasks, which access Tier 1 data using GRID or batch access.

3. ReCaS for KM3NeT: GRID VO, Storage, Computing, and Database Hosting

The contributions of ReCaS to computing in KM3NeT encompass a wide variety of services.

It is remarkable to notice that the Oracle database server for the KM3NeT detector in Italy was set up and has been hosted so far within the ReCaS facilities in the site of the University of Napoli. The database is designed to contain not only the documentation and routine duty management data for the detector; it will also be the source of all detector configuration data to run the data taking, and will receive and store monitoring and logging information from the detector management system (including the so-called "slow control" data). As the database is not physically in the shore station of the detector, network interruptions may occur, and a proper caching system in the shore station is envisaged to allow disconnected operation in case of trouble. The system is expected to store about 1 TB/year in fully relational format, running Oracle Enterprise Database Server 12c on an Intel ×64 8-core machine with 5 TB online storage and 17 TB backup storage on NAS. The database also has the task of providing a full mirror copy of the data stored in the Oracle database server hosted at CCIN2P3 in Lyon, including KM3NeT collaboration management data, component production and testing data also for other KM3NeT sites. Access to the database is not direct, but occurs through a dedicated, lightweight web server based on custom software that provides additional security levels, resource optimization and execution of optimized queries mapped onto usual HTTP URL syntax. The use of HTTP makes integration within the GRID infrastructure (see below) easy.

ReCaS helped KM3NeT achieve the important milestone of GRID access. Indeed, the KM3NeT Virtual Organisation (VO) for GRID was registered thanks to ReCaS management and the first GRID-enabled site was the ReCaS facility at the University of Napoli. KM3NeT jobs can be launched on the GRID by using the ReCaS access point and middleware for scheduling and job management. As many as 300 cores (roughly 10 HS06 each) are available to KM3NeT from ReCaS sites in non-pledged mode. This computing power satisfies the experimental needs of data calibration and first processing/reconstruction. In light load periods, it is possible to use the ReCaS facilities for data reprocessing and simulation to augment the computing power provided by CNAF. Working in the context of the GRID, it is worth noticing that ReCaS can also share the load of other European computing centers supporting KM3NeT, e.g., CCIN2P3.

The ReCaS infrastructure at the University of Napoli also provides high-performance "rolling" storage for current data. A total capacity of about 350 TB in RAID6 configuration is implemented by a dedicated network storage system with 10 Gbps full duplex speed with internal Fiber channel connections and external Ethernet access. This area will receive data from the KM3NeT shore station in Capo Passero and process them for calibration and event reconstruction. The output is buffered locally while it is transferred to CNAF for final storage.

4. Conclusions

ReCaS has given KM3NeT relevant contributions for the successful operation of the experiment. Database hosting, VO management, computing power and local storage are all provided by ReCaS. Their availability plays a major role in the computing model of KM3NeT. In addition to the remarkable machine resources, it is important to mention also the contribution of ReCaS personnel, whose skill and competence is a most valuable asset that helps everyday activity as well as long-term planning and computing model tuning and evolution.

Acknowledgments

We wish to express our grateful appreciation for the long-sighted, friendly and supportive cooperation of Prof. Leonardo Merola and Prof. Guido Russo of ReCaS. We acknowledge that all that has been achieved so far would not have been possible without the competence and commitment of Dr. Vania Boccia of ReCaS.

References

1. Available at http://www.km3net.org.
2. M.G. Aartsen *et al.*, (IceCube Collaboration), First observation of PeV-energy neutrinos with IceCube, *Phys. Rev. Lett.* **111**, p. 021103 (2013).

3. E. Kh. Akhmedov, S. Razzaque and A. Yu. Smirnov, Mass hierarchy, 2–3 mixing and CP-phase with Huge Atmospheric Neutrino Detectors, *J. High Energy Physics.* **2**, p. 82 (2013).
4. S. Adrian-Martinez *et al.*, (KM3NeT Collaboration), Deep sea tests of a prototype of the KM3NeT digital optical module, *Eur. Phys. J. C.* **74** (9), p. 3056 (2014).

Chapter 18

The ReCaS Infrastructure to Simulate Field Theory on Fuzzy Disk

F. Lizzi*,†,‡ and B. Spisso*

*INFN, Sezione di Napoli, Napoli, Italy

†Dipartimento di Fisica, Università di Napoli Federico II, Napoli, Italy

‡fedele.lizzi@na.infn.it

In this chapter, we present a first implementation in the GRID framework, using as a test-bed the ReCaS/S.Co.P.E. infrastructure, of a set of simulations in non-commutative geometry. Such calculations (Monte Carlo-like simulations) were presented in a previous paper using local resources. Our goal here is to open the way to the GRID paradigm to such models. The fuzzy disk is a discretization of the algebra of functions on the two-dimensional disk using finite matrices. Unlike other discretizations, this preserves the action of the full rotation group. We define a φ^4 scalar field theory on it, solve the theory, and analyze numerically 3 different limits for the rank of the matrix going to infinity.

1. Introduction

Field theory on non-commutative spaces (for a review see Ref. 1) is an interesting tool to study the properties of non-commutative geometry.[2–4] Here we will not discuss the motivations for a non-commutative geometry. Among the applications of non-commutative geometry are fuzzy spaces. The original fuzzy space is the fuzzy sphere,[5] but other spaces have been built[22–29] (for a review see Ref. 6). Field theory on fuzzy spaces are a way to study field theory on non-commutative spaces on a finite setting. The fact that the approximation is based on matrices makes them ideal for numerical studies. The most studied case is the fuzzy sphere, for a review see Refs. 7 and 8. Most of the interest for these investigation is the presence of a limit for which the rank of the matrices grow, and at the same time the size of the space increases, thus recovering in the limit a non-commutative plane. By non-commutative plane, we mean the algebra generated by the non-commuting variables x and y satisfying the commutator

$$[x, y] = \mathrm{i}\theta. \tag{1}$$

This *non-commutative geometry* is usually described by a non-commutative $*$ product among the fields, such as the Groenewold–Moyal product.[9,10] As we shall see, the fuzzy disk approximation turns the functions to finite matrices, and hence amenable to numerical simulations. The object of simulations is a quantized φ^4 scalar field theory approximated by $N \times N$ matrices investigating the behavior under 3 different limits, giving rise to the non-commutative plane, the commutative plane, and the commutative disk. We are interested in particular to the phase transitions of the theory as we change the parameters of the action.

The purpose of this chapter is to show a first implementation of the GRID paradigm, using the ReCaS/S.Co.P.E. infrastructure, in this field of theoretical research. The implementation of the GRID paradigm, on the simulation already carried out on a previous paper,[19] allowed us to increase the simulations complexity (aka the matrix rank and the parameters sets) in order to gather better results and even new results (eigenvalues analysis) in the same amount of time compared to the previous implementation. In this paper, we will deal only with the implementation scheme that can be used and we postpone the presentation of the numerical results in a future work. The submissions were achieved using the glite middle-ware, in particular, the parametric job tool.

The chapter is organized as follows. In the next section, we introduce the fuzzy disk and its properties. The real scalar field theory on the fuzzy disk will be presented in Sec. 3. The simulation and the implementation are in Sec. 4. Finally, the results are summarized and discussed in the conclusions.

2. The Fuzzy Disk

The construction of the fuzzy disk[15–18] is based on a quantization of the function on the plane, i.e., the association to a function an operator on an infinite-dimensional Hilbert space *F*. Considering the Hilbert space generated by the countable basis $|\psi_n\rangle$, with n positive or zero, and the operator $\hat{a}$ and its Hermitian conjugate $\hat{a}^\dagger$, acting on the basis as

$$\begin{aligned} \hat{a}|\psi_n\rangle &= \sqrt{n\theta}\,|\psi_{n-1}\rangle, \\ \hat{a}^\dagger|\psi_n\rangle &= \sqrt{(n+1)\theta}\,|\psi_{n+1}\rangle, \end{aligned} \tag{2}$$

with the commutation rule:

$$\left[\hat{a},\hat{a}^\dagger\right] = \theta 1, \tag{3}$$

we recognize the usual creation and annihilation operator of a single harmonic oscillator, with a slightly unusual normalization.

Considering a function on the plane in terms of the conjugate variables $\bar{z}$, z, with $z = x + iy$, we define a quantization map which associates to a function, an operator $\hat{f}$, according to the following rule:

$$f(z,\bar{z}) = \sum_{m,n=0}^{\infty} f_{mn}^{Tay}\bar{z}^m z^n \Leftrightarrow \hat{f} = \sum_{m,n=0}^{\infty} f_{mn}^{Tay}\hat{a}^{\dagger m}\hat{a}^n. \tag{4}$$

Here and in the following, we will not discuss the issues of convergence of the series, we assume the coefficients are of rapid decrease. A more rigorous construction, based on the coherent states of the Heisenberg group and on a Weyl–Wigner map can be found in Ref. 17.

This quantization associates an operator to a function, the inverse map is

$$f = \langle z | \hat{f} | z \rangle, \tag{5}$$

with $|z\rangle$ the usual coherent states $\hat{a}\,|\,z\rangle = z\,|\,z\rangle$. The function associated to the operator is called the *symbol* of the operator.

After quantization, if the plane using the map (4) and its inverse we define the subalgebras of finite $(N+1)\times(N+1)$ matrices considering the functions with a truncated expansion, more precisely taking in account only the expansion terms if both n and m are smaller than a given integer N. Keeping θ fixed, the limit $N\to\infty$ the truncated algebra tends to the non-commutative plane. This is obtained with the help of the projection operator

$$\hat{P}_\theta^{(N)} = \sum_{n=0}^{N} |\psi_n\rangle\langle\psi_n|. \tag{6}$$

At this operator corresponds the function:

$$\hat{P}_\theta^{(N)}(\bar{z}, z) = \sum_{n=0}^{N} \langle z|\psi_n\rangle\langle\psi_n|z\rangle = e^{-\frac{z\bar{z}}{\theta}} \sum_{n=0}^{N} \frac{(z\bar{z})^n}{n!\theta^n} = e^{-\frac{r^2}{\theta}} \sum_{n=0}^{N} \frac{r^{2n}}{n!\theta^n}. \tag{7}$$

This sum can be expressed in terms of incomplete gamma function and gamma function obtaining a radial function:

$$P_\theta^{(N)}(r,\varphi) = \frac{\Gamma(N+1, r^2/\theta)}{\Gamma(N+1)}. \tag{8}$$

In the limit $N \to \infty$, if θ is fixed and nonzero, the symbol (8) converges point-wisely to the constant function $P_\theta^{(N)}(r,\varphi) = 1$, recovering the non-commutative plane. Otherwise, if the product $N\theta$ is fixed, equals to a real constant R^2, the limit for $N \to \infty$ of (8) becomes

$$P_\theta^{(N)}(r,\varphi) \to \begin{bmatrix} 1 & r < R \\ 1/2 & r = R \\ 0 & r > R \end{bmatrix} = 11_d(r), \tag{9}$$

where the convergence still is point-wise. This limit converges to a step function in the radial coordinate r, in other words, the characteristic function of the disk on the plane. Given the algebra $\mathcal{A}$ generated by $\hat{a}$ and $\hat{a}^\dagger$, a sequence of sub algebras $A_\theta^{(N)}$ can be defined by

$$\mathcal{A}_\theta^{(N)} = \hat{P}_N \mathcal{A} \hat{P}_N. \tag{10}$$

The effect of this projection on a generic function is

$$f_\theta^{(N)} = \langle z | P_\theta^{(N)} \hat{f} P_\theta^{(N)} | z \rangle = e^{-\frac{z\bar{z}}{\theta}} \sum_{m,n=0}^{N} f_{nm} \frac{z^n \bar{z}^m}{\sqrt{m!n!\theta^{n+m}}}, \tag{11}$$

namely a truncation of the series expansion. The full algebra A is isomorphic to an algebra of operators, however the previous relations show that $A_\theta^{(N)}$ is isomorphic to $\mathbb{M}_{N+1}$ the algebra of $(N+1)$ rank matrices, on each sub algebra $A_\theta^{(N)}$, the symbol (8) is then the identity matrix in every $\mathbb{M}_{N+1}$. The name fuzzy disk derives from the cutoff; for a fixed N every function with the same first $(N+1)^2$ terms are mapped in the same matrix losing all the information of the higher orders. In addition, the symbols of the fuzzyfied function are still defined outside the disk of radius $R^2 = N\theta$, but they are exponentially damped outside the disk.

It is possible to define a fuzzy Laplacian, where the eigenstates will be the analogous of the standard Bessel function. The fuzzy

Bessel are a base for $\mathbb{M}_{N+1}$ and their symbols tends to the standard Bessel for $N \to \infty$. Turns out that the fuzzy Laplacian takes the form

$$\nabla^2 \hat{f} = \frac{4}{\theta^2} P_\theta^{(N)} \left[\hat{a}, \left[P_\theta^{(N)} f P_\theta^{(N)}, \hat{a}^\dagger \right] \right] P_\theta^{(N)}. \tag{12}$$

It can be proved that the eigenvalues of this Laplacian to converge for $N \to \infty$ to the spectrum of the standard Laplacian defined on a disk with Dirichlet homogeneous boundary conditions on the edge. In the fuzzy approximation, the continuum eigenfunctions $e^{in\varphi} J_{|n|}\left(\sqrt{\lambda_{|n|}}, k^r\right)$ are represented by matrices $\mathbb{M}_{N+1}$, but the max possible n are fixed by the dimension of the fuzzyfication $n \leq N$.

3. Real Scalar Field Theory on the Fuzzy Disk

The previous construction allows us to define a fuzzyfied version of a field theory on a disk. For the real scalar case with Euclidean signature, with the φ^4 potential, the action is given by

$$S(\varphi) = \int \left(\varphi \nabla^2 \varphi + \mu \varphi^2 + \frac{\lambda}{4} \varphi^4 \right) d^2 z, \tag{13}$$

with ∇ the Laplacian on the disk, μ and λ are mass and interaction parameters, respectively. We choose this particular model due to the well studied plane case. It is known[20,21] that the diagrammatic expansion of this theory has only one divergent diagram, the so called tadpole diagram and it is renormalizable. Using the fuzzification procedure, the action (13) can be approximated by:

$$\hat{S}_N(\hat{\varphi}) = \pi \theta \operatorname{Tr}\left(\frac{4}{\theta^2} \hat{\varphi} \left[\hat{a}, \left[\hat{\varphi}, \hat{a}^\dagger \right] \right] + \mu \hat{\varphi}^2 + \frac{\lambda}{4} \hat{\varphi}^4 \right). \tag{14}$$

In this action, $\hat{\varphi}$ are finite-dimensional Hermitian matrices and the product between the field became the standard matrix multiplication, being a finite matrix model can be approached numerically using

Monte Carlo techniques, a method currently in use for other fuzzy spaces like the sphere.

We can rescale the operators â, $\hat{a}^\dagger$ (13) as $a = \sqrt{\theta}\,\hat{a}, a^\dagger = \sqrt{\theta}\,\hat{a}^\dagger$ in order to extract the θ dependence and recast the action making θ manifest

$$\hat{S}_N(\hat{\varphi}) = \pi \operatorname{Tr}\left\{4\hat{\varphi}\left[\hat{a},\left[\hat{\varphi},\hat{a}^\dagger\right]\right] + \theta\left(\mu\hat{\varphi}^2 + \frac{\lambda}{4}\hat{\varphi}^4\right)\right\}. \tag{15}$$

It is useful to define the potential part of the action given by

$$S_V(\varphi) = \pi\theta \operatorname{Tr}\left(\mu\varphi^2 + \frac{\lambda}{4}\varphi^4\right). \tag{16}$$

From now, to simplify notations we will omit the hats and the subscript N, as all fields are matrices. In the simulation, we will use another equivalent choice of the parameters often used for the potential part, namely,

$$S_V'(\varphi) = \pi\theta r \operatorname{Tr}\left(\mu'\varphi^2 + \frac{\varphi^4}{4}\right). \tag{17}$$

The two choices are related by

$$\begin{aligned} r &= \lambda, \\ r\mu' &= \mu. \end{aligned} \tag{18}$$

The fuzzy disk arise from the relation $R^2 = \theta N$, where θ is the parameter of non-commutativity and R is the radius of the disk. Introducing in the non-commutativity parameter a dependence on the matrix size N and performing different limits we have 3 different cases:

- $N \to \infty$ with $R^2 = N\theta$ fixed. In this case, we recover the commutative disk of radius R. In the following, we will take $R = 1$.
- $N \to \infty$ with $R \to \infty$ and θ fixed. In this case, we recover the non-commutative plane. We will take $\theta = 1$ and therefore $R^2 = N$.

- $N \to \infty$, $\theta \to 0$ and $R \to \infty$. In this case, we recover the commutative plane. In the following, we will have $\theta = 1/\sqrt{N}$ which implies $R^2 = \sqrt{N}$, in the limit $N \to \infty$ we recover the commutative plane.[a]

One of the main aims of the simulation will be to find the phase diagrams for the fuzzy field theory.

4. Simulations

Using Monte Carlo method, we will produce a sequence of configurations $\{\psi_j\}_{j=1,2}, \ldots, T_{\text{MC}}$ and evaluate the average of the observables over the set of configurations. These sequences of configurations, called Monte Carlo chain, are representatives of the configuration space at given parameters. In this settings, the expectation value is approximated as

$$\langle O \rangle \approx \frac{1}{T_{\text{MC}}} \sum_{j=1}^{T_{\text{MC}}} O_j, \tag{19}$$

where O_j is the value of the observable O evaluated in the j-sampled configuration, ψ_j, $O_j = O[\psi_j]$. The internal energy is defined as

$$E(N, \mu, \lambda) = \langle S \rangle. \tag{20}$$

The specific heat takes the form

$$C(N, \mu, \lambda) = \langle S^2 \rangle - \langle S \rangle^2. \tag{21}$$

These quantities correspond to the usual definitions for energy

$$E(N, \mu, \lambda) = -\frac{1}{\mathcal{Z}} \frac{\partial \mathcal{Z}}{\partial \beta} \tag{22}$$

---

[a]It is possible recover the commutative plane taking $\theta = N^{\alpha}$, where α is a real number varying between $-1 < \alpha < 0$, our choice is just dictated by simplicity.

and specific heat

$$C(N,\mu,\lambda) = \frac{\partial E}{\partial \beta}, \tag{23}$$

where $\mathcal{Z}$ is the partition function. Another important observable used to look for a phase transition is finite-volume susceptibility defined as:

$$\chi = \langle \mathrm{Tr}(\varphi)^2 \rangle - \langle \mathrm{Tr}(\varphi) \rangle^2. \tag{24}$$

The phase transitions are located at the values where the previous quantities show a peak when varying the parameters. The simulations were conducted using a Metropolis algorithm with the jackknife and binning methods, in order to evaluate the errors on the expectation values (19). The usual update algorithm, used to generate a new configuration, changes all the entries of the independent matrices

$$\varphi_{ij} \rightarrow \varphi_{ij} + a_{ij}, \quad \text{with } i, j = 1, \ldots, N, \tag{25}$$

where N is the matrix size and a_{ij} are random real numbers. The new proposed configurations are judged by the Metropolis algorithm and the probability of accepting the new configuration $\psi_{ij} + x$ is calculated as $\min(e^{-\delta S}, 1)$. In order to reduce the computation time, instead a global matrix change, we have changed just 1 matrix entry for each Monte Carlo step. The drawbacks of this approach are mainly two: the complication of the code and the dependence of the correlation time τ in N. In fact, 2 configurations will share at least one coefficient until N^2 Monte Carlo steps, thus, these correlations introduce an increase in the correlation time increasing N and presumably τ is proportional to N^2. However, using this optimization and the implementation of parallel computing, we were able to execute our simulation with good precision and in reasonable time. The initial conditions of the Markov chain are chosen randomly in the configuration space, although we used 2 general types; hot initial conditions, which are configurations far from the minimum, and cold start conditions, which correspond to configurations close to the minimum.

5. Running on the GRID

The high grade of parallelism allow us to process several different configurations of the model potentially, completely in parallel. Besides, each Monte Carlo can be parallelized furthermore. The first result was obtained using local resource (a 10 nodes 4-core cluster) and OpenMP with the following characteristic numbers for 1 week of computation:

- Computation time: 1 week;
- Number of parameters sets used: 600;
- Maximum matrix rank achieved: 20.

This first implementation of the GRID paradigm, in this field of theoretical research, allowed us to push the complexity of the simulations (aka the matrix rank and the parameters sets), gathering much better result and even new results (eigenvalues analysis) in the same amount of time spent in the previous simulations:

- Computation time 1 week;
- Number of parameters sets used: 1,200;
- Maximum matrix rank achieved: 40;
- total Number of jobs: 5,000.

All the submissions were achieved using a parametric job at together in a suitable wrapper script. Parametric job type allows to submit bulk of jobs as a single job, and then Work Load Manager breaks the parametric job into many single jobs and submit them separately to the Computing Elements on the "father" job behalf. Upon submission, every sub-job will be associated with an individual identifier (job ID), and besides that a common job ID will be assigned to the whole set of jobs. This common ID is used to list status or retrieve output of all jobs at once.

Since the simulation depends on a set of parameters, the executable takes more than one argument and it is needed to submit large number of jobs that take different arguments.

```
#gridfuzzy.jdl
JobType = "Parametric";
Executable = "gridfuzzy.sh";
StdOutput = "gridfuzzy_out_PARAM_.txt";
StdError = "gridfuzzy_err_PARAM_.txt";
Parameters = 500 ;
ParameterStart = 1;
ParameterStep = 1;
Arguments = "gridfuzzy arguments.txt _PARAM_";
InputSandbox = {"gridfuzzy.sh", "gridfuzzy", "arguments.txt"};
OutputSandbox ={"out_PARAM_.tgz","gridfuzzy_out_PARAM_.txt"};
RetryCount = 10;
```

Fig. 1. An example of jdl file used.

```
#gridfuzzy.sh
#!/bin/bash
exec_file = $1
arg_file=$2
N=$3
arguments=$(head -n $N $arg_file | tail -n 1)
chmod +x $exec_file #sets executable
tar -czvf out$3.tgz  *.out
rm -f *.out
exit 0
```

Fig. 2. The wrapper file used.

All arguments set are contained in one text file (arguments.txt), so that each line contains an arguments set for a single run. At the time of the submission, using the jdl file in Fig. 1 and the wrapper in Fig. 2, 500 jobs will be submitted and variable PARAM (in the jdl file) will take values from 1 to 500. For each value of PARAM, the script gridfuzzy.sh will be executed performing the following operations: extract line with number PARAM from file arguments.txt and execute the simulation using previously extracted line as argument set. This implementation uses only the parameters-based parallelization obtaining excellent results, although these models offer different levels of parallelism:

- Parameters-based parallelism;
- Mote Carlo-based parallelism;
- Matrix-based parallelism.

After a first phase called thermalization, which is necessarily serial, a single Monte Carlo can be split in various parallel chains. A "father" job, after the thermalization phase, should be able to start new jobs by itself, using its results as inputs, and then gather the results of the new jobs. In this case, we have a set of jobs that are related, in a sense that output of some jobs will be used as input for another job. In order to take advantage of this type of parallelism, the second implementation will use a particular job class, supported by glite-WMS, named Direct Acyclic Graph (DAG) job. Translating grid to graph terminology, grid jobs would translate to graph nodes, and job dependencies would be graph directed edges. A graph representing a DAG job has to be acyclic (no circular dependencies allowed), otherwise, there would be no job to start from. As for other types of grid jobs, it is needed to describe a DAG job in a jdl file.

5.1. *Conclusions and Prospective*

Since these simulations have a high grade of parallelism, the use of the GRID allows us to process all considered different configurations of the model potentially completely in parallel. This first implementation of the GRID paradigm, using the ReCaS/S.Co.P.E. infrastructure, in this field of theoretical research, allowed us to push the complexity of the simulations in such a way to obtain a much better result and even new results in the same amount of time spent in the previous simulations. The numerical simulations conducted reveal 3 different phases: uniform and disordered phases already present in the commutative scalar field theory and a non-uniform ordered phase as non-commutative effects. It is important to notice that this implementation can be adapted to several other numerical simulations on high energy theoretical models (especially if based on a Monte Carlo approach): such as the Quantum Chromo-Dynamics and in general to the lattice and matrices field theories.[11–14] This implementation can be furthermore optimized. For example, implementing the matrices parallelism using the matrix × matrices parallel algorithm will be very useful to access to resource like GPU, which are very efficient for such tasks. This further optimization will push the maximum matrix rank

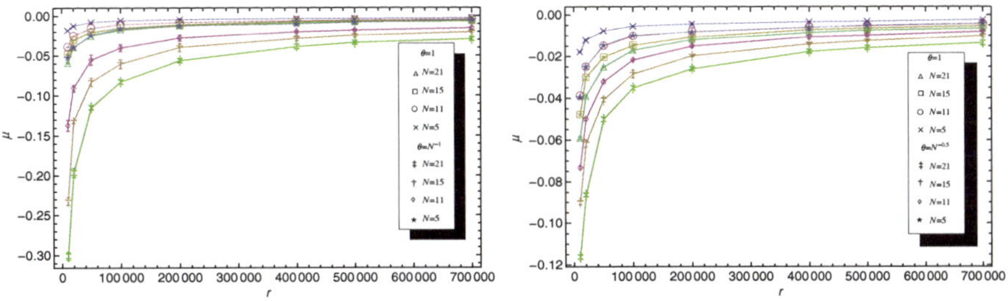

Fig. 3. Plot of the comparisons of transition curves between the non-commutative plane limit and commutative disk limit (left) and between the non-commutative limit and commutative plane limit (right) for $N = 5, 11, 15, 21$.

reachable, this is very desirable from a physical point of view, since N is connected to the approximation of the model, therefore, to the so-called, continuum limit (Fig. 3).

Acknowledgments

This work has been realized, thanks to the use of the ReCaS/S.Co.P.E. computing infrastructure at the Università di Napoli *Federico II*. A particular acknowledgement goes to Vania Boccia, Luisa Carraciuolo and Silvio Pardi who supported this job with several useful suggestions.

References

1. R. J. Szabo, Quantum field theory on noncommutative spaces, *Phys. Rept.* **378**, p. 207 (2003) [hep-th/0109162].
2. A. Connes, *Noncommutative Geometry*, Academic Press, New York (1984).
3. G. Landi, *An Introduction to Noncommutative Spaces and Their Geometries*, Lecture Notes in Physics Monographs, Vol. 51, Springer–Verlag, Berlin (1997).
4. J. M. Gracia-Bondia, J. C. Varilly and H. Figueroa, *Elements of Noncommutative Geometry*, Birkhäuser (2000).
5. J. Madore, The Fuzzy sphere, *Class. Quantum Grav.* **9**, p. 69 (1992).

6. A. P. Balachandran, S. Kurkcuoglu and S. Vaidya, Lectures on Fuzzy and Fuzzy SUSY Physics, World Scientific Singapore (2007).
7. J. Medina, Fuzzy scalar field theories: numerical and analytical investigations, arXiv:0801.1284 [hep-th].
8. J. Medina, W. Bietenholz, F. Hofheinz and D. O'Connor, Field theory simulations on a fuzzy sphere: An alternative to the lattice, *PoS LAT* **2005**, p. 263 (2006) [hep-lat/0509162].
9. J. E. Moyal, Quantum mechanics as a statistical theory, *Proc. Cambridge Philos. Soc.* **45**, p. 99 (1949).
10. S. Galluccio, F. Lizzi and P. Vitale, Twisted noncommutative field theory with the Wick–Voros and Moyal products, *Phys. Rev. D* **78**, p. 085007 (2008) [arXiv:0810.2095 [hep-th]].
11. B. Ydri, Impact of Supersymmetry on emergent geometry in Yang–Mills matrix models II, *Int. J. Mod. Phys. A* **27**, p. 1250088 (2012) [arXiv:1206.6375 [hep-th]].
12. R. Delgadillo-Blando, D. O'Connor and B. Ydri, Geometry in transition: a model of emergent geometry, *Phys. Rev. Lett.* **100**, p. 201601 (2008) [arXiv:0712.3011 [hep-th]].
13. B. Spisso and R. Wulkenhaar, A numerical approach to harmonic noncommutative spectral field theory, *Int. J. Mod. Phys. A* **27**, p. 1250075 (2012) [arXiv:1111.3050v4 [math-ph]].
14. B. Spisso, A numerical approach to harmonic non-commutative spectral field theory, [arXiv:1111.2871v1 [hep-th].
15. F. Lizzi, P. Vitale and A. Zampini, The Fuzzy disk, *J. High Energy Phys.* **0308**, p. 057, (2003) [hep-th/0306247].
16. F. Lizzi, P. Vitale and A. Zampini, The Beat of a fuzzy drum: Fuzzy Bessel functions for the disk, *J. High Energy Phys.* **0509**, p. 080 (2005) [hep-th/0506008].
17. F. Lizzi, P. Vitale and A. Zampini, From the fuzzy disk to edge currents in Chern–Simons theory, *Mod. Phys. Lett. A* **18**, p. 2381 (2003) [hep-th/0309128].
18. F. Lizzi, P. Vitale and A. Zampini, The fuzzy disk: A review, *J. Phys. Conf. Ser.* **53** p. 830 (2006).
19. B. Spisso and F. Lizzi Numerical analysis of the φ^4 scalar field theory on the fuzzy disk, *Int. J. Mod. Phys. A* **27**, p. 1250137 (2012) arXiv:1207.4998v2 [hep-th].
20. S. Minwalla, M. Van Raamsdonk and N. Seiberg, Noncommutative pertur- bative dynamics, *J. High Energy Phys.* **0002**, p. 020 (2000) [hep-th/9912072].

21. M. Chaichian, A. Demichev and P. Presnajder, Quantum field theory on noncommutative space-times and the persistence of ultraviolet divergences, *Nucl. Phys. B* **567**, p. 360 (2000) [hep-th/9812180].
22. X. Martin, A Matrix phase for the φ^4 scalar field on the fuzzy sphere, *J. High Energy Phys.* **0404**, p. 077 (2004) [hep-th/0402230].
23. M. Panero, Numerical simulations of a non-commutative theory: The Scalar model on the fuzzy sphere, *J. High Energy Phys.* **0705**, p. 082 (2007) [hep-th/0608202].
24. F. Garcia Flores, X. Martin and D. O'Connor, Simulation of a scalar field on a fuzzy sphere, *Int. J. Mod. Phys. A* **24**, p. 3917 (2009) [arXiv:0903.1986 [hep-lat]].
25. T. R. Govindarajan, S. Digal, K. S. Gupta and X. Martin, Phase structures in fuzzy geometries, arXiv:1204.6165v1 [hep-th].
26. C. R. Das, S. Digal and T. R. Govindarajan, Finite temperature phase transition of a single scalar field on a fuzzy sphere, arXiv:0706.0695v2 [hep-th].
27. M. Panero, Quantum field theory in a non-commutative space: Theoretical predictions and numerical results on the fuzzy sphere, *SIGMA* **2**, p. 081 (2006) [hep-th/0609205].
28. C. R. Das, S. Digal and T.R. Govindarajan, Spontaneous symmetry breakdown in fuzzy spheres, *Mod. Phys. Lett. A* **24**, p. 2693 (2009) [arXiv:0801.4479v2 [hep-th]].
29. S. Digal and T. R. Govindarajan, Topological stability of broken symmetry on fuzzy spheres, *Mod. Phys. Lett. A*, **27**, p. 1250082 (2012) [arXiv:1108.3320v2 [hep-th]].

Chapter 19

A Dashboard for ALICE Activity in Bari Tier 2

D. Elia*, A. Franco* and G. Vino*[†,‡]

**Istituto Nazionale Fisica Nucleare, Sezione di Bari, via E. Orabona 4, 70126 Bari, Italy*

[†]*Dipartimento Interateneo di Fisica, Università degli Studi di Bari, via E. Orabona 4, 70126 Bari, Italy*

[‡]*gioacchino.vino@ba.infn.it*

The performance and status monitoring of the ALICE activity at the *Istituto Nazionale Fisica Nucleare* (INFN), BARI ALICE Tier 2 are performed using MonALISA, Zabbix and Torquemon. The presented dashboard allows to plot on the same interface all the interested metrics coming from the three monitoring levels, together with a more customizable graphical interface.

1. Monitoring ALICE–Bari

The Istituto Nazionale Fisica Nucleare (INFN)–BARI computing farm provides computational resources to different European Organization for Nuclear Research (CERN) experiments, including ALICE.[1] It is a member of the ALICE Computing Grid as Tier 2 center. For 2015, INFN–BARI dedicated 9100 HS06 CPUs and 1,060 TB of disk space. The computational resources are managed using Torque,[2] an advanced and widely used open source batch system. The local performance and status monitoring for ALICE is performed on 3 levels: the application level service monitored by Monitoring Agents using a Large Integrated Services Architecture (MonALISA), the Torque status by Torquemon and the operating system level monitored using Zabbix.

Torquemon is a tool created in Bari. It extracts all information on the PBS activity like jobs, queues, or the batch server status and stores them in a local Structured Query Language (SQL) database with an optimized schema. The graphical interface, a web page written in Hypertext Preprocessor (PHP), allows to select a time interval that is used to require data from the database. With the returned values, the graphs are created and inserted in the web page.

MonALISA has been developed over the last 4 years by Caltech.[3] It is designed as an ensemble of autonomous multi-threaded, self-describing agent-based subsystems which are registered as dynamic services. These agents can be used to analyze and process the information in order to provide optimization decisions.[3] Also, it is described as a fully distributed service system with no single point of failure. Like all the other services belonging to the ALICE Environment (AliEn), MonALISA is hosted on a machine with a particular service profile, called VOBox. It is used to collect several information on the site facility, like the network traffic and the jobs status and it also provides a web-based interface.

Zabbix is a free monitoring system created by Alexei Vladishev, actively developed and supported by Zabbix SIA.[4] It can be described as a semi-distributed monitoring system with centralized management. The simplest architecture includes Zabbix servers and agents, processes deployed on monitoring targets to actively monitor local resources and return the gathered data to the server. It provides a notification service, the dashboard and the API interface. Zabbix is used by INFN–BARI to monitor the CPUs, memory, disk usage, network traffic and every other kind of metric on the computing farm hosts.

2. The Dashboard

Unfortunately none of the 3 monitoring systems described so far includes a complete solution for the purpose of monitoring the overall ALICE activity, because each one is focused on a different aspect of the computing activities. The dashboard deployed by us aggregates metrics coming from the different monitoring systems in a single view. The data retrieved by the dashboard are collected together in a single place, so that a part of them can be easily forwarded to other systems: for instance, in this work a subset of metrics is exported in Zabbix.

MonALISA and Zabbix return the data from their own programmatic interfaces, respectively Web Services Description Language/Simple Object Access Protocol (WSDL/SOAP) and **Hypertext Transfer Protocol** (HTTP), while Torquemon returns the batch system data using JSON. The architecture of the system presented is shown in Fig. 1.

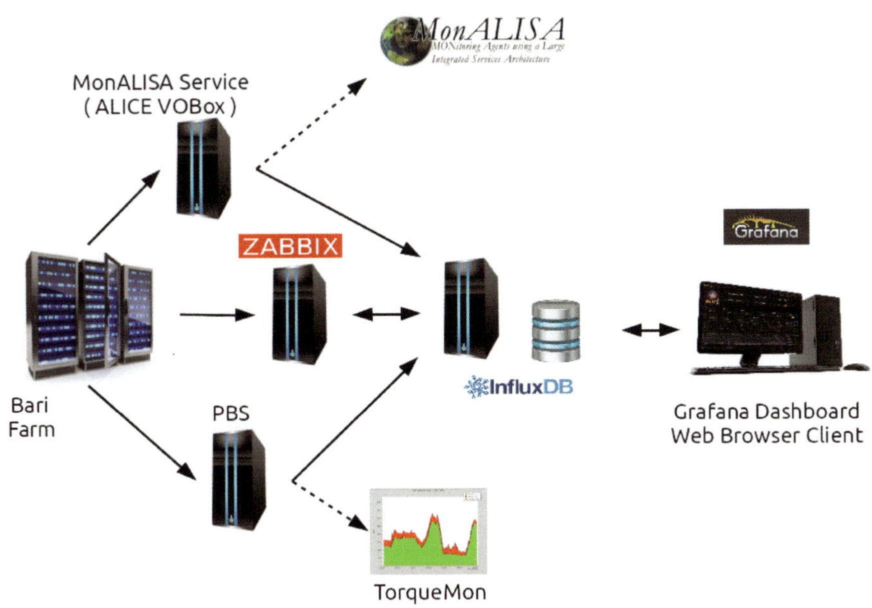

Fig. 1. Monitoring system architecture.

The system is composed of three main components:

- a database used to store all the collected data,
- a graphical interface which plots graphs and indicators,
- a component responsible of gathering the values from the three data sources and storing a part of them into Zabbix.

The chosen database is InfluxDB,[5] an open source distributed time series database with no external dependencies, written in Go. It is based on LevelDB, a key-value database. Beyond a web front-end, a HTTP interface and libraries are provided to interact with it. The real potentiality of InfluxDB is its capacity of aggregating values in time buckets on the fly without the need of doing it manually. Whilst InfluxDB is a time series database, it can be accessed using an SQL-like query language.

Another advantage of InfluxDB is the large amount of software that can be used to easily interact with it. Among them Grafana has been chosen. Grafana is a frontend for Graphite, InfluxDB and OpenTSDB with powerful visualization features for time series data.[6] It has no dependences since it's a client side application, written in Node.js, that runs in a browser: it allows to delegate the rendering task to the HTTP client. The page is published by a common HTTP server.

Grafana enables you to insert in the dashboard graphs, panels containing single values (also known as "singlestat"), text, images and HTML code.

The last component of the Dashboard collects the data from the datasources and stores them in the back-ends, InfluxDB and Zabbix. This component has been written from scratch as there is no other similar software doing the same task. A modular approach has been used to make it easier to add plugins to the system like new data sources or back-ends, to manage components and for debugging purpose. A JSON is used to store intermediate data.

Beyond InfluxDB, there is another database, called MetricDB, where only the verified ALICE metrics are stored. This solution avoids

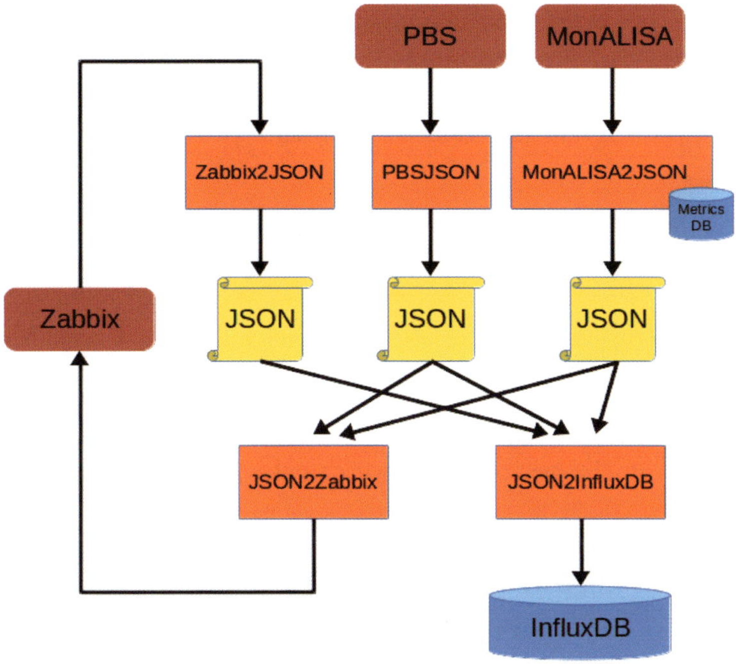

Fig. 2. Software components, the arrows show the data flows.

to check repeatedly whether a specific ALICE metric name is well written or not. Figure 2 shows the different software components.

The red blocks are data sources (Zabbix is also a back-end), the blue cylinders are the databases, the yellow ones are the JSON temporary files and the orange blocks are the written codes. The scripts *Zabbix2JSON*, *PBS2JSON* and *MonALISA2JSON* have the task to collect data from the respective data sources and save them in the temporary JSON file. MonALISA2JSON extracts the ALICE interested metrics from MetricDB before collecting data from MonALISA. *JSON2Zabbix* and *JSON2InfluxDB* extract data from the JSON file and store them in Zabbix and InfluxDB, respectively. These five codes are inserted in a script to be executed periodically to ensure new values are present in the database. The remaining 3 allow a user to interact with the databases via command line interface.

Python has been chosen as programming language because it is a Object Oriented Programming Language, has thousands of extensions, provides the InfluxDB client and can interact with the Zappix application program interface (API) and provides WSDL/SOAP library.

3. Configuration and Performances

The system has been installed on a virtual machine (VM) using Scientific Linux 6. After the creation of databases and users, all the scripts must be configured by adding username, password and the URL of the services. Next, some metrics must be added to the MetricDB and the periodical script can then be executed through a Cron job.

The last step consists of the configuration of the Grafana dashboard. The Dashboard representing ALICE activity at INFN–BARI contains one image loaded using HTML code, 12 singlestat panels and 6 graphs with 20 overall metrics. Figure 3 shows the Bari ALICE activity dashboard.

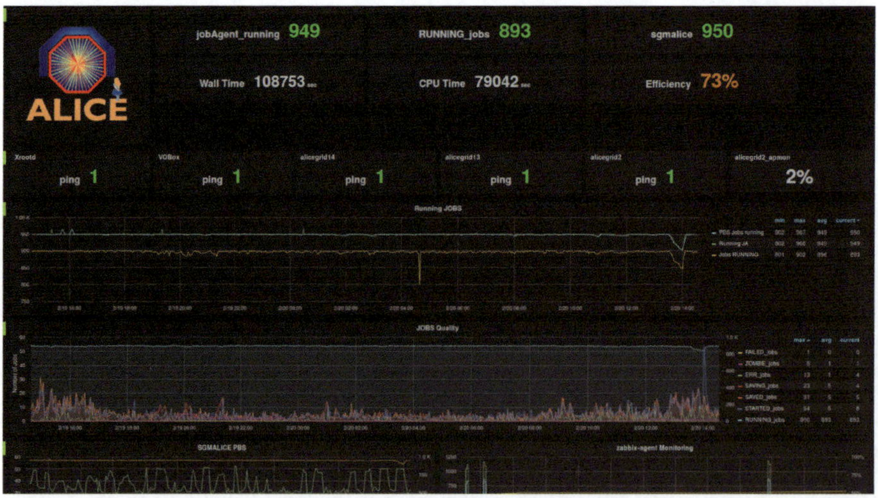

Fig. 3. Bari ALICE activity dashboard.

Table 1. Measures on InfluxDB.

Parameter	**Value**
Metrics	167
Raw values stored	5.621.167
Disk usage	101.5 MiB
Raw data query time	117.03 s
Aggregate data query time	60.23 s

Table 2. Performance of Grafana varying the visualization period.

Time period	**Aggregation time**	**Update time [sec]**
1 day	1m	2
1 week	5m	6
1 month	30m	7
5 months	3h	24

The performances of InfluxDB are measured on time spent to return values using a query and on the disk usage. Two queries are used to request all raw values and all daily aggregate values, using the mean function. The resulting measures are shown in Table 1. The overall performance is measured using the time spent from Grafana to update the dashboard for a given time period. The Grafana page is requested using the Apache server. The response time depends on the time period since the *aggregatio la domanda sullo schema aon* time is a function of it. The measured time includes the Grafana page updating time, the time spent to return the value from InfluxDB and the time required to plot graphs. Of course the overall time depends on the number of graph and metrics in the dashboard. The dashboard used for the test is that shown in Fig. 3. The measures extracted are shown in Table 2.

The values in the table show performance figures for the presented system. It is hard to find the flexibility and the customization possibilities provided by InfluxDB and Grafana in other softwares.

4. Conclusions

A new system that will be able to provide a fast dashboard for diagnostic and debugging purposes on statistical basis has been described. It provides a simple and clear interface, together with flexibility in importing data.

As local monitoring for the ALICE INFN–BARI Tier 2, the work has been presented at the Italian Tier 2s meeting in Frascati[7] where it has been proposed to extend it to the other Italian sites. Further improvements can be the implementation of agents to import data from the most popular monitoring systems and operating systems and a graphical interface capable to configure easily the overall system.

References

1. Available at http://aliceinfo.cern.ch/Public/Welcome.html.
2. Available at http://www.adaptivecomputing.com/products/open-source/torque.
3. Available at http://monalisa.caltech.edu/monalisa.html.
4. Available at https://www.zabbix.com/documentation/2.2/start.
5. Available at http://influxdb.com.
6. Available at http://grafana.org.
7. *ALICE INFN Tier 2 Workshop and Computing Board*. Available at https://indico.cern.ch/event/344910/, LNF Frascati (Italy), 18th, 19th December 2014.

Life Science

Chapter 20

Computational Approaches to the Study of Melanogenesis

O. Crescenzi

Department of Chemical Sciences,
University Federico II of Naples, Naples, Italy
orlando.crescenzi@unina.it

Melanins are the most prominent pigments in mammals, accounting for most of the colorations of human skin, hair and eyes, and are the primary determinants of racial pigmentation.

In order to address unresolved mechanistic issues in the field of melanogenesis, we have complemented state-of-the-art data obtained by experimentalists (synthesis of precursors, isolation and characterization of oligomeric intermediates, biomimetic oxidations, pulse radiolysis experiments, isotope labeling techniques, kinetic measurements, etc.), with extensive computational explorations, mostly carried out at the DFT and TDDFT level.

This integrated approach has allowed us to shed light on some hitherto unrecognized aspects of the mechanistic pathways leading to pigment formation.

1. Introduction

The term "melanin" is used in a rather broad sense to describe certain classes of black, brown and reddish pigments of widespread occurrence along the phylogenetic scale.[1,2]

In mammals, melanins represent the single most prominent pigment type, and are believed to play an important role as a defense systems against actinic damage, an aspect which is finding technological applications. Moreover, they are implicated in important disorders such as albinism, vitiligo and melanoma.

More specifically, the black-to-brown, insoluble biopolymers of dark human skin, hair, eyes, and melanomas, dubbed eumelanins, arise biogenetically by the oxidative polymerization of tyrosine *via* a range of catechol and quinone intermediates. Eumelanic polymers can also be obtained *in vitro* by oxidation of 5,6-dihydroxyindole, a key monomer intermediate in the biosynthetic pathway.

By contrast, the distinctive pigments of red human hair, dubbed pheomelanins, arise by a diversion of the eumelanogenesis pathway, leading to the formation of cysteinyldopas, followed by their oxidative polymerization *via* 1,4-benzothiazinylalanine intermediates.

The direct characterization of melanic pigments represents a formidable challenge even to the most sophisticated instrumental techniques, on account of the unfavorable physico-chemical properties (insolubility in all solvents, absence of well-defined spectral features, etc.). This state of affairs reflects in part the highly heterogeneous nature of typical melanic pigments, which contain several different types of monomeric units, at different oxidation levels, and connected through various possible linkage sites. The degree of polymerization and the extent of crosslinking are hard to quantify. Significant interunit interactions, conformational flexibility, and variable water content, all add to the complexity of the picture.

Therefore, a fertile approach to the structural and functional characterization of melanins relies on the study of model systems of smaller complexity. In particular, mechanistic studies carried out at the level of biogenetic precursors and intermediates (monomers, dimers, etc.) have proved particularly fruitful in this sense.

In the following, we will present two case studies, which highlight the potentiality of integrated experimental/computational approaches in this field. In essence, we have complemented state-of-the-art data obtained by experimentalists (synthesis of precursors, isolation and characterization of oligomeric intermediates, biomimetic oxidations, pulse radiolysis experiments, isotope labeling techniques, kinetic measurements, etc.), with extensive computational explorations, mostly carried out at the density functional theory (DFT) and time-dependent density functional theory (TDDFT) level,[3,4] to shed light on some hitherto unrecognized aspects of the mechanistic pathways leading to pigment formation.

2. Case Study 1: 5,6-Dihydroxyindole Oxidation Mechanisms

One-electron oxidation 5,6-dihydroxyindole (e.g., in the course of pulse-radiolysis experiments) in aqueous solution at pH around neutrality leads to the corresponding semiquinone. This latter decays by a second-order kinetics, giving rise to a distinctive chromophore, absorbing in the range 400–430 nm, originally attributed to the elusive 5,6-indolequinone, that would be generated by a disproportionation process.

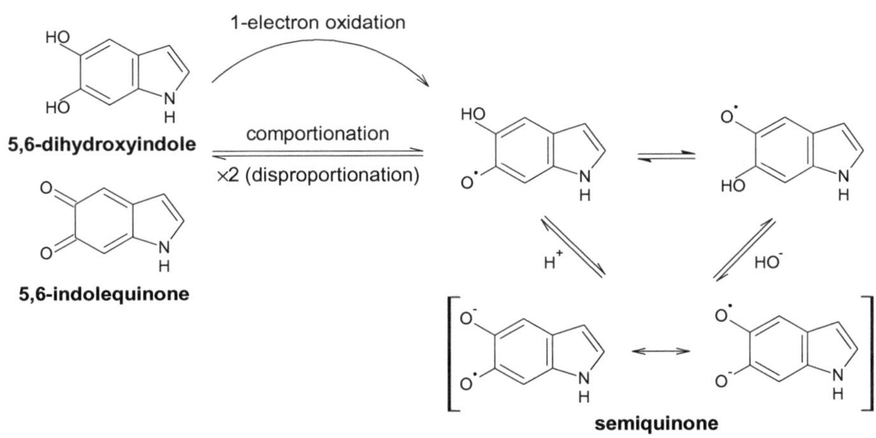

Scheme 1. The initial stages of 5,6-dihydroxyindole oxidation.

Because of the central relevance of 5,6-dihydroxyindole oxidation chemistry to eumelanin synthesis and applications, we were prompted to reinvestigate the mechanism and products of semiquinone decay.[5]

By adopting an integrated chemical/pulse radiolytic/DFT approach, and exploiting deuterium labeling as a specific mechanistic probe, we could provide evidence that 5,6-dihydroxyindole semiquinone decays mainly by a free radical coupling mechanism. This conclusion was supported by the inverse kinetic isotope effect observed with deuterated substrates, by the identification of unprecedented dihydrobiindole products, and by the good matching of experimental absorption spectra of free radical coupling products of 5,6-dihydroxyindole semiquinone with profiles computed at the TDDFT level (Fig. 1 and Scheme 2).

A detailed computation analysis of the kinetics and thermodynamics of the disproportionation equilibrium and free radical coupling of

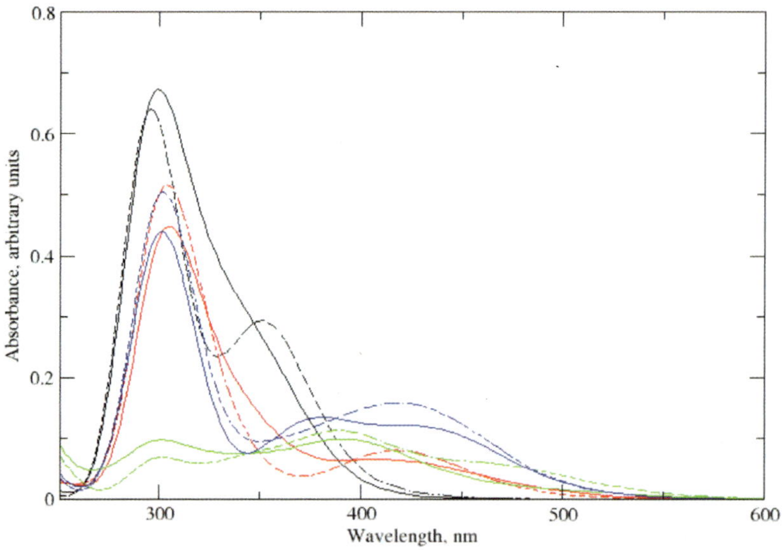

Fig. 1. Computed ultraviolet-visible (UV-Vis) spectra of dihydrobiindoles putatively involved in 5,6-dihydroxyindole radical coupling. Solid black line: 2,2′-dimer, *meso* isomer (**1**), *gauche* conformer around the inter-ring bond; dashed black: 2,2′, *dl*(**2**), *gauche*; solid red: 2,7′, 2*R*,7′*R*(**3**), *gauche+*; dashed red: 2,7′, 2*R*,7′*S*(**4**), *trans*; solid green: 7,7′, *meso*(**5**), *gauche*; dashed green: 7,7′,*dl*(**6**), *trans*; solid blue: 2,4′, 2*R*,4′*S*(7), *gauche+*; dashed blue: 2,4′, 2*R*,4′*R*(**8**), *gauche*.

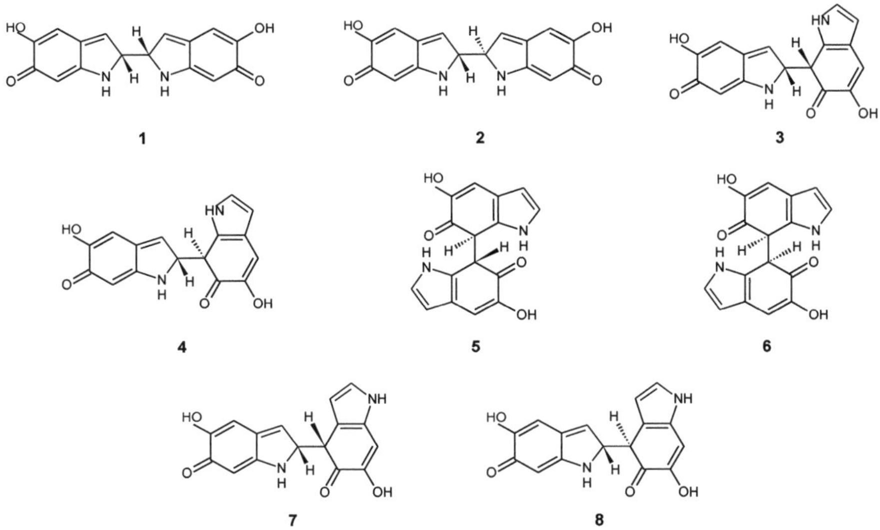

Scheme 2. Primary dimeric products putatively involved in 5,6-dihydroxyindole radical coupling.

semiquinones versus dihydroxyindole/indolequinone coupling was also performed at the DFT level.

Apart from their specific relevance to the eumelanogenesis mechanistic pathway, the results disclose an instance in which free radical dimerization of *o*-semiquinones outcompetes the classic disproportionation-driven catechol–quinone coupling, a process that may well be of broader relevance than previously believed.

3. Case Study 2: Red Hair Pigments

1,4-Benzothiazines provide the fundamental structural motif of pheomelanins. Benzothiazine units arise *in vivo* by a deviation of the usual pathway of melanogenesis, involving oxidative cyclization of cysteinyldopas (Scheme 2).

The resulting 2*H*-1,4-benzothiazine intermediates are highly unstable, and are prone to undergo oxidative self-coupling. In particular, the $\Delta^{2,2'}$-bi(2*H*-1,4-benzothiazine) skeleton characterizes the trichochromes, the pigments of red human hair (Scheme 3).

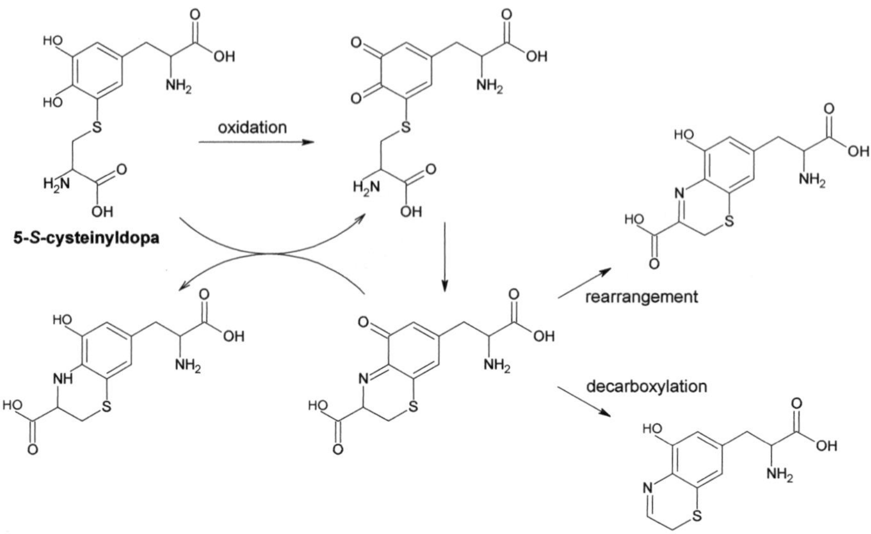

Scheme 3. Oxidative cyclization of 5-*S*-cysteinyldopas leading to the formation of benzothiazines.

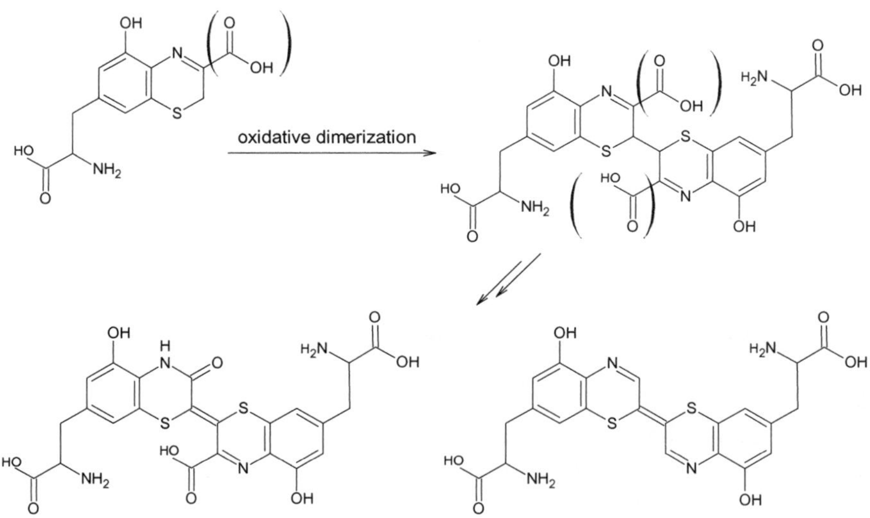

trichochromes

Scheme 4. Biogenetic origin of the trichochromes.

The system features unique photochemical and acidochromic four-state behavior, and has therefore been the object of renewed interest in view of potential bio-analytical and technological applications.[6]

However, the mechanistic course of the self-coupling reactions leading to the formation of the $\Delta^{2,2'}$-bi(2*H*-1,4-benzothiazine) system has remained largely unexplored, due to the high instability of the parent 2*H*-1,4-benzothiazine. To mitigate this problem, in a series of recent investigations we resorted to focus on the more stable 3-phenyl-2*H*-1,4-benzothiazine (see Formula) as a model compound.[7]

Again, we predicted at the DFT level of theory relative energies for different species putatively involved in the reaction pathway, and compared experimental spectra (chiefly NMR, EPR, and UV-Vis) with their computational counterparts.

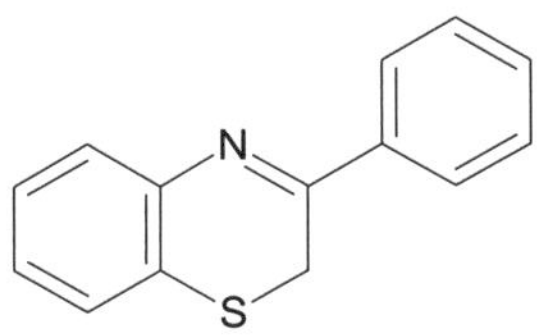

3-phenyl-2*H*-1,4-benzothiazine

We could thus provide evidence that the stable yellow form of the $\Delta^{2,2'}$-bi(2*H*-1,4-benzothiazine) from 3-phenyl-2*H*-1,4-benzothiazineis the *cis* (*Z*) isomer, rather than *trans* (*E*) isomer as previously believed. Moreover, a dark-blue species that is transiently generated under strongly acidic conditions was formulated as the *Z*-configured dication. Concerning the mechanism of formation of the $\Delta^{2,2'}$-bi(2*H*-1,4-benzothiazine) skeleton, we could further demonstrate that, under acidic conditions, the presence of peroxides or biometals (Fe(III), Cu(II), V(V) salts) induces an efficient dehydrogenative coupling of an initially formed 1,4-benzothiazinyl radical.[8]

4. Computational Approaches

All calculations were performed with the Gaussian package of programs.[9]

Structures were geometry optimized at the DFT level, with a hybrid functional (PBE0)[10] and a reasonably large basis set [6–31+G(d,p)]. Computations were performed either in vacuum, or by adoption of a polarizable continuum medium (PCM)[11–14] to account for the influence of the solution environment. In view of the faster convergence, a scaled van der Waals cavity based on universal force field (UFF)[15] radii was used, and polarization charges were modeled by spherical Gaussian functions[16,17]; non-electrostatic contributions to the solvation free energy were disregarded at this stage: these terms were accounted for in single point PCM calculations (at the PCM geometries) employing radii and non-electrostatic terms of the SMD solvation model.[18] Vibrational–rotational contributions to the free energy were also computed for selected minimum-energy structures.

UV-Vis spectra were computed in vacuum or in solution using the TDDFT approach,[3,4] with the larger 6–311++G(2d,2p) basis set, and either the PBE0 or the M06-2X functional.[19]

NMR shielding tensors were computed within the Gauge-Including Atomic Orbitals (GIAO) ansatz[20,21] at the PBE0/6–311+G(d,p) level. Computed isotropic shieldings were converted into chemical shifts using as reference the values obtained at the same level for benzene.

For computation of EPR parameters, geometry optimizations were carried out at the unrestricted DFT level, typically with the B3LYP functional[22,23] and the N07D basis set, as optimized for B3LYP.[24] Single-point calculations were then carried out with the B3LYP functional and specifically tailored basis sets, namely EPR-II or EPR-III;[25] when needed, the sets were completed for the sulfur centers with a 6–31+G(d) or 6–311++G(2d) basis, respectively.

For each species, different tautomers and when appropriate also different protonation states, were explored. In those cases where conformational enantiomers exist, a single enantiomeric series was examined. A complete conformational exploration of each individual chemical species was undertaken, in order to identify the corresponding absolute minimum. In simple cases (i.e., for monomers and double-bonded dimers), a systematic search of the potential energy surface could be carried out directly at the DFT level: in practice, a set of starting conformations was generated by a rigid scan of selected dihedrals, and then an unconstrained optimization was performed at each point.

However, for more complex structures featuring a larger number of rotatable dihedrals, the higher conformational freedom implies a combinatorial increase in the number of starting structures that must be optimized, and the straightforward approach rapidly becomes computationally unviable. For these cases, a pre-screening phase was introduced, carried out at a computationally inexpensive molecular mechanics (MM) level. In practice, each conformation of the starting set, generated as above, was initially submitted to a constrained MM (UFF) optimization, whereby all rotatable dihedrals values defining the grid were kept fixed, while all other internal coordinates were allowed to relax completely. The pool of structures thus obtained was sorted by MM energy, and all items that were more than 20 kcal mol^{-1} above the corresponding absolute minimum were discarded. The remaining low-energy structures were then processed as described above, i.e., by free optimization at the DFT level. The savings obtained by this two-stage procedure were often substantial: thus, for example, in the case of (2*R*,2′*R*)-3,3′-diphenyl-2*H*,2′*H*-2,2′-bi-1,4-benzothiazin-2-yl hydroperoxide, a putative intermediate in the oxidation of 3-phenyl-2*H*-1,4-benzothiazine, of 972 initial conformations examined at the UFF level, only 187 were passed over to the DFT optimization.

At different stages of a conformational exploration, it often happens that different starting conformations converge to the same optimized structure. An automated procedure to recognize these cases was therefore needed. Supposing a set of conformers must be screened for duplicates, each individual structure is first analyzed to produce an intramolecular Cartesian distance matrix in upper triangular form. Direct comparison between pairs of distance matrices is however complicated by the necessity to account for the possibility of chemical equivalence among different atoms (e.g., distances between a given atomic center and the 2/2′ carbons of a phenyl substituent should not be regarded as distinct when comparing distance matrices, since a

180° rotation of the phenyl ring would interconvert them without chemically altering the overall structure). As a matter of fact, the problem is even more general, since connectivity changes may well occur during *ab-initio* structural optimizations. Schemes based on the separation of atomic centers into specific equivalence classes are therefore hard to formulate in a general way in this context. The simple solution adopted consisted in removing altogether atom numbering dependence from the distance matrices, by putting in the same equivalence class all atoms of the same element. In practice, the $N(N-1)/2$ physically relevant interatomic distances (where N is the number of atoms in the molecular structure) are sorted alphabetically by the names of the elements involved in the distance; lines corresponding to the same kind of interatomic distance (e.g., all C–O distances) are further sorted by increasing interatomic distance. In this way, all arrays that describe isomers of the same chemical structures display the same ordering of lines; moreover, identical structures will give rise to identical distance arrays, quite irrespective of the initial atom numbering adopted. The sorted arrays can now be compared directly, e.g., by computing the RMS difference between corresponding elements or, alternatively, by computing an angle between the two "vectors." Structures whose distance matrices differ (in the specified sense) by less than a given threshold are regarded as coincident, and only one of them is retained for further analysis. The choice of the threshold value is hard to automate: on one hand, even genuinely identical structures will always feature a residual RMS differences, depending for e.g., on the finite convergence criteria of the structure optimization algorithm. On the other hand, the range of RMS differences spanned by a set of isomeric structures depends on such factors as the number of atomic centers, and the range of geometrical sizes and shapes covered. In practice, the choice of the threshold can be addressed empirically on a case-by-case basis, by visual examination of several pairs of structures that give rise to small RMS differences.

From a computational viewpoint, the most demanding aspect of the whole mechanistic exploration lies in the large number of DFT optimizations that must be performed — for many different chemical species putatively involved in a reaction path, and for many different conformers of each chemical structures. Fortunately, each individual optimization can be performed independently: the task is therefore "embarrassingly parallel" in character, and ideally suited for execution in a grid-computing environment.

Acknowledgments

This work was supported in part by a grant from the Italian Ministry of Education, Universities, and Research (MIUR), PRIN 2010–2011 2010FM738P project. Computational resources were provided by the SCoPE data center of the University Federico II of Naples.

References

1. G. Prota, *Melanins and Melanogenesis.* Academic Press, San Diego, CA (1992).
2. M. d'Ischia, K. Wakamatsu, A. Napolitano *et al., Pigment Cell Melanoma Res.* **26**, 616–633 (2013).
3. R. Bauernschmitt and R. Ahlrichs, *Chem. Phys. Lett.* **256**, 454–464 (1996).
4. M. E. Casida, C. Jamorski, K. C. Casida *et al.,Chem. Phys.* **108**, 4439–4449, (1998).
5. A. Pezzella, O. Crescenzi, L. Panzella *et al., J. Am. Chem. Soc.* **135**, 12142–12149 (2013).
6. A. Napolitano, L. Panzella, L. Leone *et al., Acc. Chem. Res.* **46**, 519–528 (2013).
7. L. Leone, O. Crescenzi, A. Napolitano *et al., Eur. J. Org. Chem.* **27**, 5136–5140 (2012).
8. L. Leone, O. Crescenzi, R. Amorati *et al., Org. Lett.* **15**, 4944–4947 (2013).
9. M. J. Frisch, G. W. Trucks, H. B. Schlegel *et al., Gaussian 09, Revision A.02.* Gaussian Inc., Wallingford, CT (2009).
10. C. Adamo and V. Barone, *J. Chem. Phys.* **110**, 6158–6169 (1999).
11. S. Miertus, E. Scrocco and J. Tomasi, *J. Chem. Phys.* **55**, 117–129 (1981).
12. M. Cossi, G. Scalmani, N. Rega *et al., J. Chem. Phys.* **117**, 43–54 (2002).

13. G. Scalmani, V. Barone, K. N. Kudin *et al., Theor. Chem. Acc.* **111**, 90–100 (2004).
14. J. Tomasi, B. Mennucci and R. Cammi, *Chem. Rev.* **105**, 2999–3093 (2005).
15. A. K. Rappé, C. J. Casewit and K. S. Colwell *et al., J. Am. Chem. Soc.* **114**, 10024–10035 (1992).
16. D. A. York and M. Karplus, *J. Phys. Chem. A* **103**, (1999).
17. G. Scalmani and M. J. Frisch, *J. Chem. Phys.* **132**, 1–15 (2010).
18. A. V. Marenich, C. J. Cramer and D. G. Truhlar, *J. Phys. Chem. B* **113**, 6378–6396 (2009).
19. Y. Zhao and D. G. Truhlar, *Theor. Chem. Acc.* **120**, 215–241 (2008).
20. R. Ditchfield, *Mol. Phys.* **27**, 789–807 (1974).
21. K. Wolinski, J. F. Hilton and P. Pulay, *J. Am. Chem. Soc.* **112**, 8251–8260 (1990).
22. A. D. Becke, *J. Chem. Phys.* **98**, 5648–5652 (1993).
23. P. J. Stephens, F. J. Devlin and C. F. Chabalowski *et al., J. Phys. Chem.* **98**, 11623–11627 (1994).
24. V. Barone, P. Cimino and E. Stendardo, *J. Chem. Theory Comput.* **4**, 751–764 (2008).
25. V. Barone, In *Recent Advances in Density Functional Methods*, Part I (Ed. D. P. Chong), World Scientific, Singapore (1996).

Chapter 21

Polymer Models of the Chromosomes in the Nucleus of Cells

A. M. Chiariello*, S. Bianco*, A. Piccolo*, C. Annunziatella*, M. Barbieri*, A. Pombo‡ and M. Nicodemi*,†,§

**Dipartimento di Fisica, Università degli Studi di Napoli Federico II, Via Cinthia, Napoli, Italy*

†Istituto Nazionale Fisica Nucleare, Napoli, Italy

‡Max-Delbrück Centre for Molecular Medicine, Robert-Rössle Strasse, Berlin-Buch 13092, Germany

§nicodem@na.infn.it

Chromosomes have a complex organization in the space of the nucleus of cells. Understanding their structure is a key open issue in molecular biology. Quantitative models from polymer physics, investigated by extensive Monte Carlo and molecular dynamics simulations, have been developed to find the principles of chromosome folding and its function. Here, we provide a short review of recent progress in such an important research field.

1. Introduction

In the cell nucleus of higher organisms, such as mammals, chromosomes have a complex, far from random spatial organization.[1–3] A well-described example is the co-localization of the *Xist* locus on the 2 X chromosomes in female mammalian cells during X-inactivation, but a number of other cases have been reported. Chromosome organization serves vital functional purposes, disruptions being linked to a variety of diseases, including cancer (see, e.g., Refs. 1–3 and references therein). New technologies, such as 'chromosome-conformation-capture' based methods,[3] are opening the way to probe the folding state of chromatin at a genomic level. Along with previous techniques such as FISH, these approaches are providing the first detailed exploration of the maps of contact in the cell nucleus.

Chromosome 3D organization appears to extend across spatial scales (Fig. 1). They occupy separate, yet interacting chromosomal territories[1,4] where long-range chromatin interactions are functionally important. Each chromosome is partitioned in mega base pair-long domains enriched for internal contacts, known as topologically associated domains (TADs),[1,5,6] while long stretches of chromatin interact with the nuclear lamina (defining 'lamina-associated domains', LADs),[3] and with a variety of other functional compartments, such as the nucleolus.[1] A spectrum of

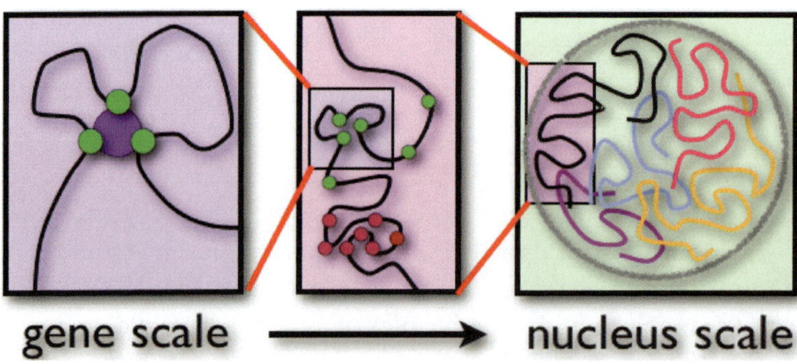

Fig. 1. Chromosome organization at different scales.

Note: Chromosomes have different levels of organization in space in the cell nucleus. Examples (from left to right): co-expressed gene co-localization at Transcription Factories; Topological Associated Domains (i.e., DNA domains with increased levels of intra- associations); chromosomal territories.

molecular factors mediates chromosomal interactions. For instance, Transcription Factories are 50 nm-wide clusters of about half a dozen polymerases that promote proximity between different, co-transcribed genes and transcription units.[7–10] Splicing factors, the machinery that splices nascent transcripts into messenger RNA, accumulate in splicing speckles, which are often associated with active genes, while repressed chromatin associates with heterochromatic regions or Polycomb bodies.[1] Early replication origins are also clustered in replication factories, which stably reassemble in consecutive cell cycles.[11]

While the understanding of chromatin folding has inspired models for many years,[12–15] the increasing level of details exposed by recent experimental advances and the complexity of the observed patterns has renewed the need to develop quantitative models, especially from polymer physics. Here we review a few recent developments in this important research field of the rapidly growing frontier between Physics and Life Sciences.

2. Polymer Models of Chromosomes

The standard model of polymer physics describing a free polymer, i.e., a polymer experiencing only self-avoidance effects, is the Self-Avoiding Walk (SAW). A SAW polymer folds spontaneously in a random, dynamically changing conformation (Fig. 1). This causes a full intermingling of polymers in a mixture by entropic forces. Yet, in case polymers also experience a strong self-attraction force, they can produce discrete chromosome territories.[16]

Under different conditions, entropy might also favor territorial separation between different chromosomes: models of free polymers folded into clustered loops can hardly find the space to penetrate into each other due to an effective entropic repulsion.[14,17–19] These models help bridge the initial picture of chromosome territories fully separated by channels devoid of DNA,[1,20] and the discovery of their intermingling.[4,21] Entropy, however, cannot be the only force behind chromosome organization as it fails to explain the variety of specific, functional contacts (e.g., enhancer–promoter interactions), and the domain structure of chromosomes (e.g., LADs, TADs).

This view has been reinvigorated by the development of 'chromosome conformation capture' (3C) technologies,[22–24] such as Hi–C,[22,25,26] which have provided the first semiquantitative measures of chromosomal interactions at a genomic scale. Hi–C contact matrices have shown, for instance, that the average interaction probability between pairs of genomic loci decreases with their genomic separation, approximately with power-law decay in the 0.5–7Mb range.[22]

The contact probability and its power law exponent, α, have been shown to be different in different organisms, cell types and chromosomes.[23] For instance, human embryonic stem cells (ESCs) have $\alpha \sim 1.6$,[23] human interphase lymphoblastoid cells have $\alpha \sim 1.1$,[22] while in metaphase HeLa cells $\alpha \sim 0.5$.[27] The exponent α reported in different species also varies widely: in yeast $\alpha \sim 1.5$,[29] in Drosophila $\alpha \sim 0.7$–0.8,[28] in mouse ESCs the *Xist* locus has $\alpha \sim 0.7$–0.9.[6,30] Even in a given cell system, different chromosomes can have different exponents.[23,31] For instance, in human lymphoblastoid cells, the gene-poor chromosome 18 has α around 1.08,while the gene-dense chromosome 19 has α close to 1.3.[23] These experiments have contradicted the idea that a single universal architecture, originally envisaged in the Fractal Globule (FG) state,[22] might recapitulate chromosome architecture. The FG state[22] is a transient, highly unstable[32] conformation traversed, for example, by a free polymer expanding from a highly confined, non-entangled state. The FG has $\alpha = 1$ and also fails to describe the formation of TADs or LADs, as well as microscopy data of interlocus distances.[23]

An alternative scenario is proposed with the Strings & Binders Switch (SBS) model (Fig. 2a)[23,33,34]: in this view, chromatin folding derives from interactions with other nuclear elements, such as molecular complexes that promote looping (e.g., contacts between co-expressed genes at transcription or replication factories) and associations with nuclear landmarks, such as the lamina.

In the simplest version of the SBS model, a chromatin filament is modeled as a "string", described by the golden standard of polymer physics, the SAW. Specific binding sites across the string have affinity to interact with diffusing binders that can crosslink distant binding sites forming loops. The SBS model can also be expanded to consider

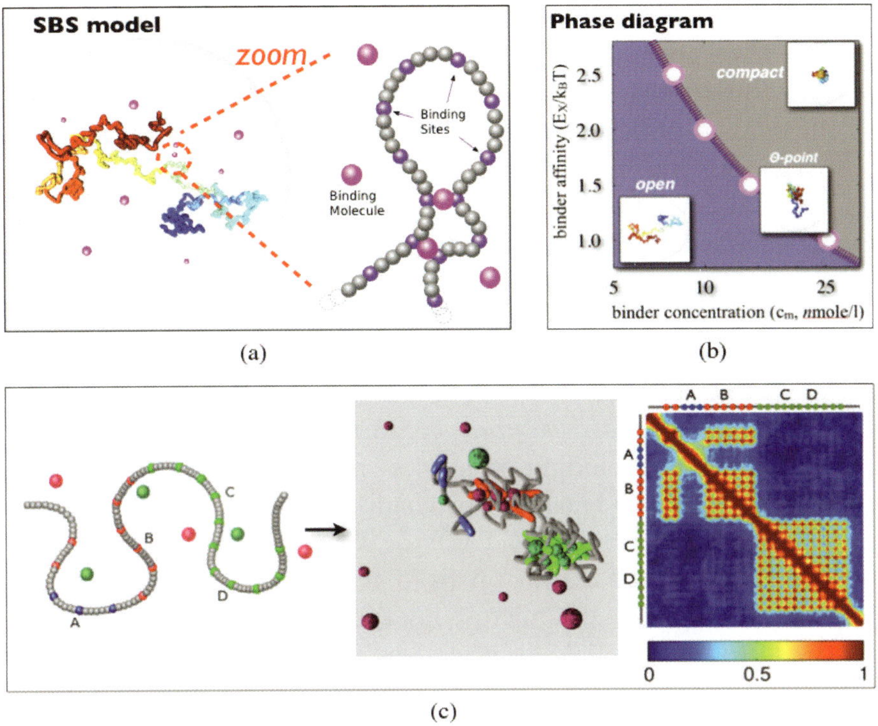

Fig. 2. The strings and binders switch (SBS) model.

Note: (a) In the SBS model,[23,33] chromosomes are represented by self-avoiding-walk (SAW) strings. They have binding sites for diffusing binders, which can produce loops. The SBS model parameters are binder concentrations, c_m, and their affinity for the polymer binding sites, E_X.
(b) The model thermodynamic phases correspond to its different, stable conformational classes: there is a phase where the polymer folds in a random open conformation (in the universality class of the free SAW) and a phase where it spontaneously folds into a compact closed conformation. At the phase transition point, there is the Θ-point state. Conformational modifications can be controlled switch-like by crossing the phase boundary through changes in binder affinity or concentration, with no need of parameter fine-tuning.
(c) Within the SBS model, the formation of chromatin domains and looping naturally emerge by the specialization of binding sites and binders. The corresponding contact matrices are close to those found experimentally by Hi–C methods.

many polymers, different binder species and interactions with nuclear landmarks.[35,36] In the model, the position of the binding sites along the polymer reflects biological information, while the concentration and affinity of binders can change under varying conditions,[23] as resulting, for instance, from the up-regulation of a corresponding gene or from chemical modifications of a chromosome sequence.

The SBS model, under simple and general settings, has been used to illustrate universal aspects of chromatin folding in quantitative terms, as deriving from polymer scaling properties: (a) three major folding classes exist (open, closed and Θ-point; Fig. 2b), corresponding to stable emergent phases of polymer physics (the Fractal Globule state is one of the many possible transient states of the SBS model); (b) conformational changes can be sharply controlled via thermodynamic phase transitions, by simple strategies, e.g., protein up-regulation or chromatin modifications, which transduce, e.g., (*analog*) transcription factor levels into (*digital*) conformational switches; (c) that randomly diffusing binding molecules can establish and dynamically change, by thermodynamics mechanisms, architectural patterns, such as territory formation, TADs or LADs, and the looping out of large stretches of chromatin from territories (Fig. 2c); (d) that population and single cell microscopy and Hi–C data, such as contact probabilities and spatial distances, can be rationalized in a single framework.[23] The results of the SBS model are confirmed by similar findings in related, simplified models, such as the Dynamic Loop (DL) model.[14,37]

The picture of chromatin depicted by the SBS model is that chromosomes are a complex mixture of differently folded regions according to local specific factors, which can self-organize across spatial scales by general physical mechanisms. A combination of single molecular factors can result in specificity of binding at different loci or domains, along with other molecular mechanisms, such as supercoiling[30] and plectoneme formation.[38]

3. Discussion

Models such as the SBS, informed with biological specificities, can also be employed to study specific genomic loci (e.g., the *Xist* locus)[36] to discriminate different biological scenarios and thus help in identifying the molecular determinants of chromatin folding. Such approaches could deliver very important applications to cases of medical relevance as they have the potential to predict chromatin interaction sites that best explain the available data. Polymer models also have the potential to enable reconstructions of the 3D shape of specific chromosomes to enhance our understanding of gene regulation and other nuclear processes.[39,40]

In summary, polymer physics, informed with and checked against experimental data, is substantially contributing to depict the first quantitative picture of the molecular mechanisms shaping chromosome folding. We expect that in the near future, further developments in modeling, experimental progresses and more detailed biological information, could push even further our comprehension of chromosome three-dimensional architecture, hopefully advancing also the diagnostic, treatment and management of genomic diseases.

Acknowledgments

We acknowledge the use of SCoPE super computing facilities of the University of Naples *Federico II.*

References

1. C. Lanctot, T. Cheutin, M. Cremer *et al.*, Dynamic genome architecture in the nuclear space: Regulation of gene expression in three dimensions, *Nature Rev. Genet.* **8**, pp. 104–115 (2007).
2. T. Misteli, Beyond the sequence: Cellular organization of genome function, *Cell*, **128**, pp. 787–800 (2007).
3. W. A. Bickmore and B. van Steensel, Genome architecture: Domain organization of interphase chromosomes, *Cell* **152**, pp. 1270–1284 (2013).
4. M. R. Branco and A. Pombo, Intermingling of chromosome territories in interphase suggests role in translocations and transcription-dependent associations, *PLoS Biol.* 4, pp. e138 (2006).
5. J. R. Dixon, S. Selvaraj, F. Yue *et al.*, Topological domains in mammalian genomes identified by analysis of chromatin interactions, *Nature* **485**, pp. 376–380 (2012).
6. E. P. Nora, B. R. Lajoie, E. G. Schulz *et al.,* Spatial partitioning of the regulatory landscape of the X-inactivation centre, *Nature* **485**, pp. 381–385 (2012).

7. A. Pombo, D. A. Jackson, M. Hollinshead *et al.*, Regional specialization in human nuclei: Visualization of discrete sites of transcription by RNA polymerase III, *EMBO J.* **18**, pp. 2241–2253 (1999).
8. C. S. Osborne, L. Chakalova, J. A. Mitchell *et al.*, Myc dynamically and preferentially relocates to a transcription factory occupied by Igh, *PLoS Biol.* **5**, p. e192 (2007).
9. C. Ferrai, S. Q. Xie, P. Luraghi *et al.*, Poised transcription factories prime silent upa gene prior to activation, *PLoS Biol.* **8**, p. 1000270 (2010).
10. S. Schoenfelder, T. Sexton, L. Chakalova *et al.*, Preferential associations between co-regulated genes reveal a transcriptional interactome in erythroid cells, *Nat Genet* **42**, pp. 53–61 (2010).
11. D. A. Jackson, A. Pombo, Replicon clusters are stable units of chromosome structure: Evidence that nuclear organization contributes to the efficient activation and propagation of s phase in human cells, *J Cell Biol* **140**, pp. 1285–1295 (1998).
12. J. Langowski, Polymer chain models of DNA and chromatin, *Eur. Phys. J. E. Soft Matter* **19**, pp. 241–249 (2006).
13. M. Emanuel, N. H. Radja, A. Henriksson *et al.*, The physics behind the larger scale organization of DNA in eukaryotes, *Phys. Biol.* **6**, p. 025008, (2009).
14. M. Tark-Dame, R. van Driel and D. W. Heermann, Chromatin folding — from biology to polymer models and back, *J. Cell Sci.* **124**, pp. 839–845 (2011).
15. M. Barbieri, A. Scialdone, A . Gamba *et al.*, Polymer physics, scaling and heterogeneity in the spatial organisation of chromosomes in the cell nucleus, *Soft Matter* **9**, pp. 8631–8635 (2013).
16. G. Kreth, J. Finsterle, J. von Hase *et al.*, Radial arrangement of chromosome territories in human cell nuclei: A computer model approach based on gene density indicates a probabilistic global positioning code, *Biophys. J.* **86**, pp. 2803–2812 (2004).
17. D. Marenduzzo, C. Micheletti and P. R. Cook, Entropy-driven genome organization, *Biophys. J.* **90**, pp. 3712–3721 (2006).
18. M. Bohn, D. W. Heermann and R. van Driel, Random loop model for long polymers, *Physical review. E, Statistical, nonlinear, and soft matter physics* **76**, p. 051805 (2007).
19. A. Rosa and R. Everaers, Structure and dynamics of interphase chromosomes, *Plos Comp. Biol.* **4**, p. e1000153 (2008).
20. T. Cremer, G. Kreth, H. Koester *et al.*, Chromosome territories, interchromatin domain compartment, and nuclear matrix: An integrated

view of the functional nuclear architecture, *Crit Rev Eukaryot Gene Expr.* **10**, pp. 179–212 (2000).
21. M. R. Branco, A. Pombo, *Chromosome organization: New facts, new models, Trends Cell Biol.* **17**, pp. 127–134 (2007).
22. E. Lieberman-Aiden, N. L. van Berkum, L. Williams *et al.*, Comprehensive mapping of long-range interactions reveals folding principles of the human genome, *Science* **326**, pp. 289–293 (2009).
23. M. Barbieri, M. Chotalia, J. Fraser *et al.*, Complexity of chromatin folding is captured by the strings and binders switch model, *PNAS* **109**, pp. 16173–16178 (2012).
24. J. Dekker, K. Rippe, M. Dekker *et al.*, *Capturing chromosome conformation. Science* **295**, pp. 1306–1311 (2002).
25. B. Tolhuis, R. J. Palstra, E. Splinter *et al.*, Looping and interaction between hypersensitive sites in the active beta-globin locus, *Mol Cell* **10**, pp. 1453–1465 (2002).
26. C. D. Rodley, F. Bertels, B. Jones *et al.*, Global identification of yeast chromosome interactions using genome conformation capture, *Fungal Genet. Biol.* **46**, pp. 879–886 (2009).
27. N. Naumova, M. Imakaev, G. Fudenberg *et al.*, Organization of the mitotic chromosome, *Science* **342**, pp. 948–953 (2013).
28. T. Sexton, E. Yaffe, E. Kenigsberg *et al.*, Three-dimensional folding and functional organization principles of the Drosophila genome, *Cell* **148**, pp. 458–472 (2012).
29. Z. Duan, M. Andronescu, K. Schutz *et al.*, A three-dimensional model of the yeast genome, *Nature* **465**, pp. 363–367 (2010).
30. F. Benedetti, J . Dorier, Y. Burnier *et al.*, Models that include supercoiling of topological domains reproduce several known features of interphase chromosomes, *Nucleic Acids Res.* 2013.
31. R. Kalhor, H. Tjong, N. Jayathilaka *et al.*, Genome architectures revealed by tethered chromosome conformation capture and population-based modeling, *Nature Biotech.* **30**, pp. 90–98 (2011).
32. R. D. Schram, G. T. Barkema and H. Schiessel, On the stability of fractal globules, *J. Chem. Phys.* **138**, p. 224901 (2013).
33. M. Nicodemi, B. Panning and A. Prisco, A thermodynamic switch for chromosome colocalization, *Genetics* **179**, pp. 717–721 (2008).
34. M. Nicodemi and A. Prisco, Thermodynamic pathways to genome spatial organization in the cell nucleus, *J. Biophys.* **96**, pp. 2168–2177 (2009).
35. A. Scialdone and M. Nicodemi, Diffusion-based DNA target colocalization by thermodynamic mechanisms, *Development* **137**, pp. 3877–3885 (2010).

36. A. Scialdone, I. Cataudella, M. Barbieri *et al.*, Conformation regulation of the X chromosome inactivation center: A model, *PLoS Comput. Biol.* **7**, p. e1002229 (2011).
37. M. Bohn and D. W. Heermann, Diffusion-driven looping provides a consistent framework for chromatin organization, *PloS One* **5**, p. e12218 (2010).
38. T. B. Le, M. V. Imakaev, L. A. Mirny *et al.*, High-resolution mapping of the spatial organization of a bacterial chromosome, *Science* **342**, pp. 731–734 (2013).
39. M. A. Umbarger, E. Toro, M. A. Wright *et al.*, The three-dimensional architecture of a bacterial genome and its alteration by genetic perturbation, *Mol. Cell* **44**, pp. 252–264 (2011).
40. T. Nagano, Y. Lubling, T. J. Stevens *et al.*, Single-cell Hi-C reveals cell-to-cell variability in chromosome structure, *Nature* **502**, pp. 59–64 (2013).
41. L. A. Parada, P. G. McQueen and T. Misteli, Tissue-specific spatial organization of genomes, *Genome Biol.* **5**, p. R44 (2004).

Chapter 22

Analysis, Tuning and Implementation of a Hippocampal CA1 Microcircuit

S. Cuomo*,‡, P. De Michele* and A. Galletti†

* *University of Naples Federico II, Naples, Italy*
† *University of Naples Parthenope, Naples, Italy*
‡ *salvatore.cuomo@unina.it*

The implementation of neural network models enables to reproduce complex biological phenomena. The tuning of a large number of biological parameters and synaptic mechanisms of several different cells that belong to a model is a very critical issue. In this work, we present the effects of increasing cAMP Response Element Binding protein (CREB)-dependent transcription on the storage and recall processes in CA1 microcircuit of the hippocampus, and we describe how parallel tools can be adopted to simulate biological behaviors experimentally observed. Here, we resort to a parallel computing strategy in order to achieve efficient and reliable simulations.

1. Introduction

Computational neuroscience is full of papers describing the implementation of single cell or neural network models, which reproduce specific biological phenomena. In general, these works highlight the biological results and rarely describe the computational aspects to be taken into account for the model implementation. From a computational point of view, it is clear that the building up of a neural network model is very expensive, even more than single neuron model. Accordingly, in the neuroscience literature it is easier to find models of single neuron rather than neural networks. In fact, comparing the results carried out by a search on Mod elDB (`http://senselab.med.yale.edu/ModelDB/`), a reference database for computational neuroscience models, we can observe that the number of neural network models is almost half compared to those of single neurons (265 versus 416). In this paper, we refer to the biological neural network model described in Ref. 1 and published in Ref. 2, for studying and understanding the effects of increasing cAMP Response Element Binding Protein (CREB) dependent transcription on the processes of memory storage and recall, also focusing on the computational aspects and solutions needed to properly tune the overall biological parameters characterizing this network model.

2. A Hippocampal CA1 Microcircuit

To understand the process of memory formation and its underlying processes is crucial to devise effective therapies for memory-related pathologies like Alzheimer's Disease. Although it is well known that encoding of facts and events first takes place in the hippocampus brain region, in Ref. 3 the molecular alterations of CA1 pyramidal neurons that result from experience- or learning-dependent synaptic activation are complex and still not fully identified. Hence, in Ref. 1, we have implemented a simplified CA1 network to investigate how and to what extent different cell properties, altered by increasing the presence of the CREB protein, may contribute to improve or rescue stored memory patterns under control and pathological conditions. More in detail, in Ref. 1 a morphological model of a hippocampal CA1 microcircuit is presented. This network is based on the model

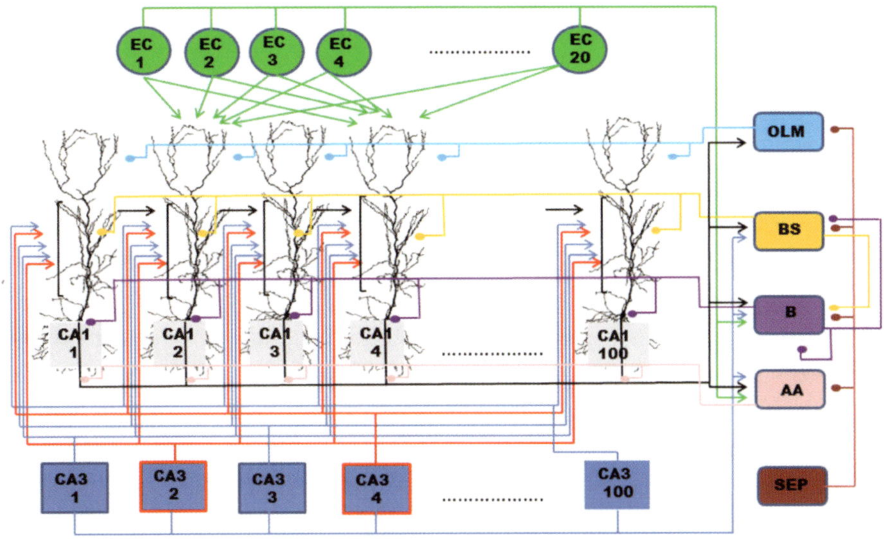

Fig. 1. Schematic representation of cell types and their connectivity; arrows and small ovals represent excitatory and inhibitory connections, respectively. EC: entorhinal cortex input; CA3: Schaffer collateral input; AA: axo-axonic cell; B: basket cell; BS: bistratified cell; OLM: oriens lacunosum-moleculare cell; SEP: Septal GABA input; active CA3 inputs are represented by a red outline.

for patterns recognition, described in Ref. 4 and implemented in Ref. 5, using several important experimental constraints. The network consists of 100 CA1 neurons with four types of inhibitory interneurons (1 OLM, 2 B, 1 BS and 1 AA), and is schematically illustrated in Fig. 1. These latter use the morphologies designed in the model implemented in Ref. 5 (ModelDB accession number 123815). Moreover, the network has three main input signals: excitatory inputs coming from EC (20 inputs), and CA3 Schaffer collateral (100 inputs) and inhibitory signals coming from the septum (10 inputs). CA1 neurons had the same morphology template used in Ref. 5 but with different distributions of ionic currents. All currents were taken from the model described in Ref. 6 (ModelDB accession number 143719), and reproduce the depolarization block observed in these neurons.

3. Implementation Details of the Model

Biological neural networks consist of several kinds of cells interconnected among them. Depending on the biological issues to be studied, it is possible to decide the level of biological details with which to define the neurons, from a computational point of view. Typically, cells representing the inputs of the network are simulated by means of electrical stimuli rather than to be implemented. Conversely, inter-neurons and output cells are biologically defined by means of the implementation of the *neuron's morphology*, *neuron's 3D spatial information* and *neuron's mechanisms* code packages. In the following, we refer to these neurons as *morphological cells*. Figure 2 reports the packages of a generic network model implemented with NEURON, assuming that this consists of K kinds of morphological cells.

As shown in Fig. 1, the microcircuit consists of $K\_TOT = 8$ different types of cells, for a total of $nCells = 235$ cells. Figure 3 illustrates

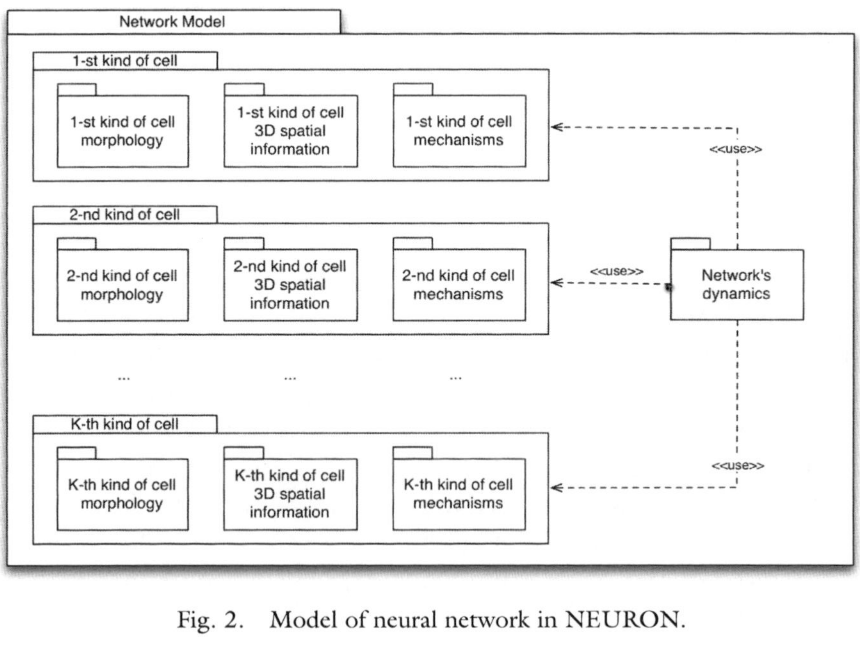

Fig. 2. Model of neural network in NEURON.

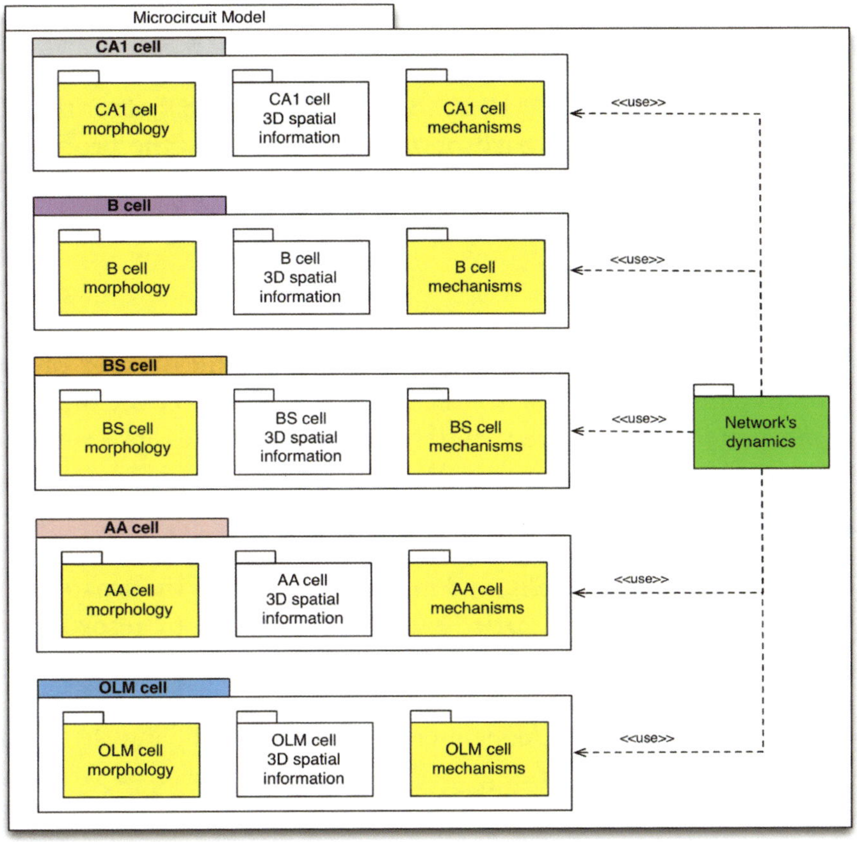

Fig. 3. The CA1 microcircuit model.

the main packages of the CA1 microcircuit. Moreover, as stated in Sec. 2, the model presented in Ref. 1 relies on the neural network described in Ref. 5. This latter is able to perform the steps of storage and recall of memory just for a single pattern. Our modification relies to redesign this model by introducing a new multi-pattern recognition strategy. Hence, given a set of patterns (*n_patterns_to_store* variable), our model is able to store and subsequently to recall all the patterns. As a result, the algorithm is divided into two main sub-algorithms: the storage and the recall schemes.

3.1. *Storage and Recall Algorithms*

Figure 4 shows the computational tree for the storage algorithm, implemented in the *neuron's dynamics* package. This algorithm is organized in 2 key steps, each of which managed by the procedure `main()`. For each pattern i ($1 \leq i \leq n\_patterns\_to\_store$) to be stored, let be

$$\mathbf{W}^{(i)} = s(\mathbf{W}^{(i-1)}, \mathbf{c}^{(i)}, \mathbf{p}^{(i)}),$$

where $\mathbf{W}^{(i-1)}$ is the weight matrix obtained from the storage of the $(i-1)$th pattern ($\mathbf{W}^{(0)} = 0$), $\mathbf{c}^{(i)}$ is a network connection vector, $\mathbf{p}^{(i)}$ is the ith pattern to be stored, and s is a function that represents the STDP rule. The weight matrix **W** covers a key role in the proper setting of the network connections.

More in detail, the first step is the network creation, which consists in the set up of the cells, and in the placement of the synapses by means of **W**. This task is carried out, only one time, by resorting to the procedures `mknet_storage()` and `acc_dist()`. In Fig. 5, the computational tree related to `mknet_storage()` is drawn. This calls `alzh()`, which simulates Alzheimer's disease by modifying the initial weight matrix $\mathbf{W}^{(0)}$, according to the decay percentage of synapses. Moreover, `mknet_storage()` calls the procedures `mkcells()` and `mkinputs()`, which create and set up the morphological and the input cells, respectively. Finally, the procedures `connectcells()`, `connectEC()`, `connectEC_2()` and `connectCA3()` are called for

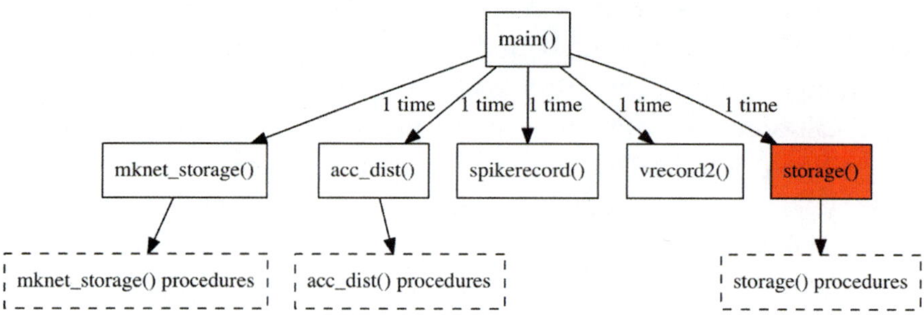

Fig. 4. The computational tree related to the procedure `main()` used in the storage algorithm.

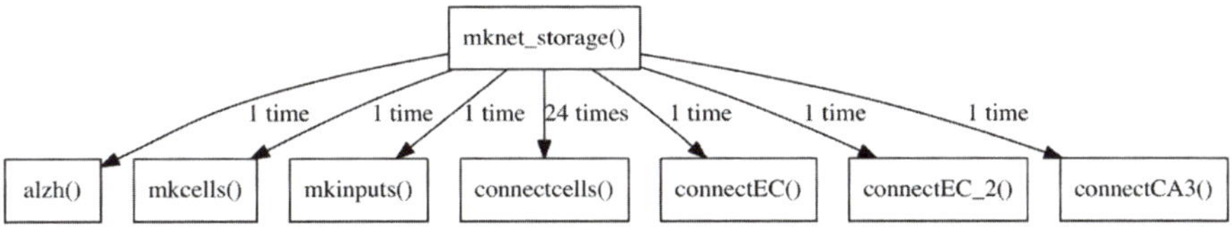

Fig. 5. The computational tree related to the procedure `mknet_storage()` used in the storage algorithm.

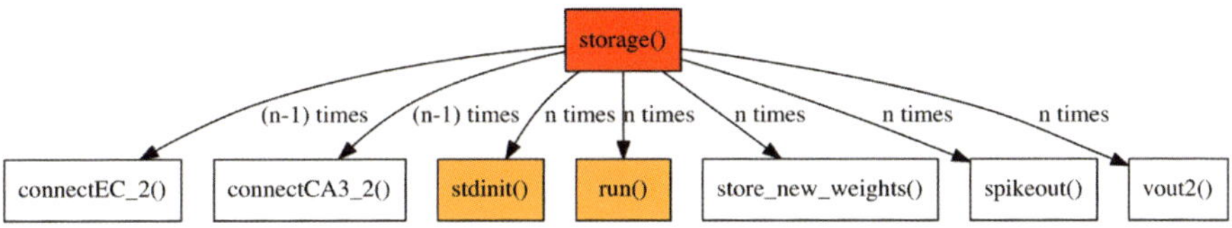

Fig. 6. The computational tree related to the procedure `storage()` used in the storage algorithm.

placing the synapses among the cells and for configuring them by using $\mathbf{W}^{(0)}$.

The second step is referred to the numerical integration of the model, where $W^{(i)}$ is updated for each *i*th pattern to be stored. In detail, `main()` calls the procedures `spikerecord()` and `vrecord2()`, which record the firing times and the potential values for each cell, respectively. Then, the storage of the patterns is completely performed by `storage()` (in the red box), which iteratively calls some sub-procedures, for each pattern to be stored. In the following, we give a brief description of the kernel procedure `storage()`, which is very computationally expensive. As we can see in Fig. 6, where for simplicity we renamed *n_patterns_to_store* with n, the procedures `connectEC_2()` and `connectCA3_2()` are called. For each pattern i (with $i > 1$), these are delegated to change the network connections by using $\mathbf{W}^{(i-1)}$, which has been updated after the storage of the previous pattern. Then, the NEURON's standard procedures (in the orange boxes) `stdinit()`, which initialize the model, and `run()`, which integrates the model, are invoked. Moreover, the procedure `store_new_weights()` is used to update $\mathbf{W}^{(i)}$. Finally, the procedures `spikeout()` and `vout2()`, which produce the output information previously recorded, are called.

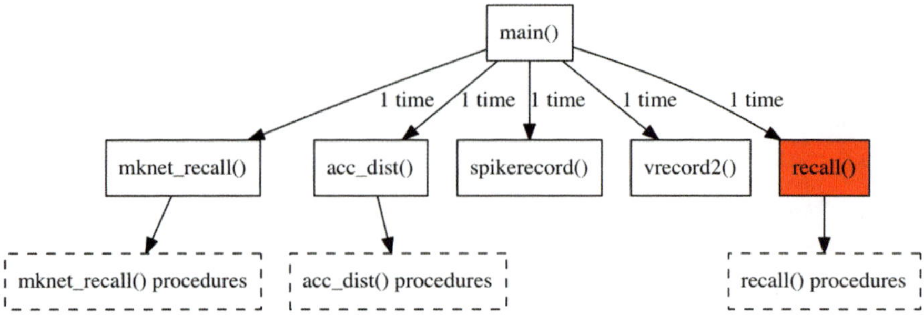

Fig. 7. The computational tree related to the procedure `main()` used in the recall algorithm.

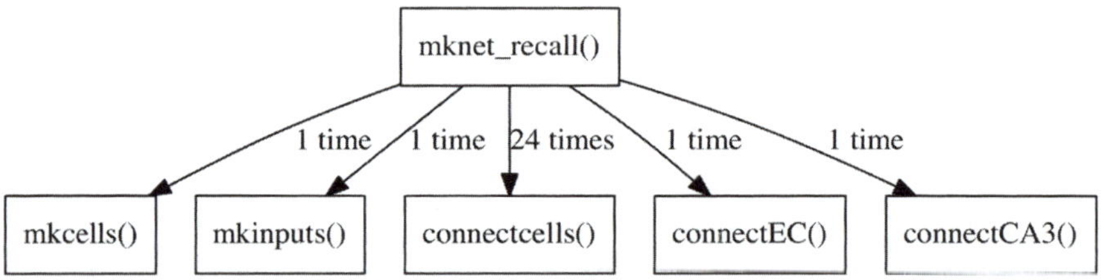

Fig. 8. The computational tree related to the procedure `mknet_recall()` used in the recall algorithm.

Figure 7 shows the computational tree for the recall algorithm, implemented in the *neuron's dynamics* code package. The scheme for the recall of a set of patterns *(n_patterns_to_recall* variable) is organized in 2 key steps, each of which managed by the procedure `main()`, as well as for the storage algorithm. Let be

$$OUT^{(i)} = r(\mathbf{W}^{(N)}, \mathbf{c}^{(i)}, \mathbf{p}^{(i)}),$$

where $\mathbf{W}^{(n)}$ is the weight matrix obtained from the patterns previously stored, $\mathbf{c}^{(i)}$ is a network connection vector, $\mathbf{p}^{(i)}$ is the *i*th pattern to be recalled. The function r evaluates the recall and gives in output $OUT^{(i)}$, which is the biological response related to the recall quality.

The first step of the recall algorithm is the network creation, as well as for the storage algorithm. In Fig. 8, the computational tree of

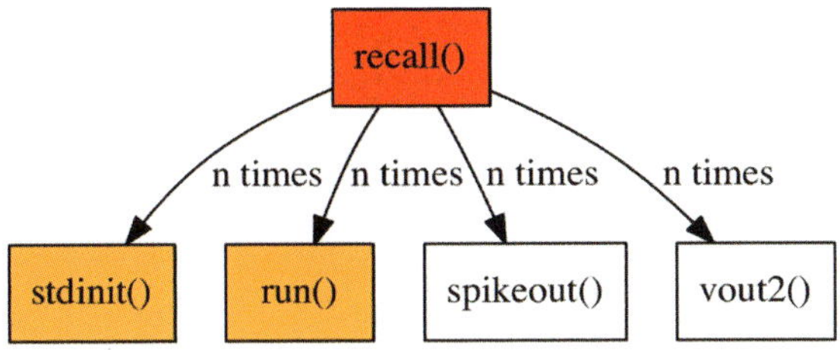

Fig. 9. The computational tree related to the procedure `recall()` used in the recall algorithm.

the procedure `mknet_recall()` is drawn. Notice that this differs from the storage version because it does not call the procedures `alzh()` and `connectEC_2()`.

The second step consists in the numerical integration of the model, performed by the procedure `recall()` (in the red box), which iteratively, for each pattern to be recalled, calls some subprocedures. This procedure represents the most expensive part of the recall Algorithm. As we can observe in Fig. 9, where for simplicity we renamed *n_patterns_to_recall* with *n*, `recall()` calls the NEURON's standard procedures `stdinit()` and `run()` (in the orange boxes), and then it resorts to the procedures `spikeout()` and `vout2()`.

4. Parallel Tools for CA1 Microcircuit Model

The tuning phase of a complex model of single neuron requires several thousand of simulations in order to properly reproduce biological behaviors. For our CA1 microcircuit, this issue emerges even more, because the number of biological parameters and synaptic mechanisms to be tuned grows with the number K of the different types of morphological neurons ($K = 5$) in the network. In fact, each kind of cell adds to the simulation runtime a different amount of processing time. This is due to the number of compartments, the types of channels in a single compartment, and the complexity and the number of synapses.[7] As a result, it is needed to resort to parallel techniques, in order to perform simulations with reasonable execution times. NEURON is able to support the parallelization of network models while maintaining, as much as possible, a separation between the network features and how the cells are distributed among all the processors. Hence, without modification, it is possible to write code that can be properly executed in any serial or parallel hardware environment, and that produces quantitatively identical results regardless of the number of CPUs.[8] More in detail, we start from the *network's dynamic* code package shown in Fig. 2 (in the green box), which reproduces the experiments on the CA1 microcircuit. Then, for each kind of cell, we properly set up the parameters of the *neuron's morphology* and the mechanisms of the *neuron's mechanisms* packages (in the yellow boxes), without changing the *neuron's 3D spatial information* package.

The parallelization of a neural network model in NEURON consists of a mapping of the network cells to the available processors (i.e., cores). Obviously, it is necessary to introduce an integer global identifier (`gid`) in order to identify any cell. The distribution of the gids (i.e., cells) on the processor is a crucial step in terms of performance. The ideal condition is that each processor has approximatively the same amount of sequential work to do. The simplest load balance approach ignores differences in individual cell processing,[7] according to the round robin (RR) strategy shown in Algorithm 1.

Here, `pc.nhost` is the number of available processors, `nCells` is the number of cells and `pc.id` identifies the *i*th processor. In Table 1, assuming that the number of available processors is N = `pc.nhost`, the mapping of cells to processors with the RR algorithm is shown.

Adopting the RR strategy, we introduced a parallel layer in the *network's dynamics* package. Here, as we can see in Fig. 10, the underline communication tool is MPI (in the blue box). The yellow boxes represent the sequential part of the code, related to the groups of neurons assigned to the same processor (in the orange box). The set

Algorithm 1 The round robin algorithm.

```
for (gid = pc.id; gid < nCells; gid += pc.nhost) {
  //create cell associated with gid
}
```

Table 1. Mapping processors-cells with round robin strategy. Processor ID: processor identifier; GID: global identifier of the cell; LID: local identifier of the cell on a specific processor.

Host ID	GID	LID
0	0, N, $2N$, ...	0, 1, 2, ...
1	1, $N+1$, $2N+1$, ...	0, 1, 2, ...
⋮	⋮	⋮
$N-1$	$N-1$, $2N-1$, $3N-1$, ...	0, 1, 2, ...

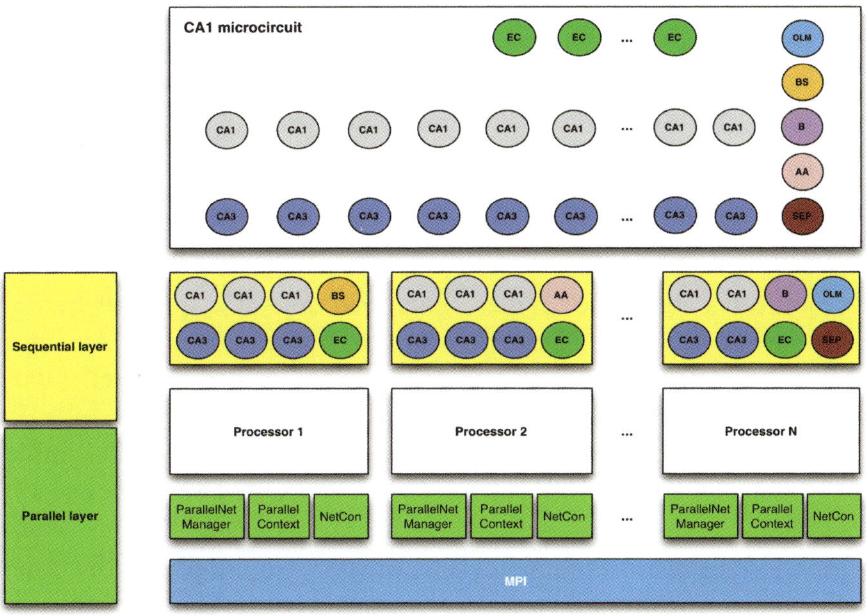

Fig. 10. Parallel framework for the CA1 microcircuit.

up of a parallel network uses the `NetCon` class as much as possible.[9] Moreover, additional parallel specific NEURON's methods from the classes `ParallelContext` and `ParallelNetManager` are required. In this way, there is no *master* host and all processors perform exactly the same program on different subsets of neurons. All together the cores *set up* the internal and external connectivity and perform the *numerical integration* on its associated cells. In the next subsections, we will present some parallel implementation of the storage and the recall algorithms.

4.1. *Data Consistency*

One of the key issues addressed to parallelize an application is the data consistency: each processor can only access data that it actually handles. For the storage and the recall algorithms, this means that there is no processor attempting to manipulate cells that it does not handle. Hence, it is very important to properly manage the communication of a spike between two cells assigned to different processors. NEURON uses an event delivery system to implement spike-triggered synaptic transmission among cells. In the simplest case on serial hardware, the connection between a spike source (pre-synaptic cell) and its target is made by instantiating an object of the class `NetCon` executing a statement of the form

```
nc = new NetCon(source, target),
```

to monitor a source for spikes. The detection of a spike throws an event which will be delivered to the `NetCon`'s target.[8] In a parallel simulation environment, the solution is to give to any cell (spike source) its own `gid` that can be referred by any processor.

The aforementioned aspects are managed by the parallel implementations of the procedures for cell connection `connectcells()`, `connectEC()` and `connectCA3()`. In Algorithm 2, the procedure `P_connectcells()` is shown. Here, the processor connects the target cells `target`, belonging to it, to the source cell `source` by means of `gid_connect()` NEURON's method (line 14 in Algorithm 2). This procedure is recalled by `mknet_storage()` (see Fig. 5) and `mknet_recall()` (see Fig. 8), for the storage and recall respectively.

More in detail, Algorithm 3 shows the parallel implementation of `mknet_storage()`, renamed as `P_mknet_storage()`, while Algorithm 4 lists the parallel implementation of the procedure `mknet_recall()`, renamed as `P_mknet_recall()`. In Algorithms 5 and 6, the parallel implementations of `main()` and `storage()` procedures are illustrated (Figs. 4 and 6). Finally, in Algorithms 7 and 8, the

Algorithm 2 A light view of the `P_connectcells()` procedure.

```
proc P_connectcells() {
  ...
  // loop over possible target cells
  for i=0, nCells-1 {
   gid = gidvec.x[i] // global id of cell
   if (gid >= first_target_cell_index && gid < last_target_cell_index) {
    ...
    // return in r the source cell index
    rs.r.discunif(first_source_cell_index, last_source_cell_index)
    source = rs.repick()
    ...
    // set up connection from source to target
    target = cells.object(i).pre_list
    nc = pc.gid_connect(source, target)
    ...
   }
   ...
 }
 ...
}
```

parallel implementations of `main()` and `recall()` procedures are shown (Figs. 7 and 9).

5. Performance Results

In this section, we report the performance results related to the parallel algorithms just described. Overall simulations are performed with NEURON (version 7.1), by using the model files available on ModelDB (`http://senselab.med.yale.edu/ModelDB/`), with identification number 151126,[2] and they are run with MPI using up to 128 cores on the S.Co.P.E. Grid infrastructure at University of Naples Federico II, Naples, Italy. The storage and recall algorithms

Algorithm 3 A light view of the `P_mknet_storage()` procedure.

```
proc P_mknet_storage() {
  ...
  /*** Alzheimer application: synapse decay ***/
  if (pc.id==0) P_alzh(alzheimer_percentage)
  pc.barrier()
  /*** Cell creation ***/
  P_mkcells() // create the morphological cells
  P_mkinputs() // create CA3, EC and SEP inputs
  ...
  /*** Network connection based on the weight matrix W^(0) ***/
  P_connectcells(nB, nEC)
  ...
  P_connectcells(nB, nOLM)
  P_connectEC()
  P_connectEC_2(pattern, 1, n_patterns_to_store) // Store the 1-st pattern...
  P_connectCA3(alzheimer_weight_matrix, 0, connection_probability) // ...with modifiable synapses
}
```

Algorithm 4 A light view of the `P_mknet_recall()` procedure.

```
proc P_mknet_recall() {
  /*** Cell creation ***/
  P_mkcells() // create the morphological cells
  P_mkinputs() // create CA3, EC and SEP inputs
  ...
  /*** Network connection based on the weight matrix W^(0) ***/
  P_connectcells(nB, nEC)
  ...
  P_connectcells(nB, nOLM)
  P_connectEC() // Restore existing pattern...
  P_connectCA3(weight_matrix, 0, connection_probability) // ...with modifiable synapses
}
```

Algorithm 5 A light view of the `P_main()` procedure for the storage algorithm.

```
proc P_main() {
  ...
  /*** Network creation and connection based on the weight matrix W^(0) ***/
  P_mknet_storage(weights_matrix, 0, connection_probability)
  P_acc_dist()
  /*** Instrumentation ***/
  P_spikerecord()
  P_vrecord2()
  ...
  /*** Numerical integration ***/
  P_storage()
  ...
}
```

Algorithm 6 A light view of the `P_storage()` procedure.

```
proc P_storage() {
  for i = 1, n_patterns_to_store {
   if (i != 1) {
    /*** New network connection based on the weight matrix W^(i-1) ***/
    P_connectEC_2(pattern, i, n_patterns_to_store) // Store the i-th pattern...
    P_connectCA3_2(weights_matrix, i-1) // ...with modifiable synapses
    }
    ...
    /*** Numerical integration ***/
    pc.set_maxstep(1)
    stdinit()
    pc.psolve(tstop)
    /*** New weight matrix W^(i) storage ***/
    P_store_new_weights(weights_matrix, i)
    ...
    /*** Output generation ***/
    P_spikeout()
    P_vout2()
    ...
}}
```

Algorithm 7 A light view of the `P_main()` procedure for the recall algorithm.

```
proc P_main() {
  ...
  /*** Network creation and connection based on the weight
matrix W^(0) ***/
  P_mknet_recall(weights_matrix, 0, connection_probability)
  P_acc_dist()
  /*** Instrumentation ***/
  P_spikerecord()
  P_vrecord2()
  ...
  /*** Numerical integration ***/
  P_recall()
  ...
}
```

Algorithm 8 A light view of the `P_recall()` procedure.

```
proc P_recall() {
  for i = 1, n_pattern_to_recall {
    ...
    /*** Numerical integration ***/
    pc.set_maxstep(1)
    stdinit()
    pc.psolve(tstop)
    /*** Output generation ***/
    P_spikeout()
    P_vout2()
    ...
}}
```

were executed on a set of 10 patterns, with an initial delay of 100 ms and 8 theta-cycles of 250 ms. Hence, the overall simulation time is 100 ms + (250 ms × 8) = 2100 ms. Notice that real experiments were performed with 50 patterns and 100 theta-cycles, for an overall simulation time equal to 100 ms + (250 ms × 100) = 25.1 s.

In order to evaluate the performance of the storage and recall algorithms, different parameters were taken into account. More in detail, **Runtime** indicates the total execution time, **Wait** represents the time spent for exchanging spikes during a simulation, and **Step** is the time spent for numerical integration. Moreover, **Setup and cell creation** represents the time needed to create the cells, **Network connection** indicate the time for connecting the network, and **New weight matrix storage** is the time for collecting and storing the values of the weight matrix. Finally, **Output** is the time for storing biological output information, and **Others** represents the time for collecting non-functional information.

Figure 11 shows that the CA1 microcircuit well scales on a single server (node) of the S.Co.P.E. infrastructure. In fact, moving from 1 up to 8 processors, we observe a huge reduction of **Runtime** from 40934 s to 3189 s. Moreover, Fig. 12 shows that the numerical integration time (i.e., **Step**) dominates the overall execution time. More in detail, with 8 cores the **Step** time is the ~95% of the **Runtime**, while the **Wait** time is the ~4%. Finally, the other execution steps (setup and cell creation, network connection, weight matrix storage and output generation) do not affect the overall time.

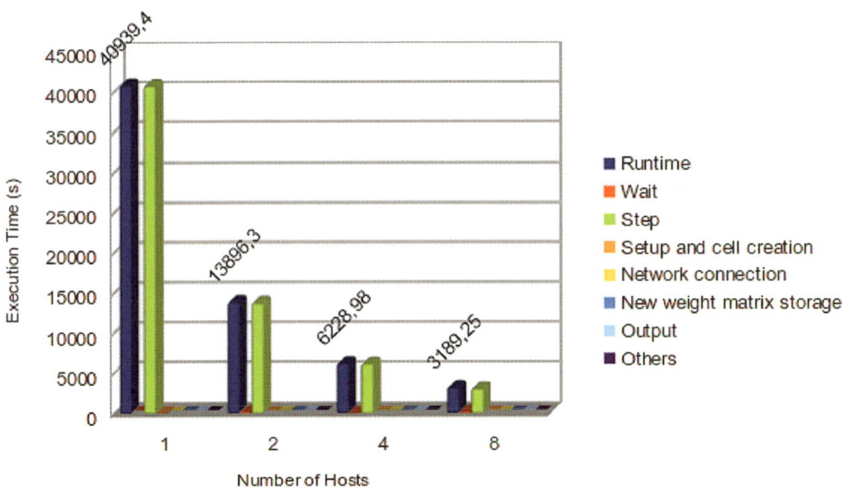

Fig. 11. Execution times for the storage of 10 patterns with 8 theta cycles moving from 1 to 8 processors.

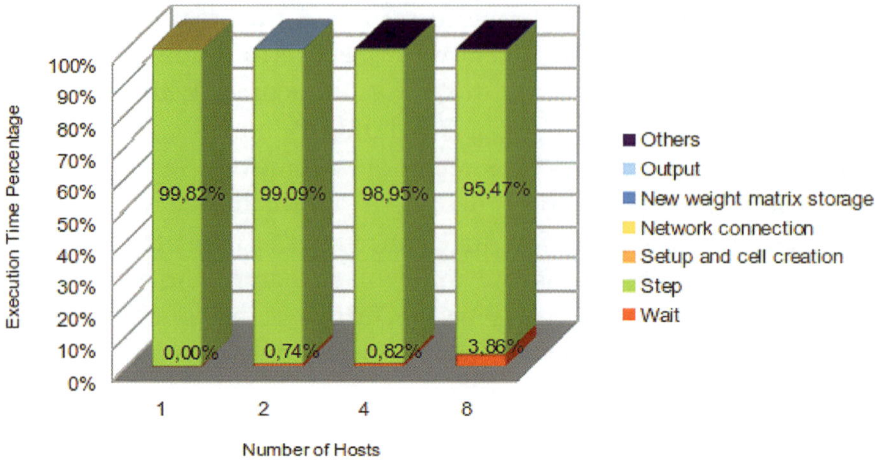

Fig. 12. Percentages of execution time steps for the storage of 10 patterns with 8 theta cycles moving from 1 to 8 processors.

Figure 13 illustrates a high worsening of the performance, moving from 8 to 128 processors. More in detail, despite the numerical integration time (**Step**) continues to decrease with doubling of the core number, the **Runtime** remains almost constant, although with slight deteriorations. In fact, as we can observe in Fig. 14, **Step** ranges from ~95% with 8 cores to 6% with 128 cores. On the other hand, there is a strong increment of the amount of time needed to synchronize the communication between two cells mapped on two different processors (**Wait**). Indeed, moving from 8 to 16 cores, we note a huge increase from ~4% to ~48%, until reaching the value ~87% with 128 processors. This phenomenon is due to the communications. In general, increasing the number of processors, the size of subnets assigned among the processors becomes small and the communication overhead for MPI calls larger. In these conditions, communication times begins to dominate the runtime.[9]

The recall algorithm has the same performance we have discussed for the storage. But, in this case, it is possible to implement the recall algorithm on a distributed architecture: in this way, each algorithm performs the recall of a single pattern on a node of the infrastructure.

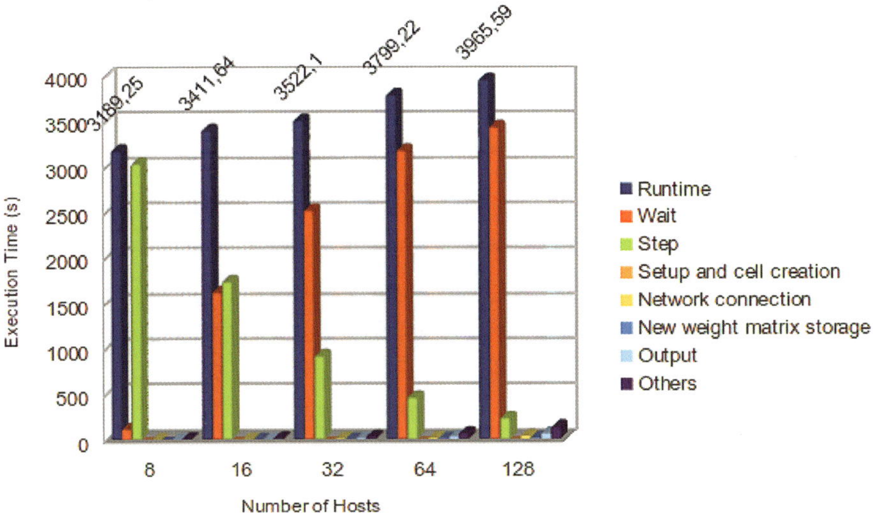

Fig. 13. Execution times for the storage of 10 patterns with 8 theta cycles moving from 8 to 128 processors.

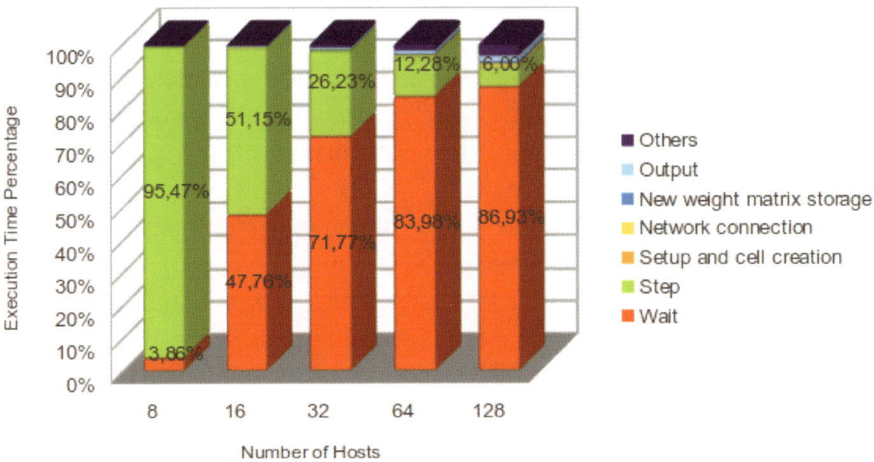

Fig. 14. Percentages of execution time steps for the storage of 10 patterns with 8 theta cycles moving from 8 to 128 processors.

Table 2. Execution times with 8 cores for the recall of 10 patterns with 16 theta cycles, distributing the patterns.

Pattern ID	Runtime	Wait	Step	#Spikes	Set and connect	Output
1	608,775	8,71033	597,987	7975	0,54125	0,61625
2	608,861	8,93715	598,457	8149	0,54125	0,6375
3	607,072	8,51312	597,083	8061	0,54125	0,64875
4	607,812	8,67061	597,683	8033	0,54125	0,635
5	607,119	9,37756	596,286	7946	0,54125	0,6275
6	607,035	8,71243	596,895	7992	0,54125	0,6325
7	607,726	8,79549	597,473	8030	0,54125	0,62125
8	607,191	8,39568	597,321	7848	0,54125	0,63875
9	608,27	10,0982	596,692	8022	0,54125	0,63625
10	608,42	9,3668	597,576	8046	0,54125	0,6325

A typical parametric execution of a distributed recall phase is characterized by the pair *(n_patterns-to-recall, n_theta-cycles*). In Table 2, the performance of the parametric execution (10 patterns, 16 theta cycles) are illustrated for an overall simulation time equal to 100 ms + (250 ms × 16) = 4100 ms for each pattern. The recall of *n_patterns_to_recall* patterns is performed in ~600 s, which is the time needed to recall just one single pattern, obtaining a substantial performance improvement: without the distribution of the patterns, the **Runtime** would have been equal to ~6000 s.

Finally, we remark that there exist alternative strategies to the RR, which allow to obtain a better load balancing. One of these is the "longest processing time" (LPT) algorithm in which one iteratively chooses the largest cell and puts it on the currently last used processor.[7] In this case, the METIS[10] graph partitioning program can be used to define a `ngid` on `nhost` partition for optimizing the load balance and for minimizing the communications.[9] We remark that we have chosen to resort to the simplest strategy of distribution of the cells (i.e., RR) for the following discussion. As observed in Ref. 9, in a context where computation time no longer scales with increasing number of CPUs, we have a large number of spikes with constant synaptic delay. In our model, there are 55% of fast spiking artificial cells (input cells), which make this point dramatically by communications point of view.

6. Conclusions

The simulation of biological neural networks is a challenging application from a computational point of view. The calibration and set up of a network require mathematical models in order to simulate the biological behavior of the different cell type and sophisticated programming environments for developing simulation codes. In practice, building up a microcircuit that mimes the real behavior of a biological neural network, with a large number of connections between its neurons, requires algorithms and communication strategies that are computationally expensive. We observed that the main problem that has to be overcome is the communication between a source cell and its target, in a parallel simulation environment. To overcome this problem, we strongly suggest to use general-purpose simulation environments that support massively parallel multi-core programming. Finally, we think that the performance analysis of the proposed microcircuit is useful in order to simulate a large number of microbiological multi-pattern recognition experiments in an acceptable computing time.

References

1. D. Bianchi, P. De Michele, C. Marchetti, B. Tirozzi, S. Cuomo, H. Marie and M. Migliore, Effects of increasing creb-dependent transcription on the storage and recall processes in a hippocampál Ca1 microcircuit, *Hippocampus.* **24**(2), pp. 165–177 (2014).
2. D. Bianchi, P. De Michele, C. Marchetti, B. Tirozzi, S. Cuomo, H. Marie and M. Migliore, Effects of increasing creb-dependent transcription on the storage and recall processes in a hippocampal ca1 microcircuit, https://senselab.med.yale.edu/ModelDB/showModel.cshtml? model=151126 (2014).
3. G. Buzski and E. Moser, Memory, navigation and theta rhythm in the hippocampal-entorhinal system, *Nat. Neurosci.* **16**(2), pp. 130–138 (2013).

4. M. Hasselmo, C. Bodelon and B. Wyble, A proposed function of the hippocampal theta rhythm: Separate phases of encoding and retrieval of prior learning, *Neural Comput.* **14**(4), pp. 793–817 (2002).
5. V. Cutsuridis, S. Cobb and B. Graham, Encoding and retrieval in a model of the hippocampal Ca1 microcircuit, *Hippocampus.* **20**, pp. 423–446 (2010).
6. D. Bianchi, A. Marasco, A. Limongiello, C. Marchetti, H. Marie, B. Tirozzi and M. Migliore, On the mechanisms underlying the depolarization block in the spiking dynamics of ca1 pyramidal neurons, *J. Comput. Neurosci.* **33**(2), pp. 207–225 (2012).
7. M. Hines, H. Eichner and F. Schurmann, Neuron splitting in compute-bound parallel network simulations enables runtime scaling with twice as many processors, *J. Comput. Neurosci.* **25**, pp. 203–210 (2008).
8. M. Hines and N. Carnevale, Translating network models to parallel hardware in neuron, *J. Neurosci. Methods* **169**(2), pp. 425–455 (2008).
9. M. Migliore, C. Cannia, W. Lyton, H. Markram and M. Hines, Parallel network simulations with neuron, *J. Comput. Neurosci.* **21**, pp. 119–129 (2006).
10. G. Karypis and V. Kumar, Multilevel k-way partitioning scheme for irregular graphs, *J. Parallel Distrib. Comput.* **48**(1), pp. 96–129 (1998).

Chapter 23

Medical Physics Applications in Bari ReCaS Farm

N. Amoroso*,†, M. Antonacci*, R. Bellotti*,†, G. Donvito*, R. Errico†, G. Maggi*,‡, A. Monaco*, P. Notarangelo*, S. Tangaro† and A. Tateo†

**INFN Sezione di Bari, Via Edoardo Orabona 4, 70126 Bari, Italy*

†*Dipartimento Interateneo di Fisica, Università degli Studi di Bari, Via Edoardo Orabona 4, 70126 Bari, Italy*

‡*Dipartimento Interateneo di Fisica, Politecnico di Bari, Via Edoardo Orabona 4, 70126 Bari, Italy*

§*nicola.amoroso@ba.infn.it*

State-of-the-art techniques of medical physics, particularly those requiring the analysis of structural magnetic resonance brain scans yield a huge amount of computational resources. Moreover, the most recent developments in big data management and sharing technologies make compelling a cross-disciplinary approach to computational neuroscience from scientists and technologists with different skills and expertize. For these reasons, Grid/Cloud technology appears to fit the requirements of such applications. To simplify the access of these technologies, the ITC INFN group has developed a lot of tools able to interact with Web and client applications like Science portal and workflow manager systems.

1. Medical Physics Applications

MRI data is very noisy and therefore arduous to analyze. First of all, noise has independent and different sources: acquisition or biological variability just to mention a few. As a consequence, this data requires the adoption of sophisticated computational strategies for multi-variate analysis. Image registration and segmentation are fundamental pre-processing steps to standardize data from multi-center studies containing even thousands of available subjects. Once data is standardized, group comparison analyses can be performed to detect significant patterns; in particular, the Bari Medical Physics Group (BMPG) has devoted its recent activity to the investigation of Alzheimer's disease (AD) patterns and fully automated disease recognition. The implemented strategies are based on the structural segmentation of regions of interest for the clinical phenotype, such as the human hippocampus (whose atrophy as a well-known bio-marker of AD), and the development of supervised classification techniques (Random Forests, Support Vector Machines, Complex Networks) to manage the huge amount of knowledge provided by magnetic resonance brain scans.

The main techniques and methodologies the BMPG has recently studied deal with three main fields of computational neuroscience: shape analysis, feature extraction and machine learning. These strategies provide a robust pipeline of analysis for neuroimaging tool and a state-of-the-art classification diagnostic tool ((for example for the AD)).

The overall pipeline for the analysis of structural brain scans require a dedicated pre-processing phase for inter-subject data normalization. Brain scans are intensity and morphologically normalized. Firstly, the removal of eventually present bias fields is performed, then brain is extracted and neck removed from each scan. Once the brain has been intensity and morphologically normalized, spatial normalization is performed. This crucial step requires different strategies, such as linear or nonlinear registrations. The pre-processing phase involves the adoption and combination of different publicly available tools which are installed on the grid/cloud computing infrastructure: FSL, ITK, FreeSurfer, just to mention a few. Once data has been normalized, the BMPG adopts several quantitative analysis approaches mainly consisting of 3 distinct frameworks:

- Shape Analysis. Fully Automated Point Distribution Model (FAPoD).[5] This algorithm segments a gross peri-hippocampal region (or other anatomical districts of interest). Its adoption improves the overall segmentation performance and reduces the computational cost of the analyses (~50% from 2 hours to about 50 minutes).
- Feature extraction. Complex networks are used to describe through community detection structural similarities between healthy controls, and subjects affected by brain diseases. These features are used along morphometric measurements[10] to detect significant patterns of the disease.[6,7]
- Machine Learning. A number of different algorithms have been explored for different classification and regression tasks, such as Random Forests, Support Vector Machines and Neural Networks (Multi-layer perceptron). The BMPG has implemented and developed these methods for ready-to-use applications on distributed grids and cloud infrastructures.[8,9]

These methodologies have proven their effectiveness in different frameworks and case studies. As a comprehensive example, the Alzheimer's disease computer aided detection system developed by the BMPG can be considered. In recent years, the study of neurodegenerative diseases and related dementias have widely conquered the attention of scientists, physicians and politicians. The study of AD requires the study of structural morpholological properties, as brain atrophy can successfully discriminate healthy controls from subjects with pathological conditions. The BMPG pipeline allows the users to automatically individuate those brain regions which play a crucial role in cognitive tasks such as the amygdala or the hippocampus. The volume of these regions can be then measured along with other interesting features such as thickness, or symmetry. Moreover, each scan can be used to properly define complex networks, i.e., mathematical

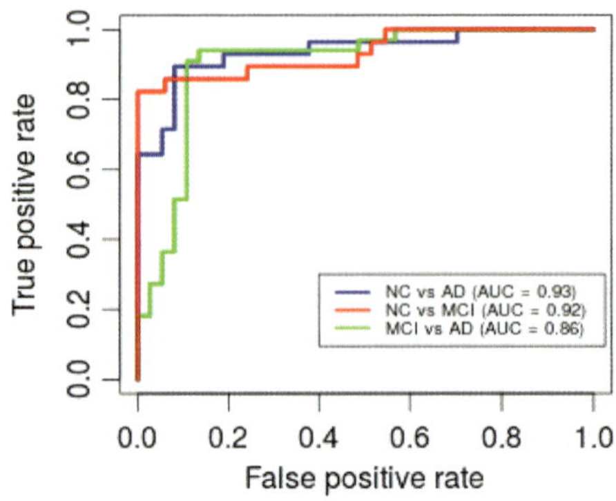

Fig. 1. The figure represents the overall performances for the one versus all classification task of Normal Controls (NC), mild cognitive impairment (MCI) and AD subjects on a set of 100 structural T1 brain scans. Performances were achieved in 5-fold cross validation, figure represents average performances on 100 cross-validation rounds.

graphs, accounting for the brain connectivity: brain regions representing the graph nodes and similarity measurements defining the edges of the network. Graph topological properties such as strength or betweenness can be also used as features of machine learning models. Finally, the BMPG has developed *ad hoc* classification strategies based on state-of-the-art techniques: Random Forests, Multi-layer perceptron, Support Vector Machines, RUSBoost and so on. These classification strategies allow to learn discriminant patterns based on the previously mentioned features. In Fig. 1, a classification performance in terms of area under the curve is represented.

The classification performances show, in this case, how healthy NC, MCI and AD subjects are accurately discriminated by the proposed methodologies. In the present case, performances were obtained in a 5-fold cross validation test led with 100 T1 structural brain scans. In particular, it is worth mentioning that these machine learning workflows (Fig. 2) are suitable for different kind of data. In particular, the BMPG collaborated with the United Nations Department of Economic and Social Affairs (DESA) for the analysis and publication of the world interest documents E-Government Survey 2012 and 2014.

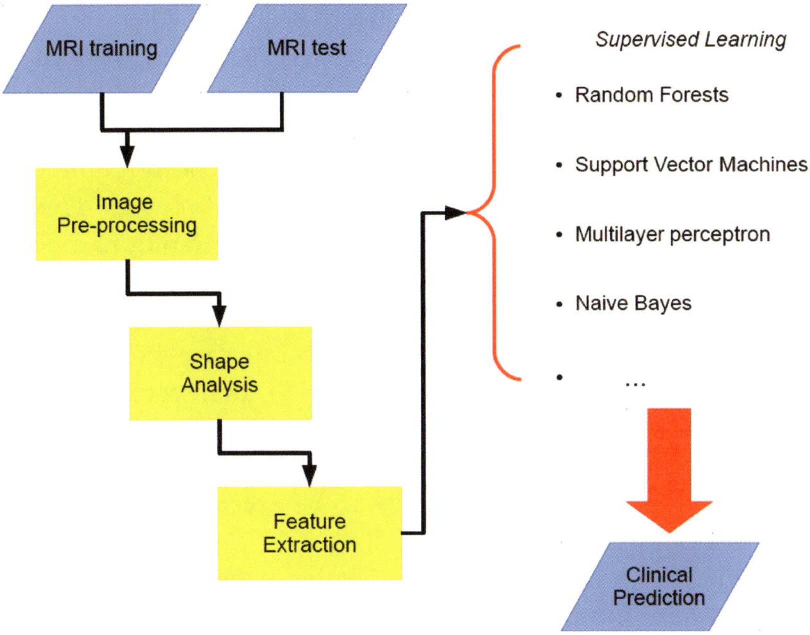

Fig. 2. The figure shows a schematic but comprehensive overview of the pipeline of analysis the Bari Medical Physics Group has implemented and adopted to tackle several tasks involving the clinical prediction of subjects affected by different brain diseases (as an example: Alzheimer's disease, Multiple Sclerosis, Schizophrenia). Moreover, this pipeline and the related software modules are available more in general for any task involving multi-variate supervised analyses and easily accessible through the applications developed for the ReCaS computer center.

2. Job Submission Tool (JST)

Medical Physics applications, particularly the ones that need to analyze MRI 3D-image, are expensive in terms of computational resources. For this reason, Grid/Cloud technology appears to fit the requirements of such applications.

To execute the medical physics applications in distributed environment, the INFN group has developed a job management tool: JST.[4] This tool is particularly useful to manage the submission and monitoring of applications, when a large number of independent executions are needed to solve a given problem. The strength of JST is the portability on different infrastructure like: the EGI grid infrastructure, dedicated servers, local batch farms and IaaS based cloud resources.

The tool takes care of the submission of all the jobs required by a given application, their monitoring and book-keeping. This JST capability hides to the user, the complexity of operating in a heterogeneous and distributed computational environment.

JST provides two WEB services interfaces: SOAP and REST. By means of this technology, it is possible to interface JST capabilities with several services and applications. The JST architecture is made of 3 macro-elements:

1. Database: contains the list of tasks to be executed;
2. FrontEnd: used to fill the TaskList via Web Services;
3. BackEnd: is an Application daemon that runs in the background and submits jobs, when there are tasks to be executed in the TaskList table.

3. Bari ReCaS Portal[1]

To improve the usability and accessibility of medical physics tools, the INFN group has realized a web interface based on the Catania Science Gateway. This interface[2] is based on Liferay framework. It allows the developers to build a simple portlet component that interacts with the JST on the back-end. In this way, the complexity of dealing with the different computational platforms is hidden by JST. This portal supports several kind of authentication, in particular, credential of social account (Facebook, Google, Twitter, etc.) and federated organizations and institutes (University, researcher institutions, etc.). In Liferay framework, new JAVA portlets were developed to support Medical Physics applications.

Through this portal, the user is able to submit a new analysis and easily check the status of the requested calculation. After the authentication, the actor can upload an MRI image to analyze and choose the required tool. When the execution is terminated, the user receives the mail notification, sent by the JST backend component. In this mail, there are the final status of the job and the link where the user can download the output results of the Medical Physics execution.

4. LONI Pipeline

JST is able to interface with workflow manager systems like LONI,[3] Taverna and Galaxy. In particular, LONI has been used by the BMPG.

The LONI Pipeline (LP) is a free workflow application primarily aimed at computational neuro-scientists. With LP, users can quickly create workflows that take advantage of all the greatest tools available in neuroimaging, genomics, bioinformatics, etc. The LP workflow application includes features that allow users to easily describe their executables in a graphical user interface. Instead of manually managing intermediate data in a script, the LP handles the passing of data between programs for the user. Once a module has been created in the LP, it can be saved into user personal library and reuse it in other workflows. LP is the principal tool used to submit request for a complete analysis of MRI-images. LP does not provide any usable plugin to exploit Torque-based systems or for the gLite/EMI Grid infrastructures. To solve this problem on Bari ReCaS Farm, a meta-scheduler based on JST has been developed.

References

1. Available at http://www.pon-recas.it/.
2. Available at https://recasgateway.ba.infn.it/.
3. Available at http://www.loni.usc.edu/Software/Pipeline
4. G. Donvito, S. Vivarion, P. Notarangelo *et al.*, The BioVeL Project: Robust phylogenetic workflows running on the GRID, EGICF2012-EMTITC2-029. (Available at http://pos.sissa.it/archive/conferences/162/129/EGICF12-EMITC2_029.pdf).
5. N. Amoroso, *et al.* Automated shape analysis landmarks detection for medical image processing, *Computational Modelling of Objects Represented in Images III: Fundamentals, Methods and Applications* **139**, 496 (2012).

6. S. Tangaro, *et al.* Automated voxel-by-voxel tissue classification for hippocampal segmentation: Methods and validation, *Physica Medica* **30**(8), pp. 878–887 (2014).
7. E. E. Bron, *et al.* Standardized evaluation of algorithms for computer-aided diagnosis of dementia based on structural MRI: The CADDementia challenge, *NeuroImage* **111**, pp. 562–579 (2015).
8. PRISMA-CAD: *Fully Automated Method for Computer-Aided Diagnosis of Dementia based on structural MRI data*, in Proc. of the MICCAI Workshop Challenge on Computer-Aided Diagnosis of Dementia Based on Structural MRI Data, 2014, pp. 16–23.
9. S. Tangaro, *et al. Active Learning Machines for Automatic Segmentation of Hippocampus in MRI* in Industrial Conference on Data Mining-Workshops, 2013.
10. S. Tangaro, *et al.* Feature Selection based on machine learning in MRIs for hippocampal segmentation, arXiv:1501.03915.

Engineering

Chapter 24

Computation of T_R-Dependent Aftershock Fragility Curves for an Existing Non-Ductile R.C. Building

M. Gaetani d'Aragona*, M. Polese, A. Prota and G. Manfredi

University of Naples Federico II, Naples, Italy
**marco.gaetanidaragona@unina.it*

In the aftermath of a damaging event, the structural safety against future earthquakes may be significantly reduced. An indirect measure of building seismic safety in its intact and damaged state is represented by the Residual Capacity (REC) index that quantifies building seismic capacity at collapse. This study investigates, by means of a detailed case study, the expected variation of REC for increasing seismic demand. The response of an existing non-ductile reinforced concrete building is simulated using a multi-degree-of-freedom (MDOF) finite element model that properly accounts for brittle failures. Different definitions of building collapse are introduced and consistent assessment of building REC are performed.

1. Introduction

The assessment of Residual Capacity (REC) of damaged buildings represent a key issue into the evaluation of building reparability. Ideally nonlinear time-history (NTH) analysis, which predicts the forces and cumulative damage in every element of the structural system, performed on multi-degree-of-freedom (MDOF) systems, would be the optimal solution for capturing building's seismic performance. On the other hand, NTH is a quite complex and time-consuming analysis, and may be justified only for the case of strategic structures. In addition, structural response of the building has to be simulated using a set of carefully selected accelerograms to overcome the sensitivity of dynamic response to the characteristics of the input motions greatly increasing the computational effort. Although simplified methods for the assessment of damaged structures behavior (e.g., Ref. 1) yield comparatively acceptable results with respect to detailed methods (e.g., Ref. 2), further investigations through detailed cases are required to analyze the realistic effects of earthquake shaking and resulting damage on the performance of non-ductile buildings.

Recent studies (Refs. 3 and 4), addressed the Aftershock (AS) collapse vulnerability of RC damaged buildings allowing to evaluate the safety variation based on the knowledge of maximum transient or residual drift. On the other hand, a faster evaluation in the post-earthquake could be achieved if AS fragilities are derived depending on the intensity of the damaging earthquake; indeed the latter information, that corresponds to T_R in a given hazard area, is quickly available soon after the Mainshock (MS).

In (4 Refs. 5–7,) a probabilistic framework finalized to the assessment of building's REC for non-ductile buildings when subjected to earthquakes of given intensity was introduced. In this paper, the framework is briefly recalled and it is applied for an existing six-story non-ductile RC building that was built in Benevento in the age '60s, presenting the results.

2. T_R-Dependent AS Fragility Curves Framework

The building capacity after an earthquake may be significantly reduced due to the spread of damage all over the building, while the probability of collapse increases. Seismic behavior of damaged buildings, and their relative seismic safety, may be suitably represented by their seismic capacity modified due to damage, the so-called REC (Refs. 8 and 9). In particular, REC_{Sa} of a building can be defined as the smallest ground motion (GM) spectral acceleration (at period T_{eq}, of the Single Degree Of Freedom (SDOF) system equivalent to the real structure) corresponding to collapse of the building.

In Refs. 5–7, a probabilistic framework for the dynamic-based assessment of REC at different return periods (T_R) was presented. This framework consists of several steps: (1) A reliable building model properly accounting for cumulative damage, as well as brittle failures in structural members is constructed. (2) Based on dynamic result of first step, the Probabilistic Seismic Hazard Analysis (PSHA) is performed for the specific site and structure, an appropriate bin of natural GMs to perform dynamic analyzes is selected and earthquake intensities at the site with a given return period T_R derived. (3) With the selected bin of accelerograms, the so-called "Back-to-back-IDA" (B2B-IDA) is used to assess the capacity of building to withstand future earthquakes after damage. In this procedure, multiple earthquake sequences MS–AS are built through suitable scaling of selected accelerograms (see Fig. 1). A first earthquake (MS) corresponding to a given T_R is applied to the undamaged structure and the response recorded; once the structure has been damaged, the IDA procedure is applied to the damaged structure in order to assess the new capacity of the building. A time gap of

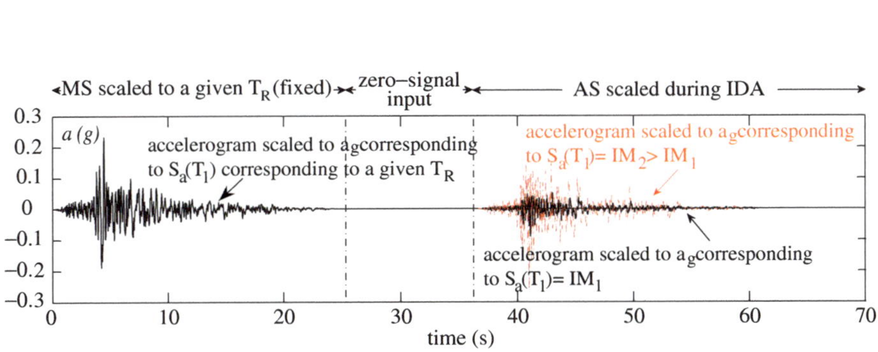

Fig. 1. Example of MS–AS sequence.

10 s between MS and AS is considered to allow the ceasing of vibration produced by the MS. Dynamic AS analysis is repeated with increasing scale factor applied to the AS record, providing IDA analysis results for MS-damaged structure, until the MS–AS sequence causes collapse. The output of this step is an IDA curve and an ultimate spectral acceleration capacity of the building for the set MS–AS–T_R.

The simulation is repeated for each combination MS–AS for all considered T_R to properly account for record-to-record variability. (4) The results of B2B-IDA are elaborated to determine fragility curves representing the REC of the building conditioned on T_R. For further details, see Refs. 6 and 7.

3. Description of the Case-Study Building Structural Model

The model frame chosen in this study is the perimeter transversal frame of a six-story non-ductile RC frame building located in Benevento, Italy.

A fixed-base two-dimensional hybrid fiber-lumped MDOF model developed using OpenSees (Ref. 10) is adopted to simulate the seismic response of the building. The frame elements are modeled using the force-based nonlinear beam–column element with distributed nonlinear fiber sections (Ref. 11). The joint behavior is modeled using a "scissor model" (Ref. 12) with pinched hysteric behavior (Ref. 13). The shear stress–strain relationship for both interior and exterior joints proposed by Jeon *et al.*[14] was employed. The model does not account for axial failure of joints. Bond-slip rotations for beams have been included modifying joint backbone as proposed in Ref. 15, while column-base bond-slip at first floor has been explicitly modeled. To account for brittle behavior in non-ductile columns, shear and axial failures in the columns are modeled using the Limit State material (Ref. 16). P-Δ effects are included. A schematic view of the frame and the adopted numerical model is shown in Fig. 2. A Rayleigh damping of 2% is assigned to the first and third modes.

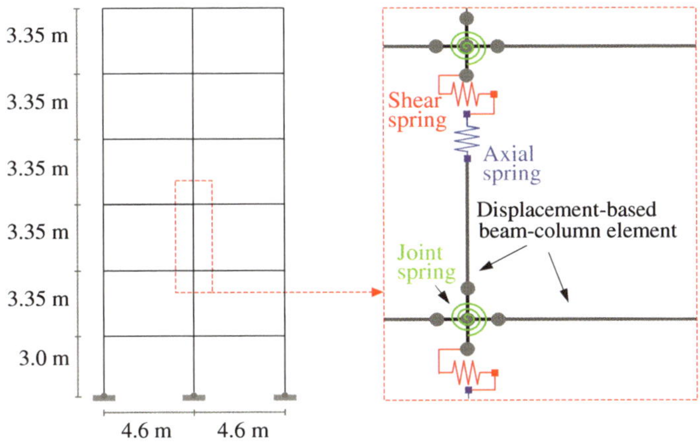

Fig. 2. Schematic view of frame and analytical model.

3.1. *Collapse Simulation*

This study considers 2 possible collapse mechanisms to capture the actual capacity in the model, according to Refs. 6 and 7: Side-sway collapse (SC) and Gravity load collapse (GLC). The SC occurs when a single storey has reached its capacity to withstand lateral loads, GLC when vertical load demand exceeds the total vertical load capacity at a given storey.

An internal algorithm monitors the dynamically varying capacity of each element and checks the GLC and SSC criteria throughout each NTH analysis to detect the collapse. A bisection algorithm is then implemented to find the collapse capacity with a precision of 0.05 g during the IDA procedure. The former algorithm, besides to supply a consistent definition of collapse, significantly reduces computing time avoiding non-convergence issues, with a similar purpose the latter is introduced to focus only on the definition of collapse capacity of the structure avoiding an excessive number of analyses.

4. Results and Discussion

Following steps indicated in Sec. 2, a bin of accelerograms was selected for the specific site and building (see Ref. 7), and the B2B-IDA was

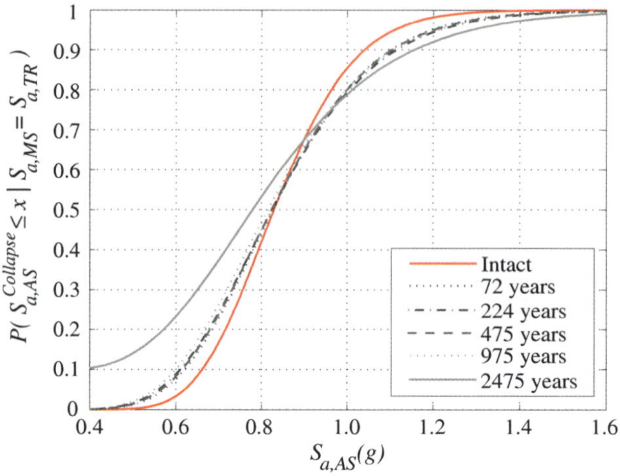

Fig. 3. T_R-dependent fragilities at collapse.

performed in order to determine AS fragility curves for increasing T_R, namely 72, 224, 475, 975, 2475 years. The REC of MS-damaged building is computed in terms of S_a based on the IDA results obtained from MS–AS sequences and here represented in terms of fragility curve at collapse. For the damaged building, the collapse fragility curve is calculated accounting for the effect of record-to-record variability on structural response. Using the set of 31 GMs applied as both MSs and ASs, a total of 961 record sequences for each selected T_R was carried out. Figure 3 illustrates the collapse fragility curves for the intact and damaged building in terms of probability of collapse conditioned on the MS for given T_R as a function of AS spectral intensity, $S_{a,AS}(T_1)$. The red curve represents the behavior of the intact building. It may be observed that the building's capacity does not change in a substantial way up to a return period of 2,475 years, for which 3 MS induced structural collapse in the structure. This is probably because the level of damage caused by MS earthquakes of increasing intensity (i.e., increasing T_R) is relatively low. In fact, local damage indicators analytically evaluated at member level for beam and columns (yielding and spalling) and joints (cracking, pre-peak, peak and ultimate strength), highlight that only for $T_R = 2475$ years a significant damage is attained, leading to a sensible shift of AS fragility curves. The median probability of collapse in 50 years is 0.052% for the intact building while it increases to 0.1% for a 72-years-MS-damaged structure and up to a maximum value of 9.9% when considering a T_R of 2,475 years.

5. Computational Effort

As evidenced above in Sec. 4, IDA is carried out on more than 4800 MS–AS sequences (961 MS–AS sequences for 5 Return periods). The process is computationally intensive, due to the use of a fiber model coupled with the Limit State Material, that allows to account for both shear and axial failure and load redistribution. In order to speed the analyses, OpenSees (OpenSees, 2014) was used on SCoPE grid at the University of Naples Federico II. Running the analysis on the SCoPE grid in parallel reduces the computation to about 7 days with respect to 240 days required on a desktop computer with 4 processors.

6. Conclusions

This paper aims at calculating REC after damaging earthquakes through a clear probabilistic framework. In particular, the REC was computed with reference to five different Return Periods, via Incremental Dynamic Analysis, for an existing non-ductile 6-story building in Benevento, Italy.

The response of the buildings was simulated by means of a detailed 2D nonlinear MDOF finite element model. The model properly account for brittle failures and captures typical non-ductile RC frames failure modes. The analytical model of building was subjected to multiple sequences comprised of 2 recorded GMs to simulate the damage produced on the building by an MS of a given Return Period.

The results show that the REC of the MS-damaged building, does not significantly change due to the low hazard level intensity at the site, which entails a small damage in the structure. Only for $T_R = 2475$, the variation in building's capacity become evident. Additional studies need to address this aspect in order to have a more accurate comprehension of the collapse sensitivity to damage and to account for earthquake polarity (Ref. 3). This study gained significant advantage from the use of a parallel computing grid that reduced significant computing time.

References

1. FEMA 306. Evaluation of Earthquake Damaged Concrete and Masonry Wall Buildings — Basic procedures manual. Federal Emergency Management Agency. Washington D.C. (1998).
2. M. Polese, M. Gaetani d'Aragona, A. Prota and G. Manfredi, Seismic behavior of damaged buildings: A comparison of static and dynamic nonlinear approach, Paper No. 1134, Proc. of COMPDYN 2013 4th ECCOMAS Thematic Conference on Computational Methods in Structural Dynamics and Earthquake Engineering, Kos Island, Greece, 12–14 June 2013.
3. M. Rahunandan, A. B. Liel and N. Luco, Aftershock collapse vulnerability assessment of reinforced concrete frame structures, *Earthquake Engng Struct. Dyn.* **44**(3), pp. 419–439 (2014).
4. A. Réveillère, P. Gehl, D. Seyedi and H. Modaressi, Development of seismic fragility curves for damaged reinforced concrete structures 15WCEE, Lisbon, Portugal, 24–28 September (2012).
5. M. Gaetani d'Aragona, M. Polese and A. Prota, *Relationship Between the Variation of Seismic Capacity after Damaging Earthquakes, Collapse Probability and Repair Costs: Detailed Evaluation for A Non-Ductile Building.* 5th ECCOMAS Thematic Conference on Computational Methods in Structural Dynamics and Earthquake Engineering, Crete Island, Greece, May 25–27, 2015.
6. M. Gaetani d'Aragona, M. Polese, K. J. Elwood, M. Baradaran Shoraka, A. Prota and G. Manfredi, *Building Performance Loss After Damaging Earthquakes: an Investigation Towards Reparability Decisions.* 12th International Conference on Applications of Statistics and Probability in Civil Engineering, Vancouver, Canada, July 12–15, 2015.
7. M. Gaetani d'Aragona, Post-Earthquake Assessment of Damaged non-Ductile Buildings: Detailed Evaluation for Rational Reparability

Decisions, ***Ph.D. Dissertation***, University of Naples Federico II, Italy, 2015.

8. P. Bazzurro, C. A. Cornell, C. Menun and M. Motahari, Guidelines for seismic assessment of damaged buildings, *13WCEE* Vancouver, B.C., Canada, 2004.
9. M. Polese, M. Di Ludovico, A. Prota and G. Manfredi, Damage-dependent vulnerability curves for existing buildings. ***Earthquake Engng. Struct. Dyn.*** **42**(6), pp. 853–870 (2013).
10. F. McKenna, G. L. Fenves and M. H. Scott, Open System for Earthquake Engineering Simulation. University of California, Berkeley, CA, 2000.
11. R. M. de Souza, Force-based finite element for large displacement inelastic analysis of frames. PhD thesis, University of California, Berkeley, United States — California, 2000.
12. S. Alath and S. K. Kunnath, Modeling inelastic shear deformation in RC beam-column joints. ***Engineering Mechanics***; ASCE, pp. 822–825 (1995).
13. L. Lowes, N. Mitra and A. Altoontash, A beam-column joint model for simulating the earthquake response of reinforced concrete frames. Pacific Earthquake Engineering Research Center, College of Engineering, University of California, 2003.
14. J. S. Jeon, L. N. Lowes, R. DesRoches and I. Brilakis, Fragility curves for non-ductile reinforced concrete frames that exhibit different component response mechanisms. ***Engineering Structures***, **85**, 127–143 (2015).
15. O. C. Celik and B. R. Ellingwood, Modeling beam-column joints in fragility assessment of gravity load designed reinforced concrete frames. ***J. Earthq. Eng.***, **12**(3), 357–381 (2008).
16. K. J. Elwood, Modelling Failures in Existing Reinforced Concrete Columns. ***Can. J. Civ. Eng.***, **31**(5), 846–859 (2004).

Chapter 25

Soft Matter in Flow and High Performance Computing

G. D'Avino*,[§], M. M. Villone*,[†], G. De Monaco*, M. De Corato*, M. Trofa*,[†], F. Greco[‡] and P. L. Maffettone*

**Dipartimento di Ingegneria Chimica, dei Materiali e della Produzione Industriale, Università degli Studi di Napoli Federico II, Piazzale Tecchio 80, 80125 Napoli, Italy*

[†]*Center for Advanced Biomaterials for Health Care @CRIB, Istituto Italiano di Tecnologia, Largo Barsanti e Matteucci 53, 80125 Napoli, Italy*

[‡]*Istituto di Ricerche sulla Combustione, Consiglio Nazionale delle Ricerche, Piazzale Tecchio 80, 80125 Napoli, Italy*

[§]*gaetano.davino@unina.it*

The progressive increase of the computing power and the development of techniques for numerical stabilization make conceivable the construction of computational codes, able to calculate *in silico* flow behavior and the rheological properties of complex fluids. In this chapter, we report on recent results in this field obtained by using the computational resources of the SCoPE-UNINA grid infrastructure.

Rheological predictions in rheometric flows for heterogeneous systems such as suspensions and emulsions, the dynamics of Brownian ellipsoids near a wall, the flow of granular materials in rotating drums are presented.

1. Introduction

The field of rheology aimed to solve, through numerical simulations, the flow of complex materials in complex geometries that is generally termed "computational rheology". Several reviews on this subject are available in literature, reporting the continuous progresses and the open challenges in the field (see, e.g., Refs. 1–3). The remarkable increase of the computational power observed in the last 30 years opened up the possibility to face difficult problems, allowing the development of new applications that could lead to a "rheometry *in silico*".

The prediction through numerical simulations of the flow behavior and the rheological properties of soft matter can be done over different dimensional scales. From a microstructural level, atomistic approaches are still limited to coarse models of polymeric systems in geometries with size comparable to molecular systems. Techniques based on Brownian dynamics allow analyzing dilute solutions of linear polymers in Newtonian solvents. These models describe mesoscale phenomena. By further increasing the characteristic dimensional scale, continuum-mechanics approaches are required. In this case, the liquid properties are modelled through constitutive equations to predict the material behavior under flow. An accurate rheological description of the complex fluids through realistic constitutive equations and the development of stable numerical algorithms made possible, in the recent years, the simulation of systems like suspensions of rigid or deformable particles in Newtonian or non-Newtonian fluids, emulsions, suspensions with non-spherical particles, granular materials. All these problems require an accurate prediction of the velocity, pressure and stress fields of the fluid to recover the bulk rheology and, in the case of suspensions/emulsions, the simulation of particles/drops moving in these fluids. The use of High Performance Computing (HPC) is, then, essential.

In this contribution, we illustrate an overview of recent results obtained by our group through the computational resources of the SCoPE-UNINA grid infrastructure. Specifically, the problems reported in this paper are: (i) linear and nonlinear viscoelasticity of solid Newtonian and viscoelastic suspensions; (ii) linear viscoelasticity of emulsions; (iii) Brownian dynamics of ellipsoids; (iv) granular materials in rotating drums.

2. Solid Particle Suspensions

We consider a suspension of solid, spherical particles and neglect both particle and fluid inertia. The equations governing the fluid dynamics of this system are the continuity and momentum balance equations, and the force and torque balances on the particle surfaces. In addition, we include a constitutive equation to model the viscoelastic suspending fluid. The equations are solved by the Finite Element Method (FEM) with triperiodic shear flow boundary conditions. A fictitious domain method is employed to handle the particle motion. More details can be found in Ref. 4.

2.1. *Small Amplitude Oscillatory Shear (SAOS)*

In this section, we present numerical simulations of suspensions subjected to an oscillatory shear flow at small amplitudes (SAOS). The aim is to show the effect of the microstructure of the rheological response of the material.

In Fig. 1a, the normalized loss modulus $G''_n = G''/G''_0$ (where G'' is the modulus of the suspension and G''_0 is the modulus of the suspending fluid) as function of the particle volume fraction ϕ is reported for a Newtonian and viscoelastic suspending liquid. Numerical simulations have been performed by starting from a initial random particle distribution (as shown in Fig. 1b). In agreement with theoretical and numerical results from the literature,[5,6] under the SAOS

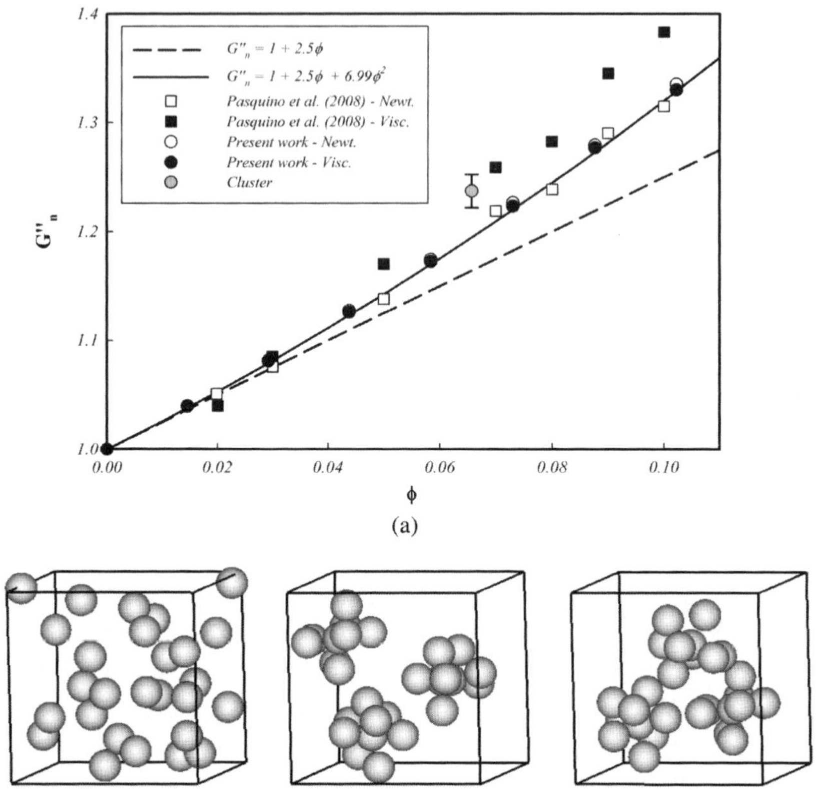

(a)

(b) (c) (d)

Fig. 1 Normalized loss modulus as a function of the particle volume fraction (a): theoretical prediction for dilute[7] (dashed line) semi-dilute[8] (solid line) systems; experimental data for Newtonian (open squares) and viscoelastic (close squares) suspensions,[9] simulations for Newtonian (open circles) and viscoelastic (close circles) suspensions with an initial random configuration (b), and with initial configurations made by clusters (c and d).

regime, the suspending medium does not alter the rheological response. Indeed, the data for Newtonian suspensions superimpose those obtained for viscoelastic suspensions. Our numerical results are in very good agreement with theoretical predictions valid for dilute[7] (dashed curve) and semi-dilute regimes (solid curve),[8] and with recent experimental data for Newtonian suspensions.[9] On the contrary, a deviation from experimental measurements in viscoelastic media is observed. To justify this discrepancy, numerical simulations for initial configurations made by aggregates (see Figs. 1c and 1d) have been performed. The computed loss modulus deviates from the one corresponding to an initial isotropic distribution (grey circle in Fig. 1a), highlighting a remarkable effect of the microstructure on the bulk rheological properties.

2.2. *Large Amplitude Oscillatory Shear*

The analysis reported in the previous section can be extended to large amplitudes Oscillatory Shear (LAOS). We select the Giesekus model as constitutive equation with the following parameters: relaxation time $\tau = 1$, viscosity ratio $\eta_s/\eta_p = 0.1$, mobility parameter $\alpha = 0.2$.

The data in Fig. 2 show the storage G′ and loss modulus G″ normalized by the corresponding values obtained in the linear regime G'_L e G''_L as a function of the oscillation amplitude γ_0 for different particle volume fractions ϕ. The oscillation frequency is $\omega = 0.2$. For increasing values of the amplitude, we observe a deviation of the moduli from the linear viscoelastic values. The entity of the deviation increases with the volume fraction. Finally, for more concentrated systems, the deviation occurs at lower amplitudes, in agreement with experimental data.[10]

3. Emulsions

The linear viscoelastic properties of a monodisperse emulsion of Newtonian drops in a Newtonian fluid subjected to SAOS flow are investigated by means of 3D Arbitrary Lagrangian Eulerian (ALE) FEM numerical simulations. Volume fractions of the suspended phase ϕ from the dilute to the concentrate regime (ϕ ~0.3), and a range of several orders of magnitude of the viscosity ratio between the drops and the matrix λ, and of the frequency of the oscillatory flow ω are considered. Interfacial slip between the drops and the matrix is also taken into account. The numerical results are compared with theoretical

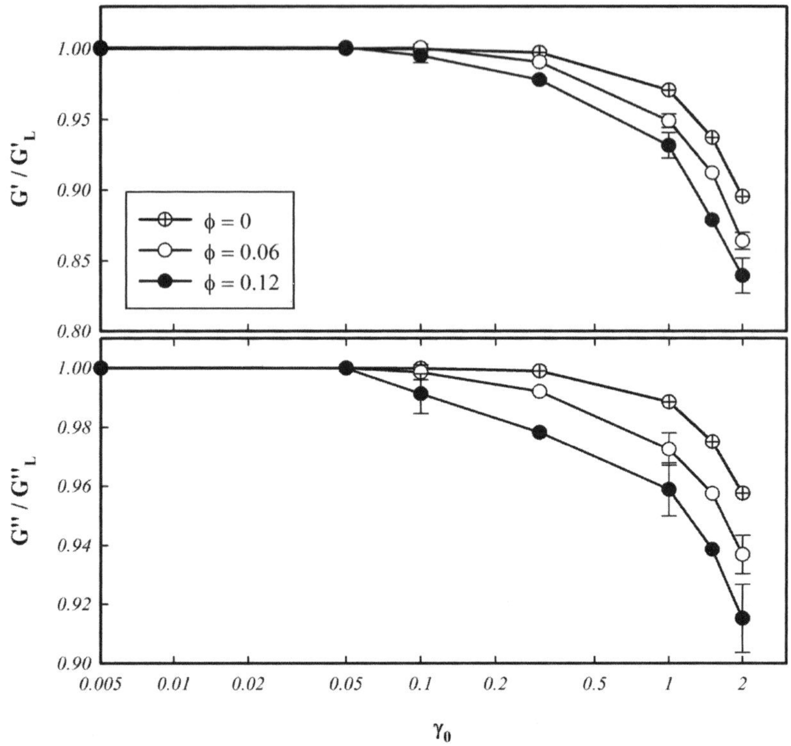

Fig. 2. Storage and loss moduli as a function of the amplitude of the oscillation for different particle volume fractions at $\omega = 0.2$.

predictions due to Oldroyd[11] for analogous systems: a quantitative agreement is found for a wide range of the parameters considered, even well beyond the limit of the dilute regime, and in the presence of interfacial slip (for details, see Ref. 12).

3.1. *Results at Fixed Frequency*

In Fig. 3, the values of the elastic modulus G'' and of the viscous modulus G'' are shown as a function of the volume fraction of the emulsion ϕ, given the frequency of the oscillations $\omega = 0.4$, and the viscosity ratio $\lambda = 1.0$. The numerical results are displayed as symbols with error bars, whereas Oldroyd predictions[11] in the same flow

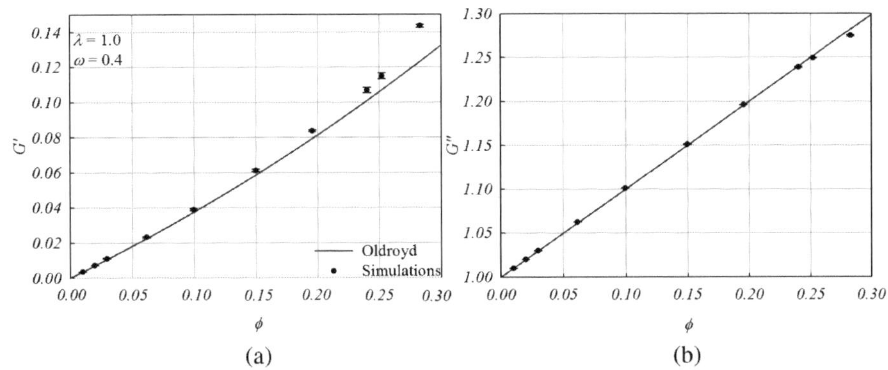

Fig. 3. Dimensionless elastic modulus G′ (a) and viscous modulus G″ (b) of a monodisperse emulsion of Newtonian drops in a Newtonian matrix subjected to SAOS flow as a function of the volume fraction ϕ. The viscosity ratio is $\lambda = 1.0$, the dimensionless frequency of the oscillations is $\omega = 0.4$, there is no slip at the drops/matrix interface. The solid lines represent Oldroyd theoretical predictions,[11] the symbols represent the numerical data.

conditions are displayed as solid curves. From Fig. 3a, it emerges that the numerical G′-values overlap Oldroyd curve in a wide range of ϕ-values, except for $\phi = 0.28$, where a discrepancy of 17% between the numerical and the theoretical data is found. In Fig. 3b, it can be seen that for G″, there is a quantitative agreement between numerical and theoretical data in the whole ϕ-range considered.

In Fig. 4, the values of G′ and G″ are shown as a function of the volume fraction of the emulsion ϕ for $\omega = 0.4$, $\lambda = 1.0$, and for two values of the interfacial slip parameter, i.e., $\alpha = 0.1$ (circles) and $\alpha = 1.0$ (triangles). It arises from Fig. 4 that, the other parameters are fixed, as the slip parameter α increases, and both the moduli decrease. For both the α-values considered, such evidences qualitatively agree with Oldroyd predictions.[11]

4. Brownian Dynamics of a Confined Ellipsoid

The anisotropic shape of an ellipsoid undergoing Brownian motion gives rise to different behaviors compared to spheres. Different diffusion coefficients arise in different directions, this anisotropy is further

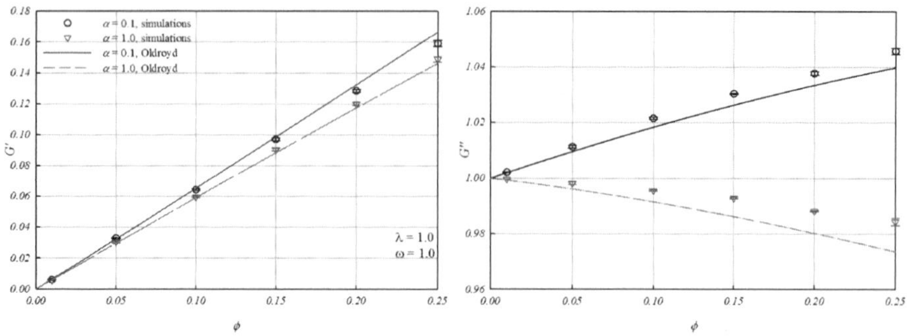

(a) (b)

Fig. 4. G′ (a) and G″ (b) of a monodisperse emulsion of Newtonian drops in a Newtonian matrix subjected to SAOS flow as a function of the volume fraction ϕ. The viscosity ratio is $\lambda = 1.0$, the dimensionless frequency of the oscillations is $\omega = 0.4$, two values of the interfacial slip parameter are considered, i.e., $\alpha = 0.1$ and 1.0. The solid lines represent Oldroyd theoretical predictions,[11] the symbols represent the numerical data.

modified by the presence of a wall[13]; the diffusion is also very sensitive to the boundary conditions at the fluid–solid interface; at small length scales no-slip condition may no more be valid on the wall or on the particle surface.

We have carried out FEM simulations to characterize the hydrodynamic interactions of a spheroid with an infinite plane wall. In particular, we have computed the mobility matrix in the range of positions and orientations covering all possible particle-wall configurations.

4.1. *Hydrodynamic Interactions with an Infinite Wall*

The presence of a wall induces several different features on the mobility coefficients compared to that of an unconfined ellipsoid. One of the most interesting is the rise of a roto-translational coupling,[13] that is to say, a force applied on the ellipsoid also causes a rotation, whereas, a torque induces a translation. In Fig. 5, a schematic picture of the problem investigated is shown. In Fig. 6, we report the mobility of an ellipsoid, with aspect ratio AR = 8, in

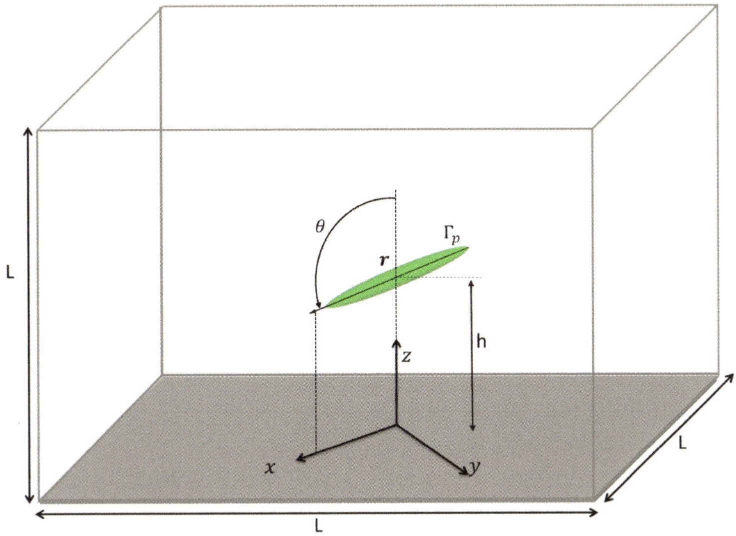

Fig. 5. Schematic representation of an ellipsoid near an infinite planar wall.

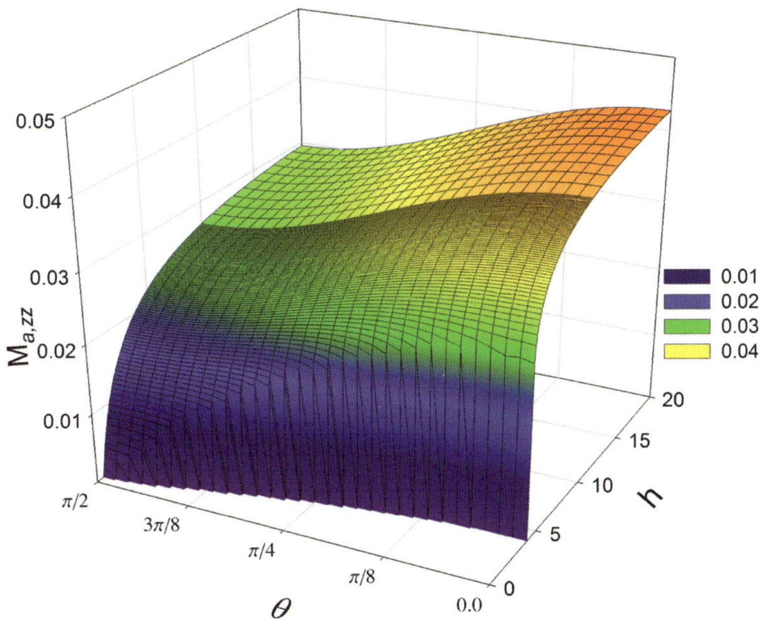

Fig. 6. Mobility coefficient of an ellipsoid, in direction normal to the wall, as a function of h and θ, a reduction is seen as the ellipsoid approaches the wall.

direction normal to the wall, as a function of the particle center-wall distance h and orientation θ. A global reduction is seen as the ellipsoid approaches the wall.

4.2. *Brownian Motion Near a Confining Wall*

The inclusion of hydrodynamic interactions gives rise to "corrective terms" in the Brownian dynamics. Those corrective terms are required in numerical simulations to recover the correct (Boltzmann) particle equilibrium distribution[14] and consist of corrective velocities in the Langevin equation.[15] In literature, these terms are typically referred to, as "drift" terms.

In Fig. 7, we report the translational "drift" velocities experienced by an ellipsoid with AR = 8, caused by the hydrodynamics interactions with an infinite plane wall.

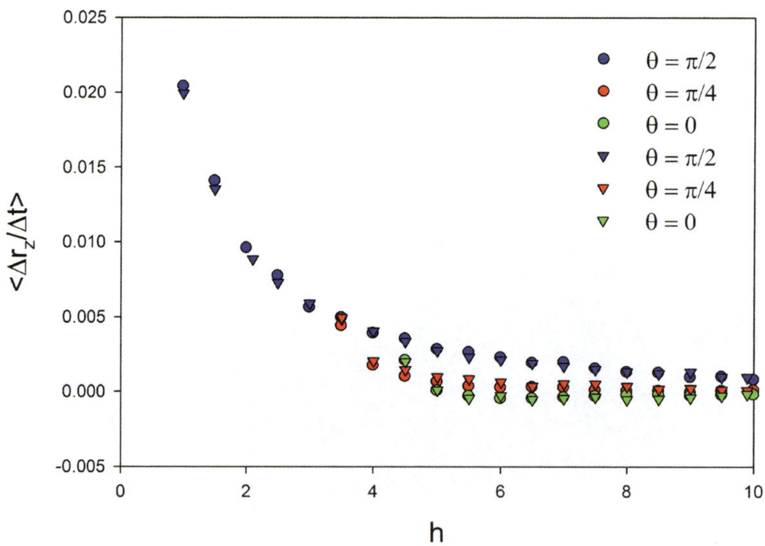

Fig. 7. Ellipsoid drift velocity as a function of h for three different angles; a positive value means that the velocity points away from the wall, a negative means the opposite.

5. Granular Materials in Rotating Drums

Granular materials are widely diffused in industry as well as in nature, but a reliable and effective description of their motion is still at a rather early stage of development. Among the benchmark problems of granular dense flow, the rotating drum is one of the most challenging, yet intriguing and technologically relevant. In proper ranges of operating conditions, granular materials inside rotating drums display a continuum motion near their free surface. The motion of those discrete systems has been studied both experimentally and through Discrete Element Method (DEM) numerical simulations; however, it can also be regarded as the flow of a continuum medium, thus allowing a continuum mechanics approach.

Here we solve the continuum dynamic equations by adopting the visco-plastic Jop–Forterre–Pouliquen (JFP) constitutive model[16] for the stress tensor, and study the continuous flow of dry grains inside half-filled, axially rotating cylinders through 3D Finite Volume simulations (FVM). Jop *et al.*[16] generalized an empirical 1D constitutive law which explicitly relies on the assumption that the (small) variation of the packing density (granular materials are known to change their arrangements if perturbed, i.e., they expand when a motion starts from a jammed configuration) can be neglected. The granular material, in whatever flow configuration, can be described as a continuum fluid having a viscosity given by:

$$\eta_g(\dot{\gamma}, p) = \left(\mu_s - \frac{\mu_2 - \mu_s}{I_0/I - 1} \right) \frac{p}{\dot{\gamma}}. \tag{1}$$

The granular viscosity depends on both the local pressure and the local rate-of-shear and μ_s, μ_s and I_0 can be considered as material properties (thus measured in independent experiments). I is a non-dimensional quantity, called the inertial number, which is uniquely dependent on the shear state of the system[17] and whose definition is:

$$I = \frac{d_p \dot{\gamma}}{\sqrt{p/d_p}}, \tag{2}$$

where d_p and r_p are the diameter and the density of a single particle.

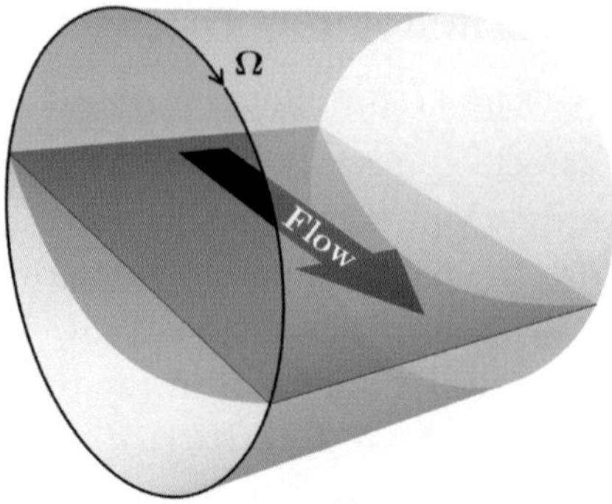

Fig. 8. Sketch of the geometry and of the flow; the shaded area is the flowing layer.

The granular constitutive equation has been tested in a fully 3D, transient, bi-phase problem with a moving interface between the granular phase and the air above it. A sketch of the flow configuration is shown in Fig. 8. We focused on the rolling regime, which is characterized by a steady granular flow, a flat free-surface and a lens-shaped flowing layer (shaded in Fig. 8).[18]

Figure 9 shows a typical result of the simulations. The orange part at the bottom is the rigidly rotating one, on top of it there is the flowing zone, which is colored by the magnitude of the velocity. The arrows represent the velocity vector field.

We investigated the effect of various flow parameters, like the particle-wall slip, the width of the cylinder, the ratio of the cylinder-to-particle diameters ratio and the angular velocity of the cylinder.

Figure 10 shows the interface between the granular phase and the air (the free surface) at various operative conditions. The inclination of the free surface increases with increasing angular velocity and decreasing aspect ratio of the cylinder, in agreement with experimental results.

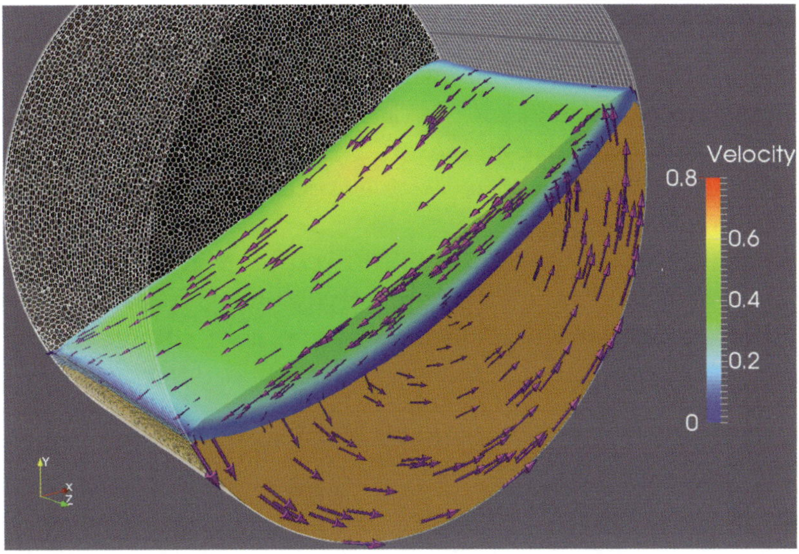

Fig. 9. A typical result of the simulations.

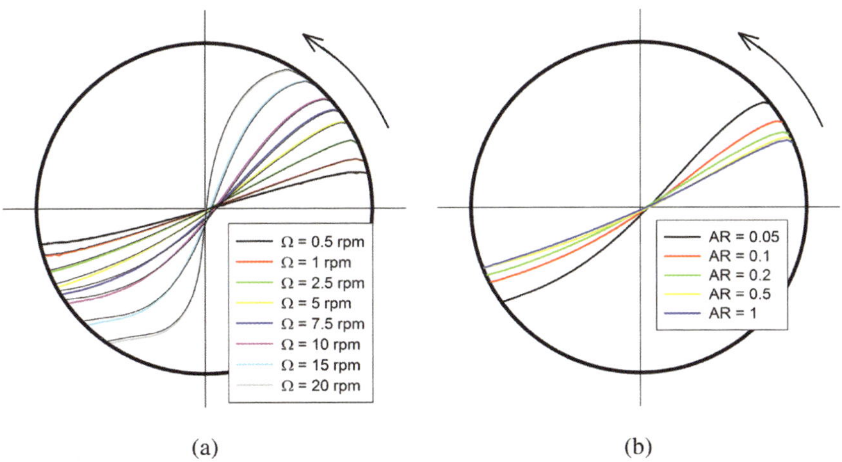

Fig. 10. Free surface at fixed slip and (a) fixed aspect ratio of the cylinder (AR = 0.1) and various angular velocities and (b) fixed angular velocity ($\Omega = 5$ rpm) and various aspect ratios.

6. Conclusions

We have presented recent simulation results on soft matter in flow. Examples are Newtonian and viscoelastic suspensions and emulsions, dynamics of Brownian ellipsoids and granular materials in rotating drums. HPC is required to successfully predict the flow behavior and the rheology of these materials. In this direction, our group has greatly taken advantage of the SCoPE-UNINA grid infrastructure. The presented results show that numerical simulations can be successfully used to realistically predict the behavior of complex materials and, in fact, reproduce experiments "*in silico*".

References

1. R. Keunings, Progress and challenges in computational rheology, *Rheol. Acta* **29**, pp. 556–557 (1990).
2. R.I. Tanner and K. Walters, *Rheology: An Historical Perspective*, Elsevier, Oxford (1998).
3. K. Walters and M.F.Webster, The distinctive CFD challenges of computational rheology, *Int. J. Num. Meth. Fluids* **43**, pp. 577–596 (2003).
4. G. D'Avino, F. Greco, M. A. Hulsen and P. L. Maffettone, Rheology of viscoelastic suspensions of spheres under small and large amplitude oscillatory shear by numerical simulations, *J. Rheol.* **57**, pp. 813–839 (2013).
5. R. Zwanzig and M. Bixon, Hydrodynamic theory of the velocity correlation function, *Phys. Rev. A* **2**, pp. 2005–2012 (1970).
6. H. M. Schaink, J. J. M. Slot, R. J. J. Jongschaap and J. Mellema, The rheology of systems containing rigid spheres suspended in both viscous and viscoelastic media, studied by Stokesian dynamics simulations, *J. Rheol.* **44**, pp. 473–498 (2000).
7. A. Einstein, Berichtigung Zu Meiner Arbeit. Eine Bestimmung der Molekuledimension, *Annl. Phys.* **339**, pp. 591–592 (1911).
8. G. K. Batchelor and J. T. Green, The determination of the bulk stress in a suspension of spherical particles to order c2, *J. Fluid. Mech.* **56**, pp. 401–427 (1972).
9. R. Pasquino, N. Grizzuti, P. L. Maffettone and F. Greco, Rheology of dilute and semidilute noncolloidal hard sphere suspensions, *J. Rheol.* **52**, pp. 1369–1384 (2008).

10. K. Hyun, M. Wilhelm, C. O. Klein, K. S. Cho, J. G. Nam, K. H. Ahn, S. J. Lee, R. H. Ewoldt and G. H. McKinley, A review of nonlinear oscillatory shear tests: Analysis and application of large amplitude oscillatory shear (LAOS), *Prog. Polym. Sci.* **36**, pp. 1697–1753 (2011).
11. J. R. Oldroyd, The elastic and viscous properties of emulsions and suspensions, *Proc. R. Soc. Lond. A* **218**, pp. 122–132 (1953).
12. M. M. Villone, G. D'Avino, M. A. Hulsen, F. Greco and P. L. Maffettone, Numerical simulations of linear viscoelasticity of monodisperse emulsions of Newtonian drops in a Newtonian fluid from dilute to concentrated regime, *Rheol. Acta* **53**, pp. 401–416 (2014).
13. J. Happel and H. Brenner, *Low Reynolds number hydrodynamics: With Special Applications to Particulate Media*, Vol. 1, Springer, Berlin (1983).
14. A. Lau and T. C. Lubensky, State-dependent diffusion: Thermodynamic consistency and its path integral formulation, *Phys. Rev. E* **76**, p. 011123 (2007).
15. H. C. ¨Ottinger, *Stochastic Processes in Polymeric Fluids: Tools and Examples for Developing Simulation Algorithms*, Springer-Verlag, Berlin (1996).
16. P. Jop, Y. Forterre and O. Pouliquen, A constitutive law for dense granular flows, *Nature*, **441**(7094), pp. 727–730 (2006).
17. F. da Cruz, S. Emam, M. Prochnow, J.-N. Roux and F. Chevoir, Rheophysics of dense granular materials: Discrete simulation of plane shear flows, *Physical Review E* **72**(2), p. 021309 (2005).
18. J. Mellmann, The transverse motion of solids in rotating cylinders — forms of motion and transition behavior, *Powder Technology* **118**(3), pp. 251–270 (2001).

Chapter 26

High Performance Computing (HPC) and Aerospace Research Activities at the University of Naples Federico II

A. De Marco*, F. Nicolosi, D. P. Coiro,
R. Tognaccini, G. Calise, P. Della Vecchia, S. Corcione,
D. Ciliberti and B. Mele

University of Naples Federico II, Napoli, Italy
**agostino.demarco@unina.it*

Research Group: A. De Marco, F. Nicolosi, D. P. Coiro, G. Calise, P. Della Vecchia, S. Corcione, D. Ciliberti, B. Mele — Aerospace Engineering Division of the Department of Industrial Engineering.

This chapter summarizes the main research activities performed by the above-mentioned research group, which have been conducted in the past 3 years with the support of the SCoPE Supercomputing Center.

1. Postprocessing of CFD Data and Aerodynamic Force Calculations

The Department of Industrial Engineering also installs, compiles and routinely runs on SCoPE, the flow solver FLOWer.[1] It has been developed at DLR (the German Aerospace Center) and it is widely being used in the industry and applied research. The FLOWer code solves the compressible three-dimensional Euler, steady and unsteady RANS equations on block-structured meshes by a central or AUSM finite volume formulation with explicit blended second- and fourth-order artificial dissipation. Time integration is carried out by an explicit hybrid multi-stage Runge–Kutta scheme with implicit smoothing and multi-grid acceleration.

Very low Mach number calculations are possible via preconditioning. Turbulence is modeled by either algebraic or transport equation models. Chimera and DES approaches are also allowed. It is written in Fortran 90 and provides very good parallel and vector performance due to the block structured approach with automatic load balancing. DII is not only using but also developing this software introducing new and advanced features, in particular, in the field of turbulence modelling.[2–4] In Figs. 1 and 2, a number of FLOWer applications are showed.

State-of-the-art expertize has been reached in particular in the post processing of Computational Fluid Dynamics (CFD) data for aerodynamic force analysis and breakdown.[5–9]

Fig. 1. Unmanned Space Vehicle (USV) designed by CIRA, Italian Aerospace Research Center (Courtesy of CIRA). Results obtained by RANS solver FLOWer on a 140 blocks structured mesh with 10 millions cells on 16–24 HPC nodes.

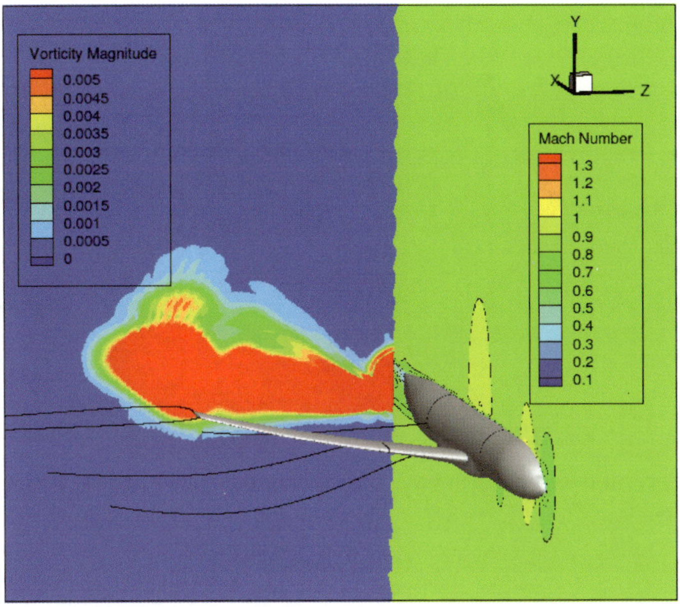

Fig. 2. CFD simulation of a transonic wing-body configuration. RANS solver FLOWer on 5 blocks structured mesh, 5 millions cells, 5 HPC nodes.

2. Vertical-Axis Wind Turbine

This research has been focused on a particular type of vertical-axis wind turbine (see Fig. 3), in which a number of inclined arms with airfoil-shaped cross-sections are mounted to connect the principal blades to their hub (see Ref. 15). While the majority of the known studies on vertical-axis turbines is devoted to the role of principal blades, in most of the cases without taking into account other parts of the wind turbine, the objective of this work has been to investigate the effect of uncommon arm geometries, such as the inclined arms.

The inclined arms are known to have a potentially beneficial role in the power extraction from the wind current but, due to the complexity of the phenomenon, the investigation on aerodynamics of this type of turbine is often impossible through analytical models, such as blade-element momentum theory. It turns out that adequate studies can only be carried out by wind tunnel experiments or CFD simulations.

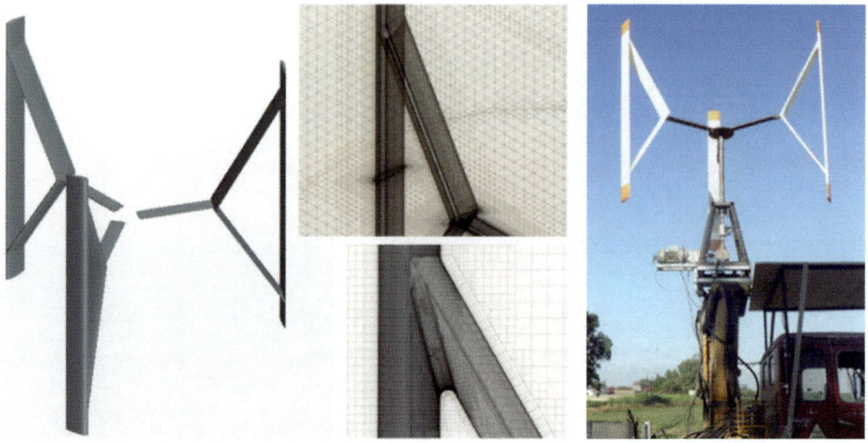

Fig. 3. A vertical-axis wind turbine. The CFD mesh (left) and the physical prototype (right).

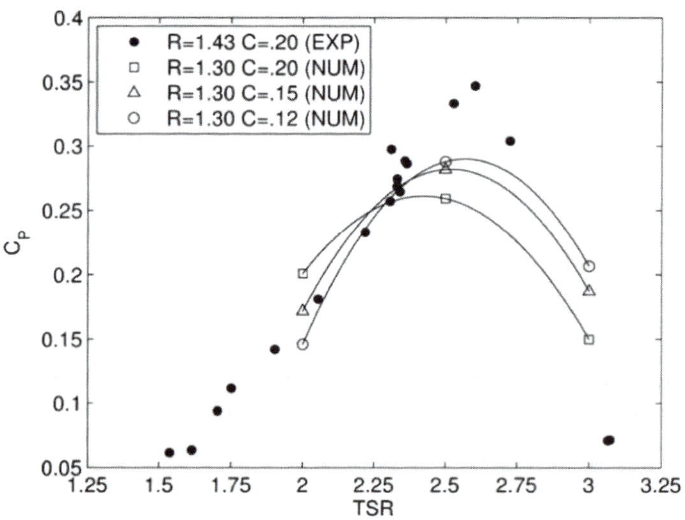

Fig. 4. Validation of experimental data. Power coefficient versus tip-speed-ratio. Grid points: 20×10^6, CPU time on 64-core: 24h/revolution.

The outcome of this research[10] has been a methodical CFD study on how inclined arms can be used on a selected wind turbine configuration to harvest additional power from the wind (see Figs. 3 and 4).

All tests are developed through *U-RANS* simulations for *3D* flow past rotating turbine. The *U-RANS* simulations have been performed by using the commercially available flow simulation software *CD-Adapco STAR CCM+ v. 9.06.009.*

3. Aircraft Aerodynamic Design

The aerodynamic analysis on the DLR-F11 high lift configuration model has been performed on the supercomputing grid infrastructure SCoPE of the University of Naples Federico II.

The model geometry is representative of a wide-body commercial aircraft, which experimental investigations at high Reynolds number have been performed at the European Transonic Wind-tunnel (ETW) for the second AIAA High Lift Prediction Workshop. The commercial CAE package, Star-CCM+ has been used to solve the Reynolds-averaged Navier–Stokes equations. Inviscid, viscous incompressible, and compressible analyses have been performed with mesh refinement. The inviscid calculations have been used to assess how far is the Eulerian prediction from experimental data. Viscous and compressible calculations have been realized using the Spalart–Allmaras turbulence model at 0.175 Mach number and 15.1 million Reynolds number. Results show that the simple Spalart–Allmaras turbulence model can predict quite accurately the stall and post-stall behavior, getting the angle of stall and underestimating the maximum lift coefficient by less than 5%. Comparisons among numerical and experimental pressure coefficients at several sections are also shown. Finally, the stall path is described. NASA High lift device workshop was concluded on June 2013. Results comparisons with workshop participants have been carried out.[11–12] Some selected results are reported in Figs. 5 and 6.

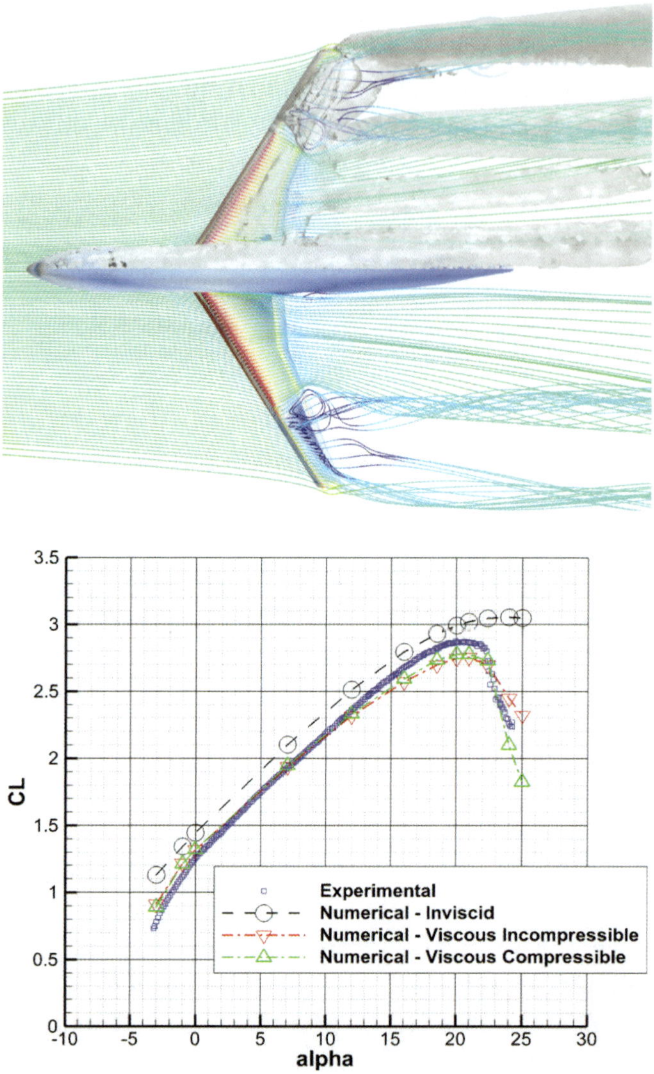

Fig. 5. Numerical simulation on a wing-body configuration (DLR F11 model) in high lift conditions (i.e., slats and flaps deployed) flap, 15 millions of polyhedral cells, Viscous/compressible flow, Spalart–Allmaras turbulence model, 12 hours per angle of attack with 128 CPUs.

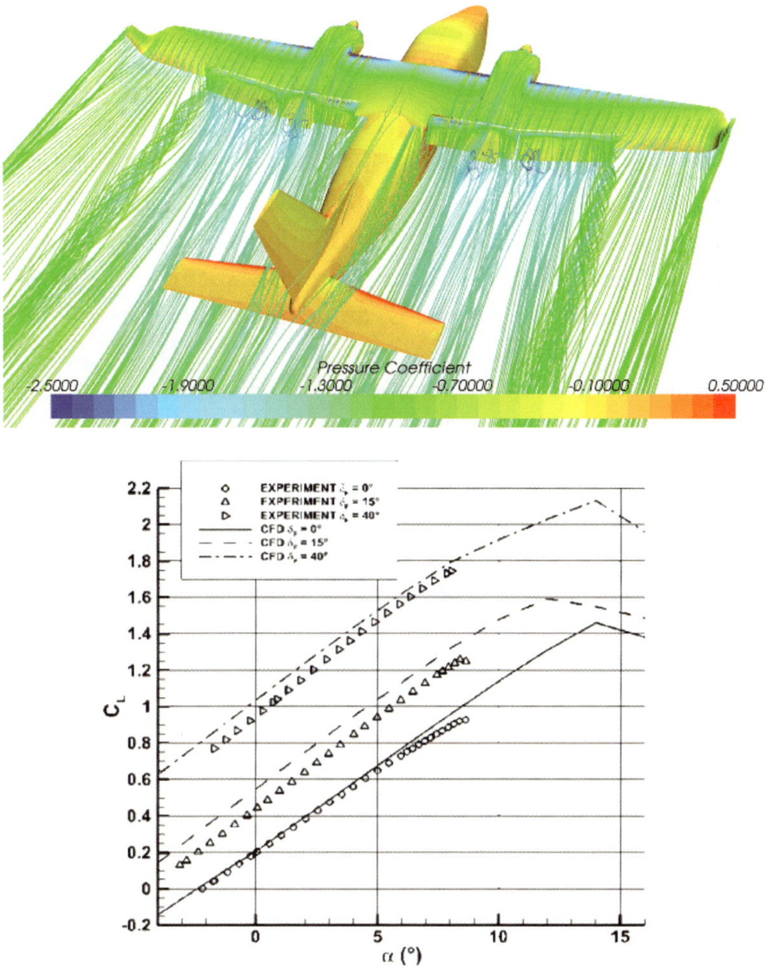

Fig. 6. Numerical simulation on a complete Commuter Aircraft with flap, 18 millions of polyhedral cells, Viscous/compressible flow, Spalart–Allmaras turbulence model, 20 hours per angle of attack with 64 CPUs.

4. Drag Reduction

Reducing aerodynamic drag and, consequently, fuel consumption is one of the most important tasks for automotive manufacturers, in order to succeed. In general, different approaches are possible: by decades, modify the overall shape of a car was the first approach to improve performances; but, in the last years, the use of active control systems to reduce drag has seemed to produce better results.[13]

Active and passive control techniques, by means of boundary-layer control, are well known in aerospace applications (see Ref. 14); this know-how and some results can be very useful for automotive industry too. This paper demonstrates that some active control systems, such as unsteady blowing or synthetic jets, are suitable in automotive engineering. Active control systems depend on a number of parameters as mass flow rate, inflow angle, position, jet frequency, etc. In this paper, CFD simulations, performed on the High Performance Computing (HPC) device available at the University of Naples, are presented. These CFD simulations have allowed analyzing the relative effectiveness of each parameter respect to drag reduction: indeed, the use of numerical analyses is really important for both the design and development of innovative devices, reducing the time and the expense for the setup and the execution of experimental tests.

The systems dealt in this work are respectively unsteady blowing system and synthetic jet system. They are different respect to the mass flow rate: the former has a nonzero mass flow rate, while the latter has a zero mass flow rate. Both systems have been compared by means of both two-dimensional and three-dimensional CFD simulations, in order to select the most effective device.

These devices were numerically tested on the so-called Ahmed body: the Ahmed body represents a shape of reference for such kind of investigations (see Fig. 7). Automotive industry has individuated that the Ahmed body represents the reference model for studying the flow behavior of a car. Indeed, small modifications of the shape of the Ahmed body, in particular of the so-called slant angle, makes it capable of reproducing the rear flow of different categories of car, from fast-backs to square-backs. This is due to its drag curve, which, up to the critical slant angle of 30° respect to the upper surface, drops off quickly. The studies performed by Ahmed have highlighted that, for slant angles slightly below 30°, the flow structure is more complicated than the flow at other slant angles: so, this study investigates the effectiveness of active control techniques in the most unlucky condition, that is for 25° of slant angle.

At first stage, the Ahmed body, with 25° of slant angle, has been analyzed by means of two-dimensional CFD simulations. These simulations have the aim to search an operative set of parameters that will also be tested later on the more expensive three-dimensional simulations. Two-dimensional computational mesh has about 3,00,000 cells and it requires about 2 days on 8 CPUs desktop device. With this mesh, steady simulations, with a simulated steady jet, have allowed to individuate:

- The critical separation point of the Ahmed body;
- The most effective position for the jet injection;
- The most effective inflow angle;
- The minimum mean jet velocity capable of reducing drag;
- The less influenced position from inflow angle deviation.

All these parameters constitute an initial guess that has to be validated with further investigations. Several unsteady simulations have been performed switching from steady jets to unsteady ones. Unsteady jets are modeled using a sinusoidal wave form, with nonzero mean for the unsteady blowing case and zero mean for the synthetic jet case; the amplitude of the sinusoidal wave is equal for both types of jets and it is equal to the minimum mean velocity found with steady simulations. While steady two-dimensional simulations have shown a drag reduction due to the steady jet, using unsteady one has shown a detrimental behavior. Indeed, even if two-dimensional simulations don't require huge time and hardware efforts and, so it is quick to obtain information about effectiveness of the chosen parameters, conversely, because the flow around a car is highly three-dimensional, successive three-dimensional simulations are fundamental.

Three-dimensional unsteady simulations have been performed using a mesh of about 14M of cells. This mesh has several refinements to catch the typical tip vortices of the 25° Ahmed body configuration and the lower pressure zone behind it. Probably, a finer mesh should be necessary to achieve the same drag value measured by Ahmed, but, using it for comparison purpose, it seemed to be well suited. For the unsteady simulations, the use of a small time step is a requirement to better fit the simulated sinusoidal wave form of the control jet. Thus, to achieve both requirements in a suitable time, these simulations are performed on the SCoPE facility: using 64 CPUs, each simulations required about 2 days for simulating about 0.4 s.

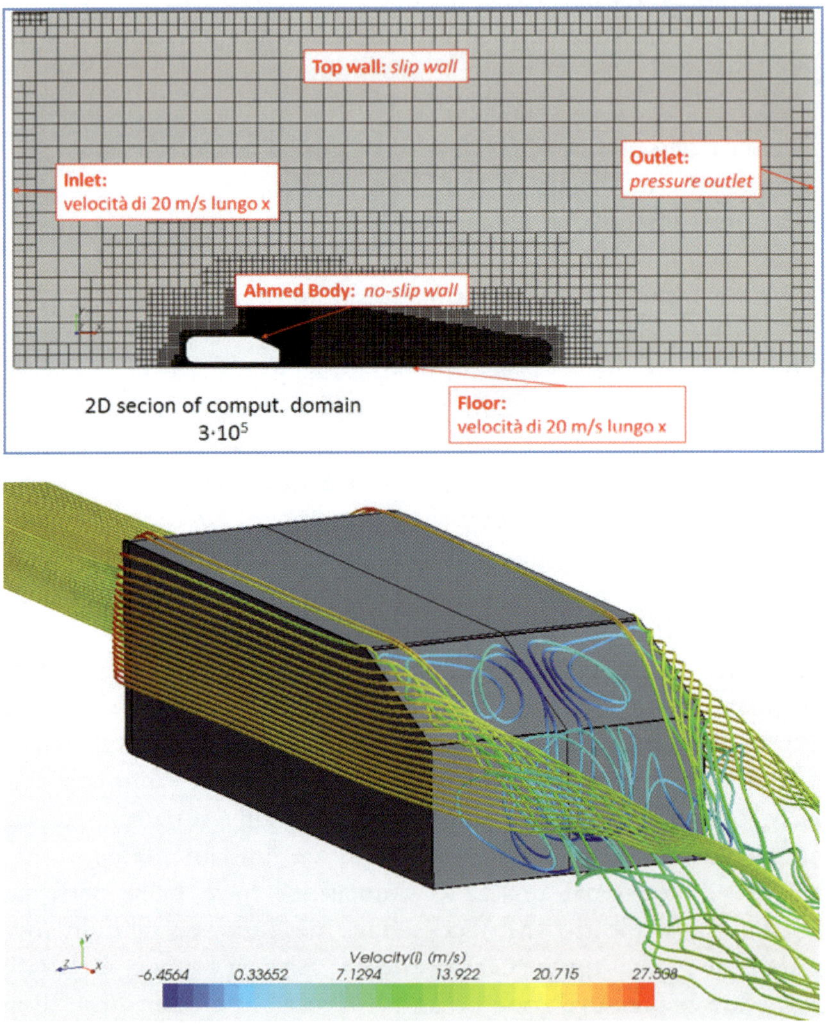

Fig. 7. Unsteady RANS simulation over a Ahmed body. The computational grid (top) and streamlines behind the slanted rear part of the body. Results obtained with a $k-\omega$ turbulence model, 15×10^6 grid cells. CPU time at 128 cores: 24 h per 0.1 s of simulated time.

In the light of this, only a small number of three-dimensional simulations have been done. Data from simulations highlight that a drag reduction is possible with such device: a drag reduction of about 11% has been obtained with the unsteady blowing jet, while a reduction of about 13% has been obtained with the synthetic jet device. For each device, the best combination of position, velocity and inflow angle, has been individuated and these set of parameters has to be compared in the light of energy consumption in order to select the most efficient device. The first experimental tests, made at Industrial Engineering Department of University of Naples, have confirmed the results obtained from CFD simulations.

Summarizing, it is possible to reduce drag of a car-shaped body by means of active control jet devices. CFD simulations can be very useful in the development and pre-test stages for investigating different engineering solution, with less expense and time, respect to experimental tests. But, to solve CFD problems, usually, a huge computational effort is needed, so HPC infrastructure represents more than a requirement. The average drag reduction of about 10%, predicted by means of CFD simulations, has been confirmed by tests. But this reduction has to be reviewed looking also to the power requirements of each device: due to the need of providing a pressurized tank for the unsteady blowing device, it seems to be less efficient than the synthetic jet device.

References

1. J. Raddatz and J. Fassbender, Block structured Navier–Stokes solver FLOWer, Vol. 89 of Notes on Numerical Fluid Mechanics and Multidisciplinary Design, Springer, Berlin, 2005, pp. 27–44.
2. P. Catalano and R. Tognaccini, Turbulence modelling for low Reynolds number flows, *AIAA J.*, **48**, pp. 1673–1685 (2010). Available at http://dx.doi.org/10.2514/1.J050067. 03/09/2016.
3. P. Catalano and R. Tognaccini, RANS analysis of the low-Reynolds number flow around the SD7003 airfoil, *Aerospace Science & Technology*, **15**, pp. 615–626 (2011). Available at http://dx.doi.org/10.1016/j.ast.2010.12.006. 03/09/2016
4. P. Catalano, B. Mele and R. Tognaccini, On the implementation of a turbulence model for low Reynolds number flows, *Computers & Fluids*, **109**, pp. 67–71 (2015).

5. M. Lanzetta, B. Mele, R. Tognaccini, Advances in aerodynamic drag extraction by far-field methods, *J. of Aircraft*, 52, 1873–1886, 2015.
6. B. Mele and R. Tognaccini, Aerodynamic force by Lamb vector integrals in compressible flow, *Phys. of Fluids*, **26**, pp. 056104–1 056104-16 (2014).
7. C. Marongiu, R. Tognaccini and M. Ueno, Lift and lift induced drag computation by Lamb vector integration, *AIAA J.*, **51**, pp. 1420–1430 (2013).
8. M. Ueno, K. Yamamoto, K. Tanaka, M. Murayama and R. Tognaccini, Far-field drag analysis of NASA Common Research Model simulation, *J. of Aircraft*, **50**, pp. 388–397 (2013).
9. C. Marongiu and R. Tognaccini, Far field analysis of the Aerodynamic force by Lamb vector integrals, *AIAA J.*, **48**, pp. 2543–2555 (2010). Available at http://dx.doi.org/10.2514/1.J050326. 03/09/2016
10. A. De Marco, D. P. Coiro, D. Cucco and F. Nicolosi, A Numerical Study on a Vertical-Axis Wind Turbine with Inclined Arms, *Int. J. Aerospace Eng.*, (2014).
11. P. Della Vecchia and D. Ciliberti, *Numerical Aerodynamic Analysis on a Trapezoidal Wing with High Lift Devices: A Comparison with Experimental Data*, XXII AIDAA Conference, Napoli (Italy), 9–12, September, 2013.
12. F. Nicolosi, S. Corcione and P. Della Vecchia, *Commuter Aircraft Aerodynamic Design: Wind-Tunnel Tests and CFD Analysis*, 29th ICAS Conference, 7–12 September 2014, St. Peterburg (Russia).
13. Ahmed, S., Ramm, G., and Faltin, G., "Some Salient Features Of The Time-Averaged Ground Vehicle Wake," SAE Technical Paper 840300, 1984, doi:10.4271/840300.
14. E.-F. Bellobuono, *Aircraft Endurance Improvement Through Turbulent Separation Control*, Tesi di Dottorato, Università degli Studi di Napoli Federico II, 2006.
15. Cucco D. (2012), Numerical analysis of aerodynamic flow around vertical-axis wind energy microgenerators on the SCoPE supercomputing framework, Tesi di Laurea, Università degli Studi di Napoli Federico II.

Chapter 27

Contribution of the High Performance Computing (HPC) in Naval Architecture Researches

A. De Marco*, F. De Luca, S. Mancini, S. Miranda, C. Pensa, R. Scognamiglio and G. Staiano

University of Naples Federico II, Napoli, Italy
**agostino.demarco@unina.it*

Research Group: A. De Marco, F. De Luca, S. Mancini, S. Miranda, C. Pensa, R. Scognamiglio, G. Staiano.

In this chapter the main research fields of the above mentioned research group of the naval section of the Department of Industrial Engineering performed by SCoPE Supercomputing Center are summarized.

1. Flettner Rotor

Flettner Rotors (FR) are spinning cylinders, which produce fluid dynamic lift by using the Magnus Effect. The Magnus force can be much greater in magnitude than the wing lifting force, given the same projected area and dynamic air pressure. The FR was used for ship's propulsion system in the 1920s. Nowadays, the FR is being seriously examined due to the necessity of renewable resources use.

Since several months, the *Dipartimento di Ingegneria Industriale* of the *Università degli Studi di Napoli "Federico II"* has started a research about preliminary analysis for marine *application* of FR through Unsteady Reynolds Average Navier–Stokes (U-RANS) simulations for 3D flow past a full-scale rotation.[3]

The aerodynamic forces, Lift (*L*), Drag (*D*) and Aerodynamic Efficiency (*L/D*) of a FR are the response variables and are influenced by the following control factors:

- Spin Ratio (SR), i.e., the ratio of circumferential-cylinder-velocity-to-free stream-velocity;
- Aspect Ratio (AR),[3] i.e., the ratio between the cylinder length and diameter;
- Endplate Ratio (De/D_0), i.e., ratio between the endplates diameter and cylinder diameter.

As the first step, a preliminary assessment of numerical setup has been conducted in comparison with experimental investigation carried out by Badalamenti and Prince at City University of London. The tested cylinder has AR = 5.1, $D_0 = 0.0889$ m, $De/D_0 = 2$.

As the second step, a statistical approach based on Design of Experiment (DoE) has been applied to the testing program of FR in order to evaluate the functional relationship and the interaction between the control factors SR, AR and De/D_0 which influence the response variables *L*, *D* and *L/D*.

All tests are developed through *U-RANS* simulations for 3D flow past rotating cylinder. The *U-RANS* simulations have been performed by using the commercially available flow simulation software CD-Adapco STAR CCM+ v. 9.06.009 (see Figs. 1 and 2).

2. Wake and Thrust Deduction Fraction

The aim of this research is the numerical evaluation of the hull–propeller interactions obtained by the commercial RANSE software CD-ADAPCO Star CCM+.

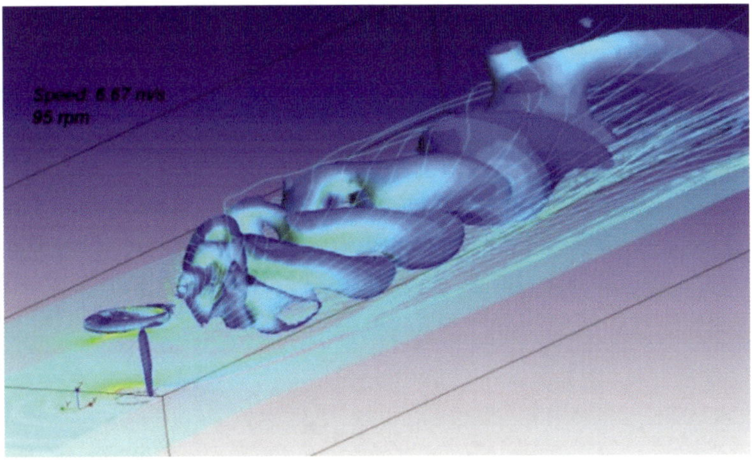

Fig. 1. FR: Vortex visualization (at SR = 1).

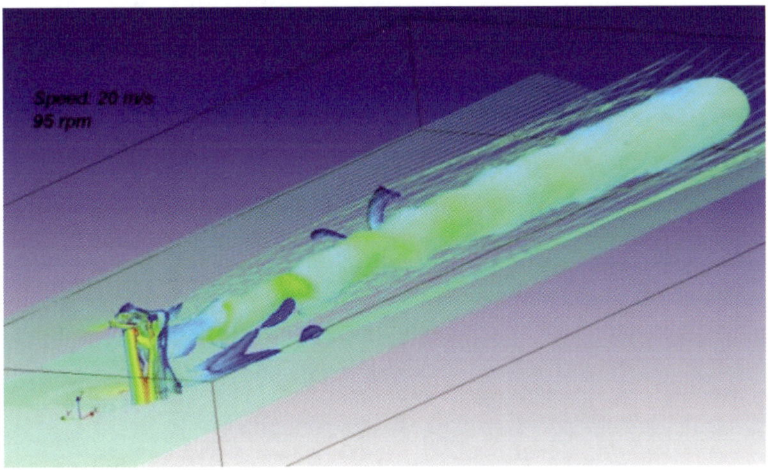

Fig. 2. FR: Vortex visualization (at SR = 3).

In literature, there is a significant lack of experimental data on wake and thrust fraction magnitude, especially regarding medium and high speed vessels. This is due to the high costs and the reduced reliability of the tests carried out with small models whose dimensions are constrained by the high Froude numbers. In spite of the high model speed, the smallness of the models induces small Reynolds numbers and therefore non-negligible scale effects on propellers dynamic (laminar flow) and on wake intensity.

The numerical analysis has been performed on the parent hull of the Naples Systematic Series.[2] In particular, 3 procedures, characterized by simulations with different degree of freedom and different relations with experimental data, have been applied (See Figs. 3 and 4). The simulations have been carried out in the speed range Fr = 0.5/1.5. This way is submitted also as an alternative to towing tank experiments on big models if these tests are not feasible.

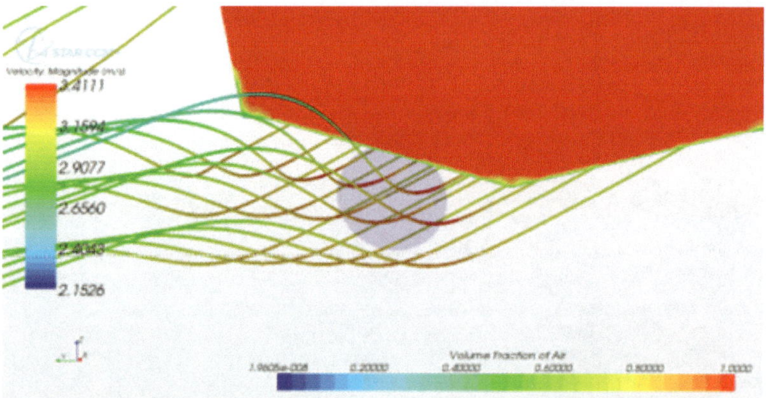

Fig. 3. Streamlines around propeller disk.

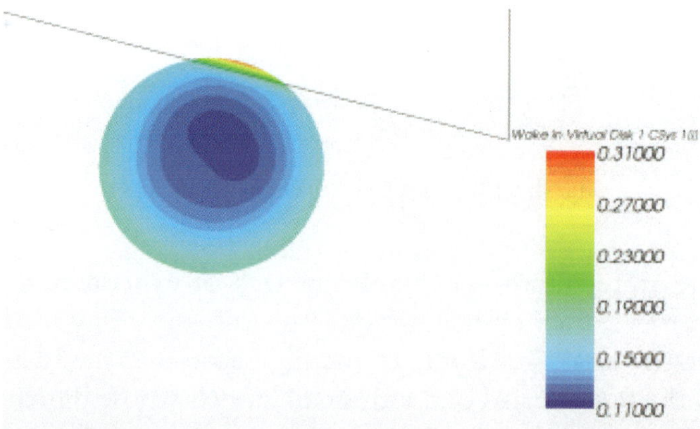

Fig. 4. Wake distribution in the propeller disk.

3. Hull Form Optimization of Units of Italian Navy

Three different units have been analyzed by numerical simulation carried out with RANSE code CD-Adapco STAR CCM+ v. 9.06. In particular, a feasibility study of a new concept of Mine Hunter Ship (MHS) has been carried out as preliminary support to towing tank test (see Figs. 5–7).

For the other 2 ships, the aim of the research has been the optimization of hull forms of new Offshore Patrol Vessels (OPV), characterized by very high maximum speeds and two different speeds

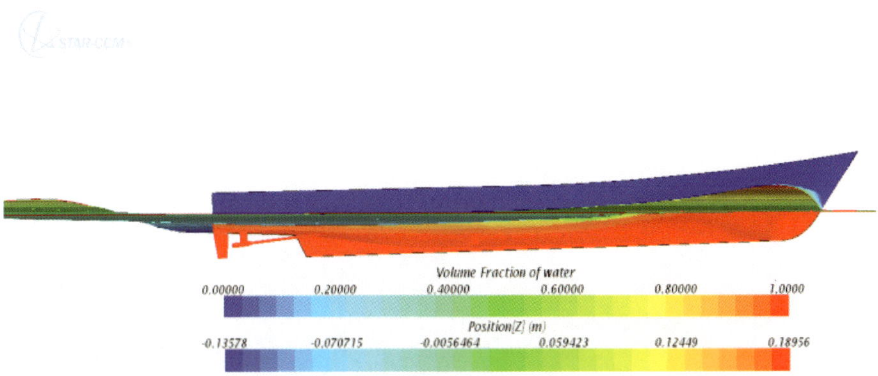

Fig. 5. MHS — no bulb configuration (at maximum speed).

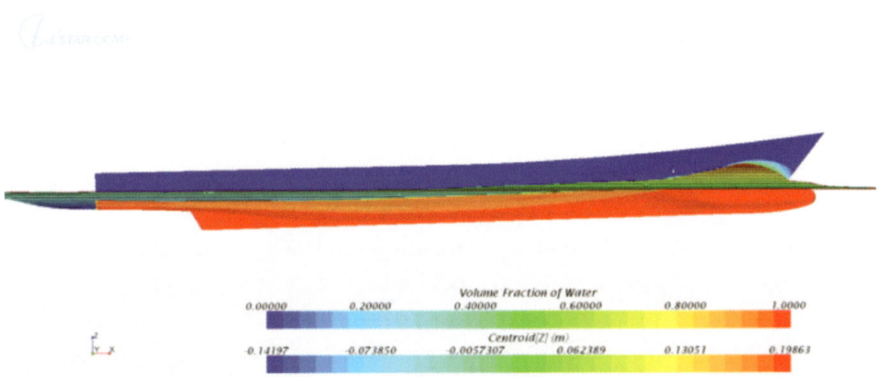

Fig. 6. MHS — bulb configuration (at maximum speed).

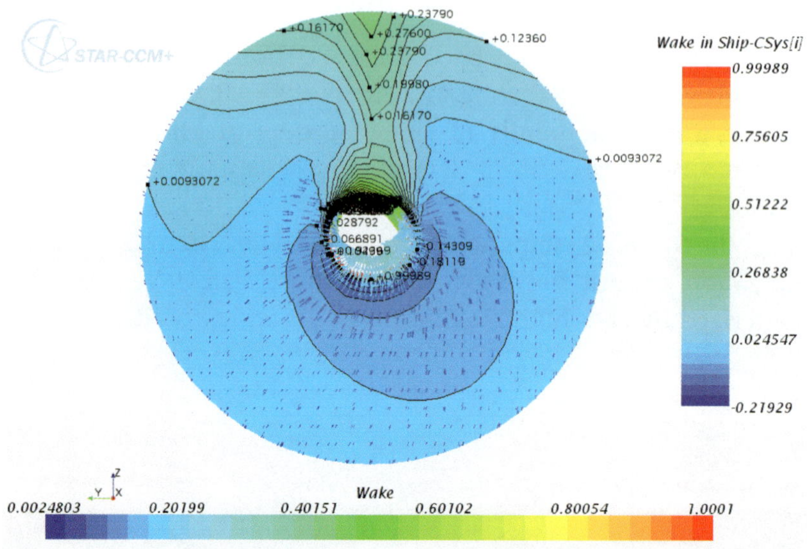

Fig. 7. MHS — Nominal wake visualization (at maximum speed).

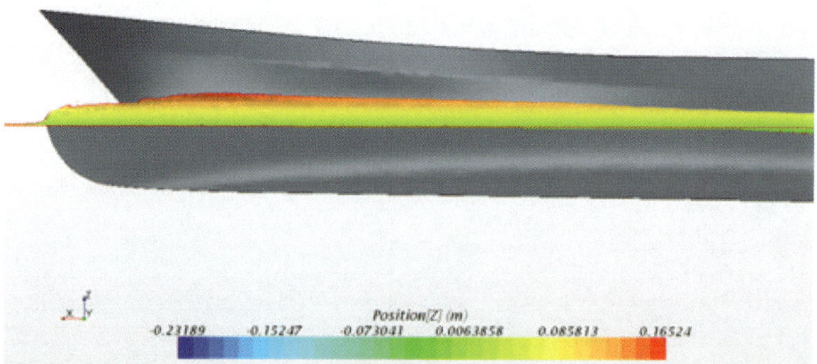

Fig. 8. Forebody wave pattern elevation (OPV unit).[1]

of interest (see Figs. 8 and 9). Simulations have been carried out by using the RANSE code CD-Adapco STAR CCM+ v. 9.06.

Nowadays, the numerical simulations are widely employed for macro-modification of hull form. To extend the applicability of this procedure to micro-modification of the form, as the first step in the study, the simulations have been compared with the towing tank data.

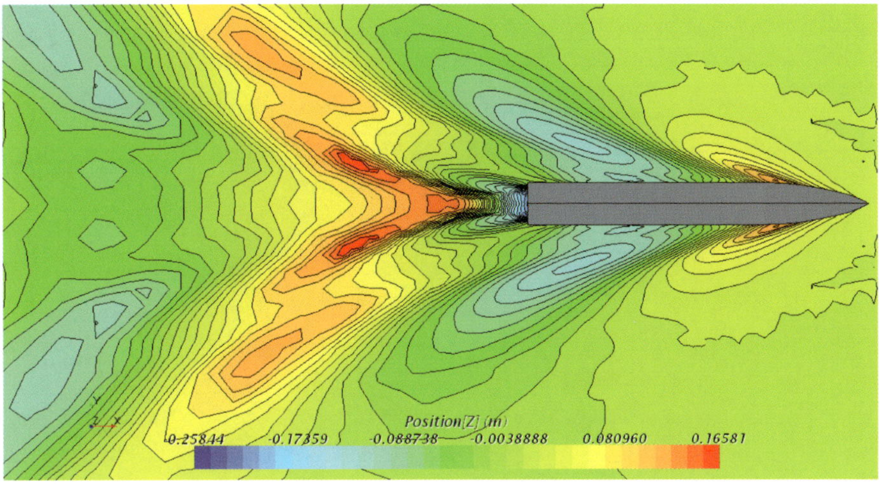

Fig. 9. Wave pattern visualization (OPV unit).

In addition, with the same aim, different mesh sizes and turbulence models have been tested.

In brief, this research is divided into the following parts:

- Towing tank test experiences;
- Choice of the more reliable numerical procedure (by validation with towing tank test);
- Optimization of the hull forms in relation to drag resistance evaluated for the two operational speeds;
- Identification of interceptor suitable size to be applied to the final hull form.

Tests and numerical simulations have been performed for a length model of 8.3 m (14.931 scale), i.e., Reynolds Number (Re) > 2.1×10^7.

4. Verification and Validation Activity for CFD Applied on Partially Hydrodynamic Sustained Hulls

Nowadays, the reliability of the CFD is suitable in the displacement hull field. For the hydrodynamic sustained ships (planing craft), the simulations are significantly less accurate (see Fig. 10).

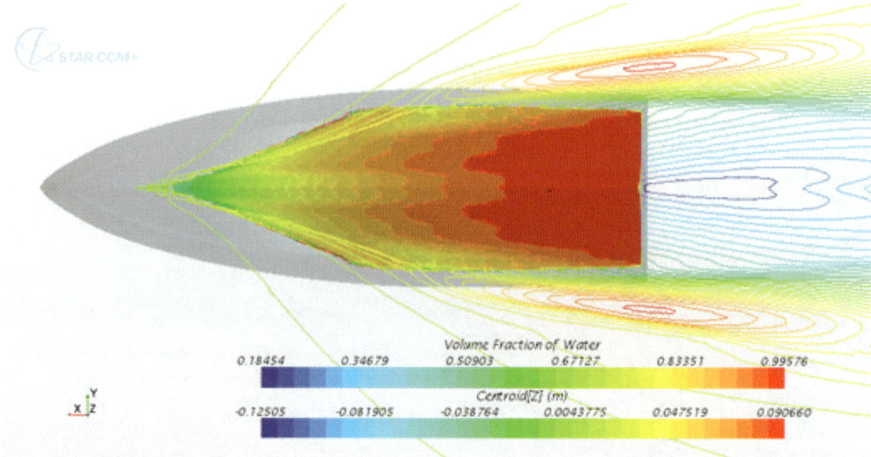

Fig. 10. Planing craft: Volume of Fraction of water and wave pattern.

As suggested by International Towing Tank Conference (ITTC), an intensive Verification & Validation activity is working in progress. The main parameters and the relatively interactions, are under investigation. In particular, the following parameters have been take into account:

- Iteration number;
- Grid density;
- Time step;
- Turbulence models;
- Overset method *versus* static grid.

The validation activity is performed according to the ITTC prescriptions.

References

1. F. De Luca, S. Mancini, A. Manfredini, C. Pensa and R. Scognamiglio, *Interceptor Device for a High-Speed Displacing Craft (Comparison Between CFD Simulation and Experimental Data)*. 18th International

Conference on Ships and Shipping Research (NAV 2015), Lecco, Italy, 2015.
2. F. De Luca, S. Mancini, C. Pensa and G. Staiano, ***Numerical evaluation (CFD) of Wake and Thrust deduction fraction of a Warped Hard Chine Hulls Systematic Series.*** 10th High Speed Marine Vehicles Symposium (HSMV 2014), Naples, Italy, 2014.
3. A. De Marco, S. Mancini and C. Pensa, Preliminary Analysis for Marine Application of Flettner Rotors. 2nd INT-NAM, Istanbul, Turkey, 2014.
4. A. De Marco, S. Mancini, C. Pensa, R. Scognamiglio and L. Vitiello, Marine ***Application of Flettner Rotors: Numerical Study on a Systematic Variation of Geometric Factor by DoE Approach.*** 6th Conference on Computational Methods in Marine Engineering (Marine 2015), Rome, Italy, 2015.

Chapter 28

Non-Equilibrium Molecular Dynamics of Oligomeric Chains

G. Ianniruberto*,‡, A. Brasiello†, G. Park* and G. Marrucci*

**Dipartimento di Ingegneria Chimica,*
dei Materiali e della Produzione Industriale,
Università Federico II, Piazzale V. Tecchio 80,
80125 Napoli, Italy
†Dipartimento di Ingegneria Industriale,
Università di Salerno, via Giovanni Paolo II 132,
84084 Fisciano (SA), Italy
‡iannirub@unina.it

We recently interpreted latest, unusual rheological data in fast extensional flows of Polystyrene (PS) melts by invoking flow-induced friction reduction effects. In order to test this hypothesis, Non-Equilibrium Molecular Dynamics (NEMD) simulations of styrene oligomers were performed to measure the oligomer mobility in fast flows. This computationally heavy task was successfully achieved by using the GRID computing infrastructure.

1. Introduction

Polymeric liquids exhibit a very complex rheological behavior, especially in fast flows as those encountered in processing applications.

Traditional rheological models based on macroscopic, continuum-level approaches are giving way to molecular models, i.e., to the so-called molecular rheology. In this context, molecular dynamics simulations are progressively becoming a very useful tool to investigate the dynamics of polymeric liquids, thanks to the increasing computational power of available computers. However, in order to deal with the long time scales relevant for polymer rheology, most simulations have been performed by introducing some level of coarse graining, i.e., by "lumping" groups of atoms into beads endowed with force fields, determined in such a way to match properties of the more detailed simulations. The lowest level of coarse-graining is achieved in the so-called united-atom simulations where hydrogen atoms are "merged" with larger atoms like carbon or oxygen.[1] At the opposite extreme, it is worth mentioning simulations where the polymer is replaced by a single bead, as in the so-called RAPID approach developed by Briels *et al.*[2] Most molecular dynamics simulations have been performed under equilibrium conditions, but in recent times, there has been an increasing interest for Non-Equilibrium Molecular Dynamics (NEMD) simulations with the aim of studying the effect of flow on polymer dynamics.[3,4]

Here, in particular, we briefly summarize our recent research activity based on NEMD simulations, aiming at investigating possible flow effects on the monomeric friction coefficient ζ. Indeed, all existing molecular theories for polymer rheology are based on the assumption that, since ζ is a local monomeric property, flows of practical interest in rheology are not fast enough to modify it. In polymers, however, the thermal relaxation rate of the molecule can be sufficiently slow that flows can induce conformation changes. In particular, when flows (or deformations) start stretching the molecules by inducing some orientational anisotropy of the monomers, there is no reason to assume that the friction coefficient remains unperturbed. This is particularly true in the case of monodisperse polymer melts where all molecules behave in the same way. In those systems, any chain is surrounded by similarly stretched chains, and this anisotropic environment may "act" on the test chain in a way that can be different from equilibrium. In particular, it is highly tempting to assume that flow-induced molecular alignment determines a reduction in the monomeric friction coefficient.

Since friction is a local property, the effect of flow on it can be sensitive to the chemical structure of the polymer. For this reason, we decided to resort to detailed, united-atom molecular dynamics simulations. The chemistry of choice was Polystyrene (PS). This was motivated by recent data on monodisperse PS melts in fast extensional flows,[5] revealing only moderate chain-stretch-induced strain hardening effects. This unexpectedly "soft" behavior of PS melts is fully compatible with the assumption that friction becomes less and less important because of flow-induced monomeric alignment. In order to gain computing time, and since monomeric friction is a local property, the simulations were limited to short polymers, i.e., to oligomers, the aim being that of proving the existence of possible flow-induced friction-reduction effects. Furthermore, we have limited so far our NEMD simulations to shear flows, which can easily be run indefinitely in time, thus allowing to determine flow effects on the molecular mobility in steady-state conditions, as described in the next section.

2. Methods

The most natural way of measuring the friction coefficient of the oligomer consists in performing the (numerical) experiment of pulling a single oligomer by applying an external force F to its center of mass (as schematically depicted in Fig. 1), and then by measuring the resulting drift velocity ν.[4] The friction coefficient ζ of the oligomer is then given by the ratio F/ν, provided the force is small enough not to perturb both the pulled and the surrounding molecules. In other words, care must be taken to ensure that the experiment be performed in the Stokes limit where ν is directly proportional to F (implying F-independent ζ values). Under equilibrium conditions, the value of the friction coefficient must also be consistent with the Einstein relationship $D = kT/\zeta$, where D is the diffusion coefficient of the molecule and kT is the thermal energy. Under flow conditions, the Einstein relationship breaks down, and both the friction and the

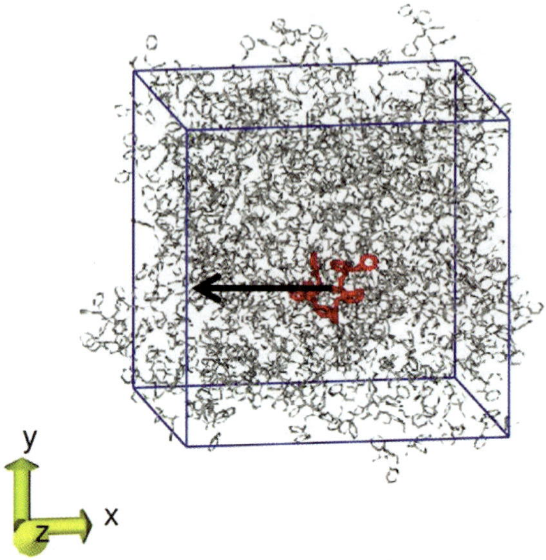

Fig. 1. Snapshot of the simulation box. The arrow represents the force pulling the red molecule.

diffusion coefficient become tensors. This implies that in order to determine the whole friction (or mobility) tensor, the pulling experiments should be independently performed along the *x*, *y* and *z* directions. Needless to say, the velocity ν of the pulled molecule must be deprived of convective contributions in the case of NEMD simulations.

2.1. *Simulation Details*

Molecular dynamics deals with the motion of material points (beads) due to forces derived from potentials.[6] The equation of motion of the beads is the following:

$$m_i \frac{d^2\mathbf{r}_i}{dt^2} = \frac{\partial U}{\partial \mathbf{r}_i}, \tag{1}$$

in which $\mathbf{r}_i$'s are the spatial coordinates of the beads, while U represents the sum of all the interaction potentials among the beads.

In the most detailed description, the beads represent single atoms while in coarser descriptions, a single bead can stand for a specific group of atoms. As mentioned above, in our case we adopted the so-called united atoms description. This is a common choice since it is a good compromise between the required level of detail and the computational efforts needed to obtain the desired information. Despite the level of description, the mass of a bead is given by the sum of the masses of the atoms accounted for by the bead itself.

The potential U, here adopted for simulations of PS melts, has the following form:

$$\begin{aligned} U &= U_{bonded} + U_{unbonded} \\ &= \sum_{bonds} \frac{k_{ij}^{b}}{2}\left(d_{ij} - d_{ij}^{eq}\right)^2 + \sum_{angles} \frac{k_{ijk}^{\theta}}{2}\left(\theta_{ijk} - \theta_{ijk}^{eq}\right)^2 \\ &+ \sum_{dihedrals} k_{ijkl}\left[1 + \cos\left(n\phi_{ijkl} - \phi_{ijkl}^{eq}\right)\right] + \sum_{i>j} 4\epsilon_{ij}\left[\left(\frac{\sigma_{ij}}{d_{ij}}\right)^{12} - \left(\frac{\sigma_{ij}}{d_{ij}}\right)^{6}\right]. \quad (2) \end{aligned}$$

The first two terms on the right-hand side of Eq. 2 are harmonic type potentials for bonds between two consecutive beads and for angles between two consecutive bonds, respectively. The third term represents dihedral potentials accounting for torsions, while the last term is the sum of the Lennard–Jones potentials related to the non-bonded interactions. In all the simulations, we adopted the TraPPE united atoms force field.[7] In order to perform simulations of NVT ensembles, Berendsen thermostat with time constant of 0.1 *ps* is then introduced. Periodic boundary conditions with a cut-off range of 1 nm are imposed. Simulations are carried out using a leap-frog algorithm via GROMACS software package,[8] with a time step of 0.001 *ps*. Boxes of about 100 styrene oligomers made up of 10 monomers (decamers) are initially minimized and then equilibrated through an NPT simulation at $T = 463K$ and $P = 1\,atm$ run for a time much longer than the oligomer relaxation time. For such purpose, a Berendsen barostat with time constant of 2 *ps* is adopted.

Compressibility is fixed at 9.5×10^{-5} bar^{-1}. Computation time on 27 processors working in parallel is 30 min/*ns*. This means that each simulation, typically lasting 100 *ns*, was performed in about 2 days.

3. Results

Figure 2 shows the typical trajectory under equilibrium conditions of a pulled molecule in terms of its coordinates x, y and z as a function of time. The force F has been applied in negative x direction, and its magnitude has been chosen with the following criterion. As mentioned above, the force should be small enough not to perturb the conformation of the pulled molecule. This is expected to occur if the characteristic time τ_c of the convective effect induced by F is much longer than the thermal (diffusive) relaxation time τ_d of the molecule. The convection time is the time required for the molecule to move over a distance equal to its size R, i.e., $\tau_c = R/\nu$ (where we recall that ν is the drift velocity), while the diffusion time is $\tau_d = R^2/D$. Now, since $\nu = F/\zeta$ and $D = kT/\zeta$, the condition $\tau_c \gg \tau_d$ is equivalent to $F \ll kT/R$. As expected, Fig. 2 shows that the force induces a systematic drift along x, while thermal fluctuations dominate the y and z

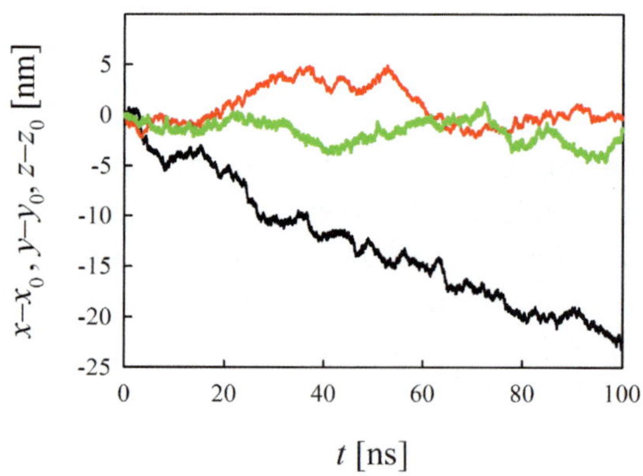

Fig. 2. Trajectory of a pulled molecule along the pulling direction x (black line), and along the orthogonal directions y and z (red and green lines, respectively).

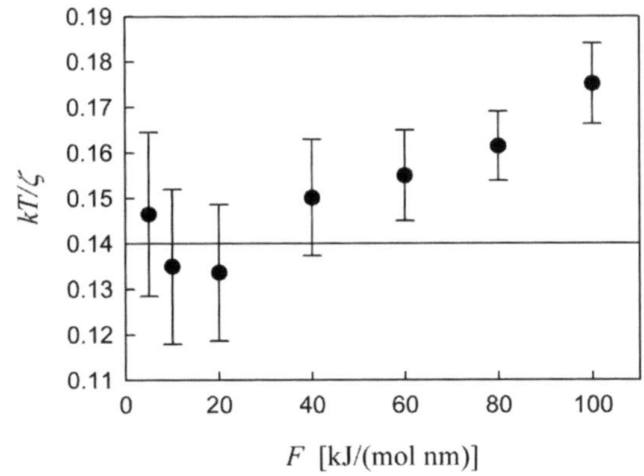

Fig. 3. Effect of the pulling force F on the friction at equilibrium. The horizontal line is the center of mass diffusion coefficient D.

components. The average slope of the function $x(t)$ determines the drift velocity v, and hence the friction coefficient ζ through the ratio $\zeta = F/v$.

Figure 3 shows the effect of the external force F on the friction coefficient, always under equilibrium conditions. The horizontal line in the same figure represents the value of the diffusion coefficient measured from the mean square displacement of the unpulled molecules. The diffusion coefficient extracted from pulling experiments is fully consistent with the Einstein relationship $D = kT/\zeta$ at low values of the force. Significant deviations are observed with increasing values of the applied force. Figures 2 and 3 also reveal that fluctuations, and hence error bars, can be comparable to the measured quantities, and even longer simulations would be required to reduce the statistical uncertainty. Needless to say, we could reduce error bars in the friction coefficient values by also performing (independent) pulling experiments on different molecules.

We now examine the effect of flow on the friction coefficient by performing NEMD simulations. As mentioned above, with the software GROMACS, we can apply a continuous shear deformation to the simulation box, thus easily reaching a non-equilibrium steady-state in which we can measure several observables, including friction

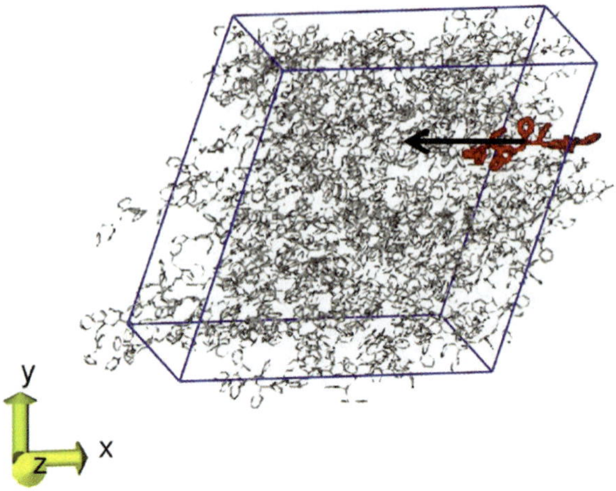

Fig. 4. Snapshot of the simulation box under shear. The arrow represents the force pulling the red molecule.

coefficients. In other words, GROMACS allows for simultaneously shearing the box and pulling molecules through external forces, as schematically depicted in Fig. 4. The rate of shear deformation $\dot{\gamma}$, often referred to as shear rate, should be high enough to induce some orientational anisotropy in the oligomers. In other words, $\dot{\gamma}$ should be significantly larger than the thermal relaxation rate whose order of magnitude is, as mentioned above, $1/\tau_d = D/R^2$. Figure 5 reports the typical trajectory of a pulled molecule under non-equilibrium conditions at a non-dimensional shear rate $Wi = \dot{\gamma}\tau_d = 100$, where *Wi* is the so-called Weissenberg number. Figure 5 clearly shows that, differently from the equilibrium case in Fig. 2, the molecule also drifts, though slightly, along the *y* direction. In other words, in the non-equilibrium case, friction becomes a tensor with nonzero off-diagonal components.

The other (and perhaps the most important) message of Fig. 5 emerges when comparing the trajectory along the *x* direction with the analogous trajectory in Fig. 2. The comparison clearly shows that under flow, the pulled molecule has a significantly larger drift along *x*, thus revealing that, when pulling along the shear direction *x*, the mobility along that same direction, equal to the reciprocal of the

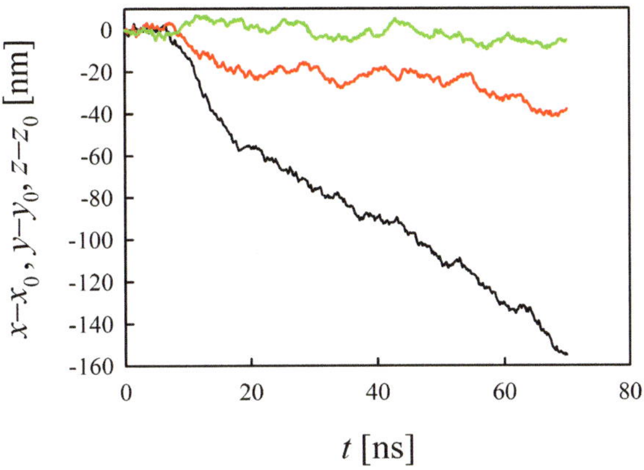

Fig. 5. Trajectory of a molecule pulled along the x direction with force $F = -10$ KJ/(mol·nm) under a shear flow with $Wi = 100$. Black line is the x coordinate; red and green lines are y and z coordinates, respectively.

xx component, ζ_{xx}, of the friction tensor, significantly increases with respect to the equilibrium case. In other words, ζ_{xx} becomes significantly smaller than its equilibrium value ζ_{eq}. It should be recalled that the trajectories in Fig. 5 were deprived from any direct convective contribution. It should also be mentioned that convection is in the positive x direction, while the pulling experiments was performed in the negative x direction. Hence, the friction reduction effect is not due to convection *per se*, but it is due to some indirect flow effect, as discussed below.

The upper panel of Fig. 6 reports ζ_{xx} values as a function of dimensionless shear rate Wi, as obtained from pulling experiments analogous to those reported in Fig. 5 for the case $Wi = 100$. In slow flows ($Wi \ll 1$), the function $\zeta_{xx}(Wi)$ approaches a shear rate independent plateau value, equal to ζ_{eq}, as expected. With increasing shear rate, ζ_{xx} significantly decreases, seemingly approaching a second lower plateau, though no definite conclusion can be drawn in very fast flows because of the statistical uncertainty. The lower panel of Fig. 6 reports the so-called monomeric order parameter S, measuring the degree of alignment along the shear direction x of the monomers of the styrene decamer. More specifically S measures the order of the unit vectors

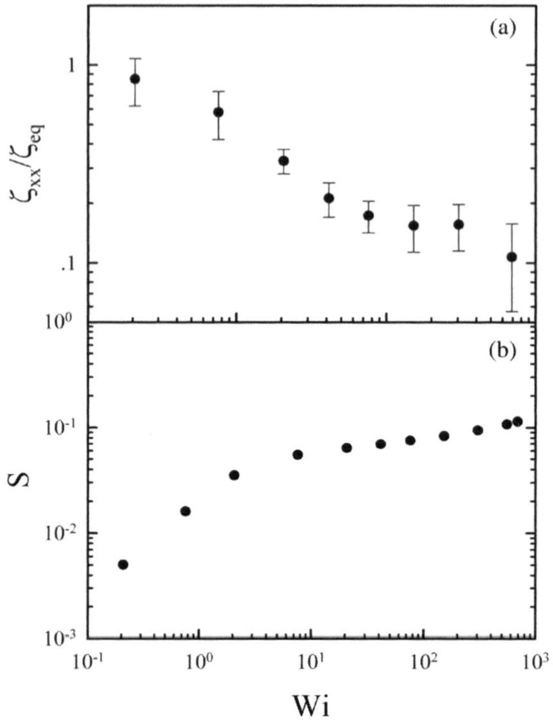

Fig. 6. Effect of shear flow on the *xx* component of the friction tensor (a) and on the monomeric order parameter (b).

specifying the direction connecting alternate carbon atoms along the decamer backbone. Starting from its equilibrium value, $S_{eq} = 0$, S first steeply increases with increasing values of the shear rate, to then approach a quasi-plateau at intermediate rates, before seemingly increasing again. From inspection of Fig. 6, it is highly tempting to link the friction reduction effect to flow-induced monomer co-alignment in shear direction. In fast flows, oligomers are strongly aligned along the shear direction (see Fig. 4), and it is "easier" to pull each of them through the "maze" of the surrounding ones. In other words, the increase in the molecular mobility is due to flow-induced nematic-like effects.

In order to confirm our interpretation, we decided to perform pulling experiments on the decamer in different matrices, made by

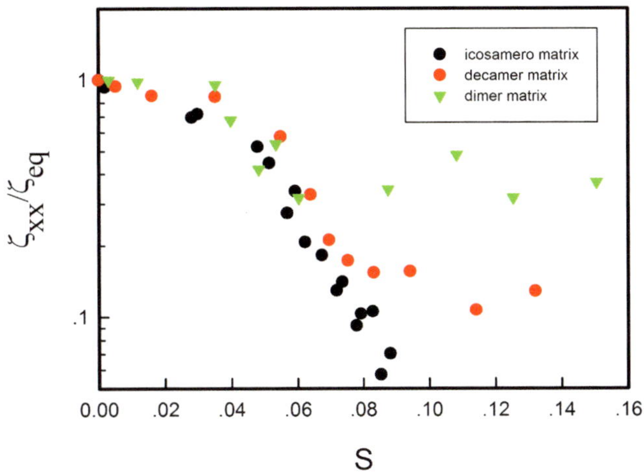

Fig. 7. Influence of the order parameter S of the probe decamer on the xx component of the friction tensor in several matrices. Error bars are omitted for clarity.

either smaller or larger molecules. More specifically, we made simulations of a single decamer in a matrix of dimers, as well as in a matrix of icosamers (oligomers with 20 monomers). Typical results are reported in Fig. 7, in terms of ζ_{xx}/ζ_{eq} as a function of probe decamer order parameter. When the decamer is pulled in a melt of dimers, no significant friction reduction effect is observed in spite of the fact that the decamer is aligning along the shear direction. This is because, the matrix is now made by fast relaxing molecules that are in fact not affected by the background flow. In other words, in order to observe significant friction reduction effects both the matrix and the probe molecule should become anisotropic.

This interpretation is confirmed by the pulling experiments performed in the matrix of icosamers, also reported in Fig. 7. The results clearly show that, when the matrix is made by longer molecules, flow has a more pronounced effect on decamer mobility with respect to pure decamer case. This is due to the fact that, since the icosamer obviously has a thermal relaxation time longer than the decamer, the icosamer matrix gets more oriented by the flow than the decamer matrix, thus allowing for a larger mobility of the probe molecule.

4. Conclusions and Perspectives

In this chapter, we have briefly reviewed our research activity based on the molecular dynamics simulations performed in recent times by using the GRID computing infrastructure at the Federico II University. Thanks to the computational and storage power made available to us, we were able to perform molecular dynamics simulations that we would not have been able to run by using our own local machines. The simulations have helped us to prove (for the first time) that in fast flows of monodisperse polymer melts the molecular mobility can be significantly enhanced because of flow-induced monomeric alignment. This phenomenon, here quantitatively measured in the case of styrene oligomers, can help understanding the unusual rheological behavior of PS melts recently observed in fast uniaxial extensional flows. Those flows are often encountered in many processing applications, and hence the results here reviewed can have a significant impact in several industrial processes.

The flow-induced friction reduction effect is expected to be very sensitive to molecular architecture, polymer chemistry, as well as polymer concentration. Rheological experiments reveal that PS melts behave differently from PS solutions, or from melts with other chemistries (methyl methacrylate, butyl acrylate, etc.). These evidences motivate further simulations, and we are currently working on polymethylmethacrylate (PMMA) melts (though still focusing on oligomers) seemingly showing even stronger friction reduction effects.[5] Due to partial charges embedded in the molecule, however, PMMA simulations are computationally more demanding, but we rely on the further increasing computational power of the GRID infrastructure. For that reason, we hope to report results on PMMA in the near future.

Acknowledgments

We acknowledge financial support from EU through the ITN SUPOLEN project (Grant No. 607937).

References

1. D. Frenkel and B. Smit, *Understanding Molecular Dynamics Simulation — From Algorithms to Applications*, Academic Press, San Diego (2002).
2. I. S. Santos de Oliveira, B. W. Fitzgerald, W. K. den Otter and W. J. Briels, Mesoscale modeling of shear-thinning polymer solutions, *J. Chem. Phys.* **140**(10), 104903 (2014).
3. C. Baig and V. A. Harmandaris, Quantitative analysis on the validity of a Coarse-Grained Model for non-equilibrium polymeric liquids under flow, *Macromolecules* **43**(7), pp. 3156–3160 (2010).
4. G. Ianniruberto, A. Brasiello and G. Marrucci, Simulations of fast shear flows of PS oligomers confirm monomeric friction reduction in fast elongational flows of monodisperse PS melts as indicated by rheooptical data, *Macromolecules* **45**(19), pp. 8058–8066 (2012).
5. Q. Huang, O. Mednova, H. K. Rasmussen, N. J. Alvarez, A. L. Skov, K. Almdal and O. Hassager, Concentrated polymer solutions are different from melts: Role of entanglement molecular weight, *Macromolecules.* **46**(12), pp. 5026–5035 (2013).
6. M. P. Allen and D. J. Tildesley, *Computer simulation of liquids*, Clarendon Press, Oxford (1987).
7. V. A. Harmandaris, N. P. Adhikari, N. F. A. van der Vegt and K. Kremer, Hierarchical modeling of polystyrene: From atomistic to coarse-grained simulations, *Macromolecules* **39**(19), pp. 6708–6719 (2006).
8. D. Van der Spoel, E. Lindahl, B. Hess, G. Groenhof, A. E. Mark and H. J. C. Berendsen, GROMACS: Fast, flexible, and free, *J. Comput. Chem.* **26**(16), pp. 1701–1718 (2005).

Chapter 29

Applicability of CFD Methods for Roll Damping Determination of Intact and Damaged Ship

E. Begovic*,‡, S. Mancini*, A. H. Day† and A. Incecik†

* *University of Naples Federico II, Napoli, Italy*
† *University of Strathclyde, Glasgow, UK*
‡ *begovic@unina.it*

This chapter presents an assessment of the roll damping of DTMB 5415 naval ship model in both intact and 2 compartments symmetric damaged scenarios. A numerical assessment of roll decay is performed by CDAdapco StarCCM+ software investigating the accuracy and efficiency of the numerical approach. The sensitivity analysis on mesh refinement is performed for damaged ship while time step and turbulence models sensitivity is investigated for the intact ship. Numerical results for decay curves, natural frequency and period of roll for intact and damaged ship are plotted against experiments, performed at the University of Strathclyde, Glasgow, to verify the precision of the numerical simulations. Obtained numerical results are shown to be reasonably accurate although the calculation time still precludes the use of Computational Fluid Dynamics (CFD) analysis as a standard design procedure.

1. Introduction

Although most vessel responses can be calculated with acceptable accuracy by potential theories in the frequency domain, this is more difficult for roll response due to the viscous damping effects, which are not negligible in roll. Roll damping plays an important role in the vessel seakeeping, which is the basis for the precise prediction of vessel motions in waves. The most common approach adopted is based on the Ikeda (1976) empirical method in which the equivalent total damping coefficient is calculated as a sum of potential, friction, eddy-making, appendages and lift contributions. The roll damping coefficient can also be obtained through a ship model roll decay tank test but there is evident lack of this approach in typical design procedures.

Very recently, use of Computational Fluid Dynamics (CFD) methods in calculating roll damping has become possible due to developments in computing power. Numerical simulation based on CFD offers the advantage of considering viscous flow, although calculations are still very time consuming, and experience of the modeling of this phenomenon is still very limited. A major problem in roll decay simulation, common to any problem of transient ship motion, is the necessity of special computational techniques such as deforming mesh, moving mesh and grid interface.

One of the first CFD assessments of roll decay was given by Wilson[7] who performed simulations for a bare hull and bilge-keel-appended surface combatant model (referred as DTMB 5512) using the software CFD Ship-IOWA. Roll decay simulations are performed for 3 cases: the bare hull at Fr = 0.138 and 0.28, and the hull with bilge keels at Fr = 0.138. Comparisons of EFD and CFD damping coefficients for the low speed case with bilge keels showed very small differences, generally less than 0.4%, while comparisons for the bare hull cases at both speeds showed larger differences for damping coefficients (up to 20%) even though the difference in time histories for the roll motion showed reasonable agreement (<4.5%).

Yang *et al.* presented simulation performed using the commercial software package Fluent-ANSYS of roll decay for the same vessel,

DTMB 5512, with initial heel angles: 5°, 10° and 15° at Fr = 0.28. The authors reported very good results in terms of damping coefficient and two examples of decay curve but no details on the method and calculation procedure are given. Yang *et al.*[2] performed numerical simulations of free decay and forced rolling at various forward speeds and amplitudes for DTMB 5512 and S60 hulls to predict ship roll damping, using a RANS solver using a dynamic mesh technique. The influences of forward speed, roll amplitude and frequency on the ship roll damping are evaluated. The authors report the difference between numerical and experimental results as 1.3–2.5%.

Handschel *et al.* (2012) applied RANS simulations to calculate roll damping coefficients of a RoPax vessel in full scale. The influence of the roll amplitude up to 35°, 3 ship speeds, the vertical position of the roll axis, and the interaction between the bilge keels and the ship hull are analyzed. Detailed validation data for a RoPax ship was not available but authors compared the numerical results with Ikeda's method. Avalos *et al.*[1] investigated a roll decay test of the middle section of an FPSO with bilge keels by the numerical solution of the incompressible two-dimensional Navier–Stokes (NS) equations. Very good results were obtained for cases with bilge keels, although sometimes the agreement for the oscillation period was not so good in the case with the larger bilge keel. The worst results in terms of damping and oscillation period were obtained for the section without bilge keels. Gao and Vassalos[4] presented results of numerical simulations of roll decay of DTMB 5415 with bilge keel in both intact and damage conditions by RANS. The comparison shows that the agreements between calculation and model test are acceptable with slightly larger period and smaller damping obtained from the calculation. Gao *et al.*[3] presented an integrated numerical method that couples a seakeeping solver based on the potential flow theory and an NS solver with the Volume of Fluid (VoF), developed to study the behavior of a damaged ship in beam seas. The integrated method was used to simulate the roll decay of a damaged Ro–Ro ferry and the ferry's motion in regular beam seas. Validation against experimental data showed that the proposed method can yield satisfactory results with acceptable computational costs.

This work continues the stream of investigation on the applicability of CFD methods for roll damping determination. The commercial software CD Adapco StarCCM+ is used for roll decay simulation of an intact and damaged bare hull DTMB 5415 model, tested by authors at the University of Strathclyde.

2. Model DTMB 5415 Geometry and Data

Roll damping was studied for the well-known benchmark naval hull form DTMB 5415, constructed in fiber glass as 1/51 scale model used in experimental campaign in Begovic *et al.*[2] The main particulars of the DTMB 5415 model are given in Table 1.

Table 1. Main Particulars DTMB 5415.

Particulars	Ship	Model 51
L_{OA} (m)	153.300	3.0
L_{PP} (m)	142.200	2.788
B_{WL} (m)	19.082	0.374
B_{OA} (m)	20.540	0.403
D (m)	12.470	0.244
T (m)	6.150	0.120
V(m3)	8424.4	0.0635
D (t, kg)	8635	63.5
C_B	0.505	0.505
C_P	0.616	0.616
C_M	0.815	0.815
KM (m)	9.493	0.186
KG (m)	7.555	0.148
GM (m)	1.938	0.038
LCG (m)	70.137	1.375
$k_{xx\text{-}WATER}$ (m)	6.932	0.136
$k_{yy\text{-}AIR}$ (m)	36.802	0.696
$k_{zz\text{-}AIR}$ (m)	36.802	0.696

Fig. 1. DTMB 5415 ship model.

Fig. 2. Damage opening of DTMB 5415.

The internal geometry of the 1:51 model was identical to that presented by Lee *et al.*[6] The model has been fitted with the 5 watertight bulkheads located as shown in Fig. 1. The damage opening shown in Fig. 2 leads to 2 compartment (3 and 4) symmetric flooding. The flooded length extended from x_1 = 65.66 m (ship scale) to x_2 = 90.02 m, corresponding to 17% of the length between perpendiculars. This extension seemed reasonable for a destroyer type of ship, as it is expected that this type of ships have to preserve all functionality with two compartments damage. Both compartments were fitted with the small tube to assure the air-flow during tests, visible on the port side of model at Fig. 2.

The exact amount of flooded water is determined from hydrostatic calculations, i.e., for the measured immersion and trim angle,

Table 2. Damaged case principal characteristics.

Particulars	Ship	Model
$L_{\text{flooded compartments}}$ (m)	24.36	0.478
B_{WL} (m)	19.458	0.382
T_{mean} (m)	7.41	0.145
Trim [+ aft] (deg)	–0.656	–0.656
D (t)	11273.8	0.083
Mass of flooded water (t/kg)	2638.9	0.019
LCG (m)	71.622	1.404
KM (m)	9.427	0.185
KG (m)	6.654	0.130
GM (m)	2.773	0.054

the displaced volume was found. All characteristics of damaged ship are reported in Table 2.

3. Numerical Setup

The commercial software CD Adapco StarCCM+ V.8.04. has been used for the calculations of roll decay curves. It is well known that the accuracy of CFD results and the calculation time strongly depends on the type of the mesh and number of cells used, and therefore meshing is optimized for the "most challenging" case, i.e., damaged ship with 19.1° initial heel. In present work, a moving mesh and grid interface have been used for modelling the roll decay phenomenon. For the interaction between the moving body and the free surface, a Chimera grid or overset mesh technique is used. To solve the time-marching equations, an implicit solver has been used to find the field of all hydrodynamic unknown quantities, in conjunction with an iterative solver to solve each time step. The software uses a Semi Implicit Method for Pressure Linked Equations to conjugate pressure field and velocity field, and an Algebraic Multi-Grid solver to accelerate the convergence of the solution.

The free surface is modeled with the two phase VoF technique. A segregated flow solver approach is used for all simulations. The Reynolds stress problem is solved by means of k-ε turbulence model.

3.1. *Mesh Generation and Sensitivity Analysis*

Four different meshes, shown in Figs. 3–6 have been used for the sensitivity analysis: 2 trimmed meshes and 2 hybrid meshes (polyhedral and trimmed). A summary of cell numbers and CPU time for 5 s simulation of model roll decay at 32 processors is given

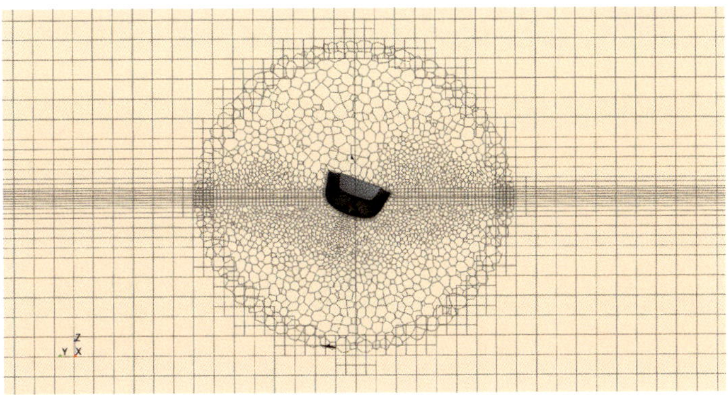

Fig. 3. Mesh Hybrid_1.

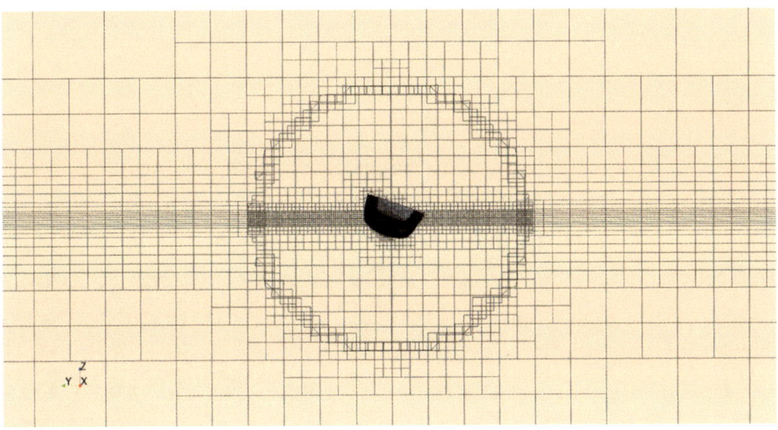

Fig. 4. Mesh Trim_1.

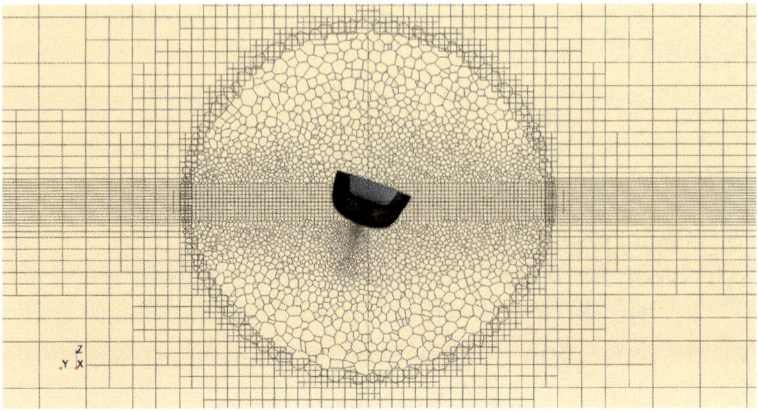

Fig. 5. Mesh Trim_2.

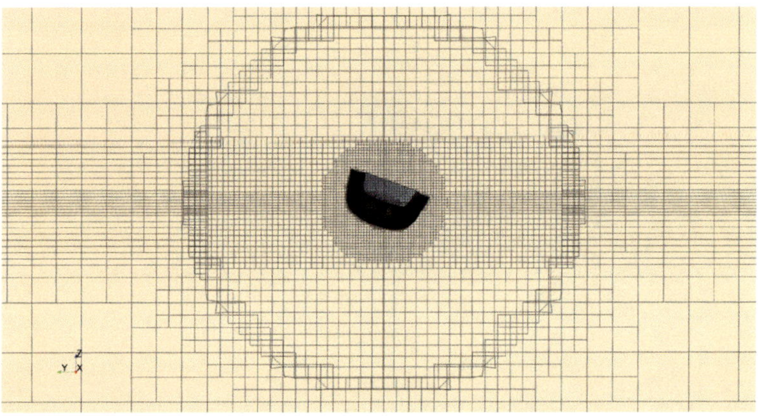

Fig. 6. Mesh Hybrid_2.

in Table 3. The obtained roll decay histories are shown in Fig. 7 indicating that the Hybrid_1 mesh gives completely incorrect results, and it was thus stopped after 3 s. It can be noted how the refinement of the free surface VoF (Hybrid_1 versus all others) in the range of the complete hull model height (not only the "seakeeping" free surface) yields significant improvement in roll decay simulation. From Fig. 7, very small difference can be noted between Trim_2 and Hybrid_2 meshes in quality of results while the computational time is extremely prohibitive for the Hybrid_2 case.

Table 3. Mesh sensitivity analysis summary.

Grid Type	No. of Cells $*10^5$	CPU Time (h)
Hybrid_1	1.194	90
Trim_1	0.709	40
Trim_2	1.476	90
Hybrid_2	2.590	192

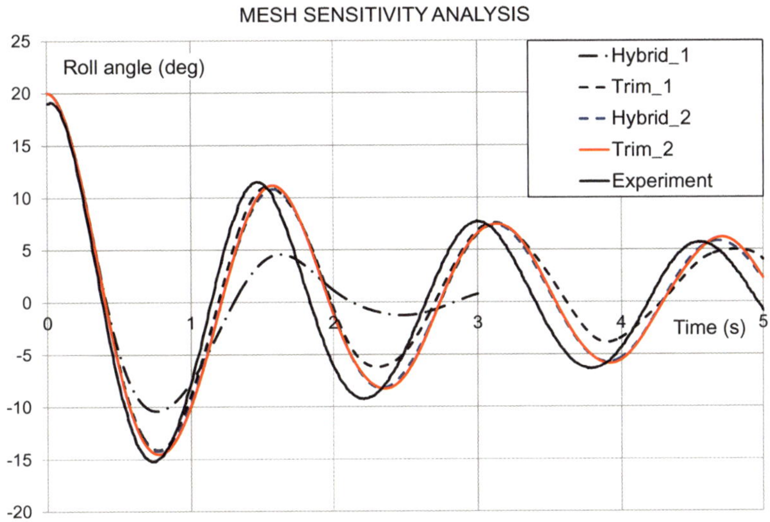

Fig. 7. Mesh sensitivity results.

Finally, in order to optimize the discretization of each region and to avoid large computational costs, a trimmed mesh of hexahedral type (shown in Fig. 5) is used. The region around the hull is finer than the far field regions as can be observed in Fig. 8.

Boundary conditions applied to each boundary are summarized in Table 4.

Once all the boundary conditions and the origin of the coordinate system at the model CG have been imposed, the last step is defining the numerical set up. The ITTC[10] "Practical Guidelines for Ship CFD Applications" recommendation for time step choice for periodic

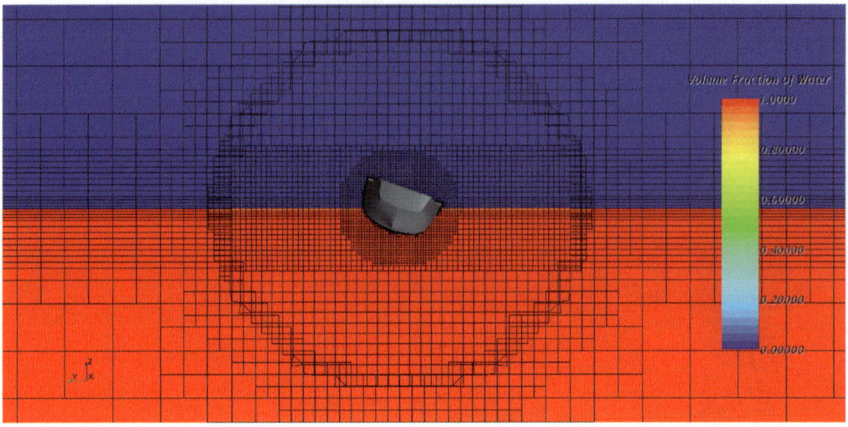

Fig. 8. Hexahedral trimmed mesh.

Table 4. Boundary conditions summary.

Inlet	Velocity inlet condition
Outlet	Velocity inlet condition
Bottom/Top	Velocity inlet condition
Sides	Pressure outlet
Hull	Wall with no-slip condition
Symmetry plane	Not existing
Overset	Boundary Interface

phenomena such as roll decay and vortex shedding is at least 1/100 of phenomenon period. The experimental roll period varies from 1.37 to 1.52 s resulting in recommended minimum of 0.015 s. Sensitivity analysis has been performed for time steps equal to 0.002 s and 0.001 s. The simulations have been performed for intact ship at 19.43° initial heel and results are given in Fig. 9.

Although the initial step of 0.002 s is one order of magnitude lower than ITTC recommended time step, it can be seen that the simulation results is not stable with this time step. Both, decay curve and roll period are improved in simulation with 0.001 s time step.

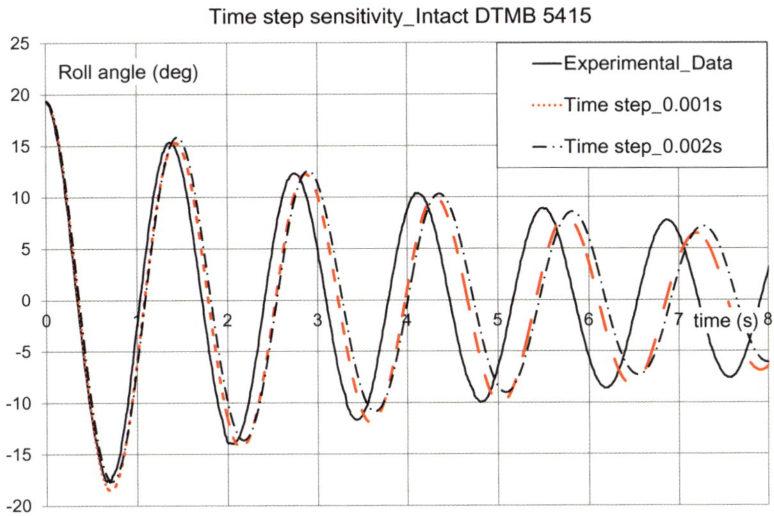

Fig. 9. Time step sensitivity.

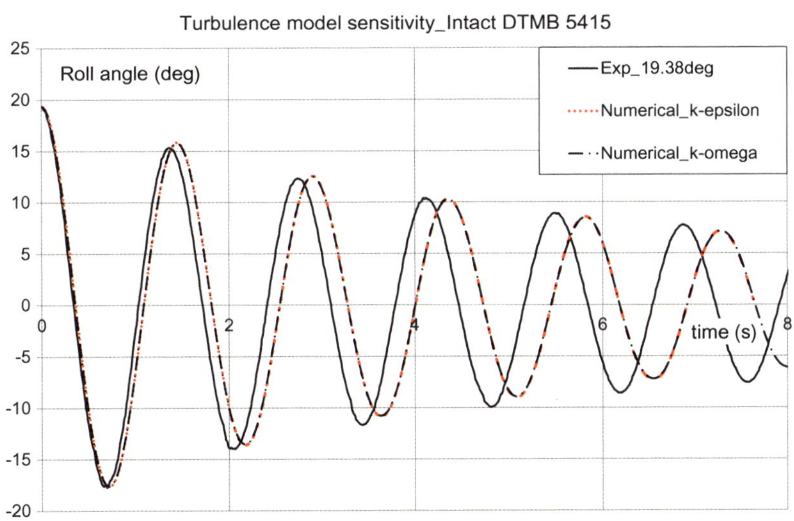

Fig. 10. Turbulence models sensitivity.

Table 5. Solver settings summary.

Convection term	Second-order
Temporal discretization	Second-order
Time-step (s)	0.001
Iteration per time step	12
Turbulence model	$k-\omega$; $k-\varepsilon$

Trying lower time step has been considered too expensive in terms of calculations costs.

Results of simulations with $k-\omega$ and $k-\varepsilon$ turbulence models are given in Fig. 10. Numerical results are within 1% difference although it is not possible to appreciate the difference between two numerical curves. Final numerical set up used for the simulations is reported in Table 5.

4. Numerical Results

4.1. *Intact Ship*

The final simulations for the intact ship have been performed for 4.00 and 28.00° initial heel. The larger angle represents a limit for mesh functionality. The lower angle gives the part of extinction curve common to all experimental decays, where none of the simulations arrived due to the necessary computing time. The mesh scene given in Fig. 11 is common for both simulations. The total number of cells is 1.24M. The calculation time depends on the turbulence model and the initial heel angle; for the $k-\varepsilon$ model using 32 processors, 1s of simulation takes about 13 hours for 4.00° and about 8 hours for 28.00° initial heel. Results compared with the experimental data are given in Figs. 12 and 13.

In both simulations, a good trend of magnitude of decay curves with higher roll period can be observed. Roll oscillation period in all simulations is 1.443 s, and does not show dependence on roll angle. With respect to experimental result of 1.369 s, this gives a difference of 5.4%.

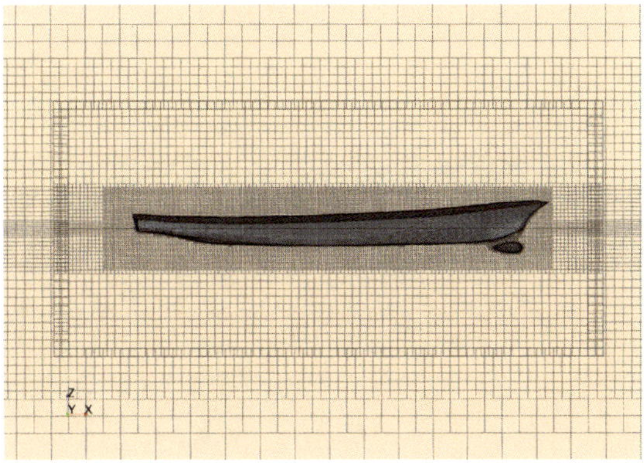

Fig. 11. Mesh Scene for Intact model.

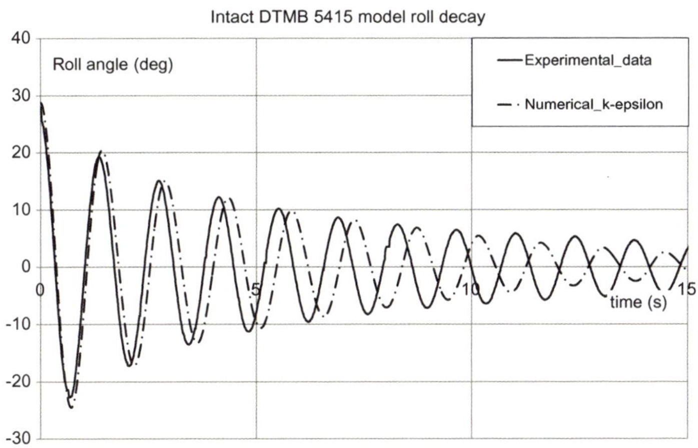

Fig. 12. Comparison of experimental and numerical results.

4.2. *Damaged Ship*

The final simulation for the damaged ship is performed for 15 s model time. Details of the mesh in the flooded compartments is shown in Fig. 14. The numerical roll decay curve compared with the experimental data for the damaged ship is given in Fig. 15.

It can be seen that oscillation period of numerical results 1.56 s is longer than of experimental, 1.518 s, leading to the difference of 2.8%.

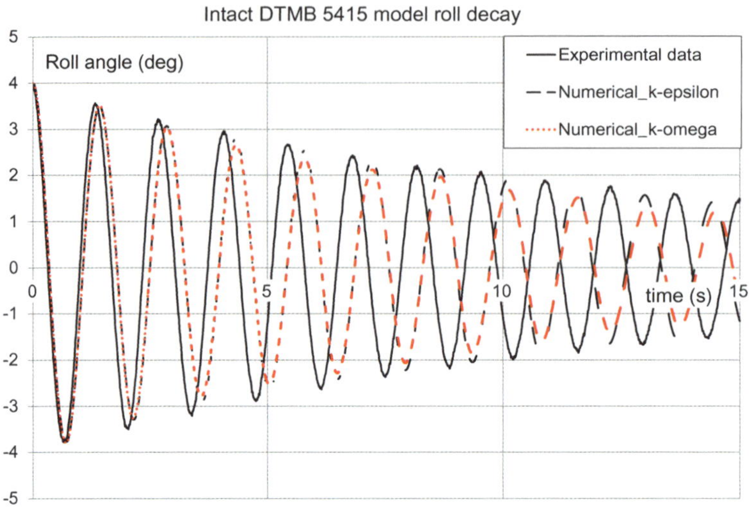

Fig. 13. Comparison of numerical and experimental results.

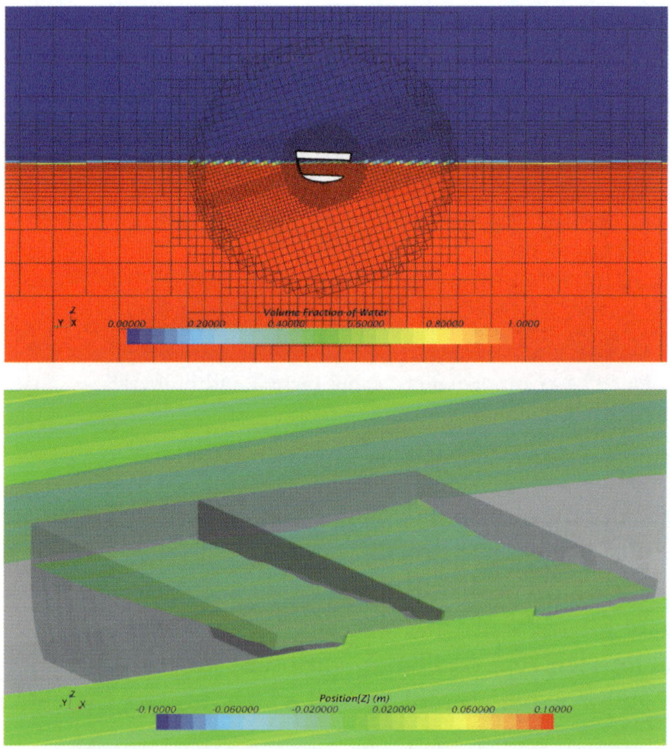

Fig. 14. Damage detail.

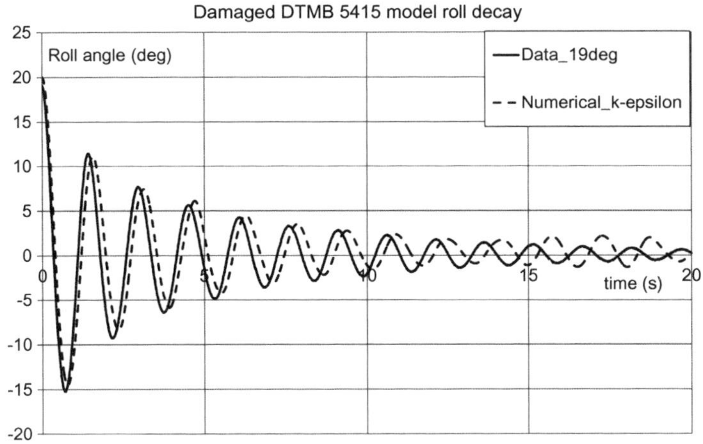

Fig. 15. Numerical versus Experimental Roll Decay.

5. Conclusions

This work focuses on the use of commercial software CD Adapco StarCCM+ RANS solver for the analysis of roll damping properties of the bare hull naval ship DTMB 5415.

Roll damping is considered through the roll decay curve prediction, which is the beginning for any further analysis of roll damping coefficients and it is directly compared with the decay curves obtained from experiments performed by the authors. Experimental results concern intact and damaged ship behavior in free roll decay starting from different angles ranging from 4° to 28° and damaged ship data can be added to Gothenburg CFD behavior workshop (2010).

Mesh sensitivity in numerical simulations is optimized for the damaged ship case considering hexahedral trimmed and hybrid meshes with different refinements, sizes and shapes. The trimmed mesh is chosen as it has the same accuracy of fine hybrid but significantly lower computational time. Obtained numerical results have reasonable damping coefficient prediction but the period of oscillations differ from experiments by up to 4%. These results are in line with those presented by Gao[3,4] and Avalos.[1] It has to be commented that numerical predictions are highly determined by the quality rather than the quantity of the mesh.

The serious challenge for the use of CFD method for damping prediction lies in the extremely high computational time required. Without considering the time necessary for the mesh generation, the calculation time of 5–6 days on 32 computers is impractical for common design practice. However there is great potential to use these simulations to generate damping coefficients numerically for flooded compartments of different geometry and to use these results to improve those semi-empirical formulae typically used in design practice, such as those based on experiments by Katayama.[5]

Acknowledgments

The authors gratefully acknowledge the availability of 32 processors at Calculation Centre SCoPE, University of Naples. Thanks to SCoPE academic staff for the given support and technical staff of Kelvin Hydrodynamic Laboratory for the assistance during experimental campaign.

References

1. G. O. G. Avalos, J. B. V. Wanderley, A. C. Fernandes and A. C. Oliveira, Roll damping decay of a FPSO with bilge keel, *Ocean Eng.*, **87**, pp. 111–120 (2014).
2. E. Begovic, G. Mortola, A. Incecik and A. H. Day, Experimental assessment of intact and damaged ship motions in head, beam and quartering seas, *Ocean Eng.*, **72**, pp. 209–226 (2013).
3. Z. Gao, Q. Gao and D. Vassalos, Numerical study of damaged ship flooding in beam seas, *Ocean Eng.*, **61**, pp. 77–87 (2013).
4. Q. Gao and D. Vassalos, *Numerical Study of the Roll Decay of Intact and Damaged Ships*, Proc. of the 12th International Ship Stability Workshop, Washington D.C., 2011, pp. 277–282.
5. S. Handschel, N. Kollisch and M. Abdel-Maksoud, "Roll Damping of Twin-Screw Vessels: Comparison of RANSE with Established Methods", Proceedings of the 11th International Conference on the Stability of Ships and Ocean Vehicles, Athens, Greece, pp. 887–897 (2012).
6. T. Katayama, M. Kotaki, T. Katsui and A. Matsuda, A Study on Roll Motion Estimation of Fishing Vessels with Water on Deck, *Journal of the*

Japan Society of Naval Architects and Ocean Engineers, **9**, pp. 115–125 (2009) (in Japanese).

7. Y. Lee, H. S. Chan, Y. Pu, A. Incecik and R. S. Dow, Global Wave Loads on a Damaged Ship, *Ships and Offshore Structures* 7(3), pp. 237–268 (2012).
8. R. V. Wilson, P. M. Carrica and F. Stern, Unsteady RANS method for ship motions with application to roll for a surface combatant, *Computers and Fluids*, **35**, pp. 501–524 (2006).
9. B. Yang, Z. C. Wang and M. Wu, Numerical Simulation of Naval Ship's Roll Damping Based on CFD, *Procedia Eng.*, **37**, pp. 14–18 (2012).
10. C. L. Yang, R. C. Zhu, G. P. Miao and J. Fan, Numerical simulation of rolling for 3-D ship with forward speed and nonlinear damping analysis, *J of Hydrodynamics*, **25**(1), pp. 148–155 (2013).
11. ITTC Recommended Procedures and Guidelines. *Practical Guidelines for Ship CFD Applications*, (2011).
12. A Workshop on CFD in Ship Hydrodynamics, Chalmers University, Gothenburg (2010).

Earth Science

Chapter 30

3D Spectral Element Model for Numerical Simulation of the Seismic Ground Motion in the Aterno Valley (Italy)

L. Evangelista*,‡, S. del Gaudio†, A. d'Onofrio†, G. Festa†, A. Santo†, I. Iervolino† and F.Silvestri†

**Istituto per l'Ambiente Marino e Costiero, IAMC-CNR*
†Università di Napoli Federico II, Napoli, Italy
‡lorenza.evangelista@cnr.it

In this chapter, we present 3D Spectral Element simulations of the 2009 *L'Aquila* earthquake aimed at defining a physics-based seismic input for site response analysis. The numerical description of the Aterno basin was built on the subsurface geological model and on the dynamic soil properties of the filling material. We adopted a kinematic description of the rupture that fits near source observations. The comparison between synthetics and recorded accelerograms shows that the simulated signals reproduce with a satisfactory accuracy the real data in the frequency range 0.1–0.7 Hz, consistently with the adopted source model. Simulations were performed by the cluster SCoPE at University of Napoli Federico II.

1. Introduction

With the development of large supercomputer facilities and workstations, physics-based source-to-site 3D numerical simulations of seismic ground motion are becoming a viable tool to construct ground shaking scenarios for earthquakes. These simulations aim at coupling the modeling of seismic source, propagation and site effects, to provide simulations of earthquakes for seismic hazard studies and earthquake engineering applications. We performed an advanced 3D numerical simulation of the 2009 Mw = 6.3 *L'Aquila* earthquake, to be used as an input in the framework of the RELUIS project (see acknowledgments).

The selected numerical tool was the code Spectral Elements in Elastodynamics with Discontinuous Galerkin (SPEED), developed at *Politecnico di Milano*[6] and designed for the simulation of large-scale seismic wave propagation problems including a kinematic description of the rupture, the propagation through a heterogeneous Earth, localized geological irregularities such as alluvial basins and topography.

2. 3D Spectral Element Model of the Aterno Basin

The area hit by the 2009 *L'Aquila* earthquake is located in an intra-Apenninic basin, elongated in NW–SE direction, crossed by the Aterno river, and surrounded by the Gran Sasso and the Velino-Sirente mountains. The bedrock consists of Meso-Cenozoic carbonate rocks, generally outcropping along the sides of the valley and on ridges located within the Aterno River basin. The maximum thickness of the Quaternary deposits is estimated as high as 500 m.

Through the collection and interpretation of geophysical and geotechnical investigations,[7] a simplified shear waves velocity model of the Aterno basin was defined. We assumed an outcropping bedrock outside the boundaries of the basin. A crustal model, based on the profile suggested by Ameri *et al.*[1] was adopted. Moreover, the S-wave velocity value of the shallower layer was reduced according to the results of site investigation (Working Group AQ-MS)[8] to mimic the impedance contrast at the bedrock.

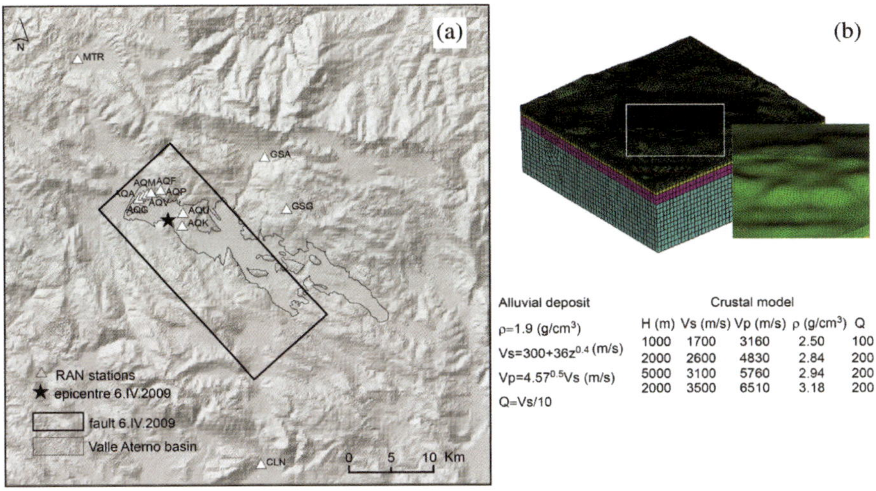

Fig. 1. 3D model of the *L'Aquila* basin: (a) area extension of the analysis domain, (b) Numerical mesh consists of 263,936 hexahedral elements.

The 3D numerical model of the *L'Aquila* basin extends for 58 km in the NS and EW directions with a maximum depth of 20 km (Fig. 1). It was discretized to provide a mesh of 263,936 hexahedral elements (Fig. 1b). The size of the elements ranges from a minimum of 133 m, within the quaternary basin, up to 400 m in the outcropping bedrock. The mesh was generated to propagate up to about 2 Hz for a spectral degree 3. The software ParMetis[5] was used to partition the mesh over the selected number of threads.

2.1. *Kinematic Source Model Along the Paganica Fault*

To represent the rupture process of the *L'Aquila* earthquake, the fault plane and the focal mechanism (strike 133°, dip 54°, rake 102°) were fixed[4] leading to a fault size of 28×20 km^2. The kinematic model is shown in Fig. 2, within the misfit between real data (black records) and synthetics (red records). The model shows a major slip patch between 5 and 10 km southwards of the hypocenter. A smaller asperity in the upper part of the fault with slip as large as 50 cm is instead responsible of the up-dip directivity observed at *L'Aquila* and GSA

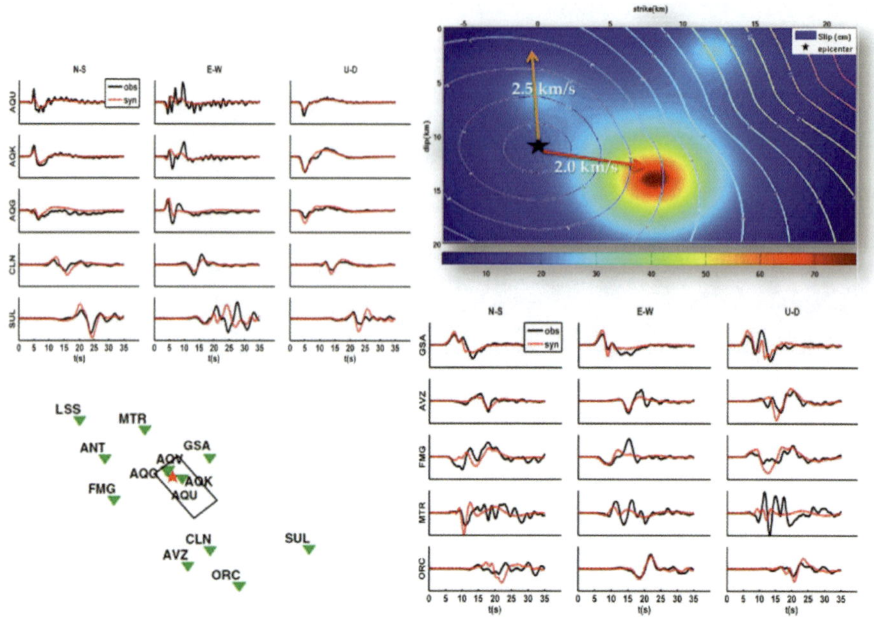

Fig. 2. Final slip distribution and isolines of rupture time from kinematic inversion of *L'Aquila* earthquake data. Comparison of real data (black records) and synthetics (red records) is also shown for some stations.

stations (see Fig. 2). To extend the kinematic model at higher frequencies, the low frequency slip was coupled with a high frequency k^{-2} distribution.[3] We found that the slip roughness does not significantly affect the spectral shape at near and far fault stations, while randomization of the rise time, with a uniform distribution and a mean of 0.75 s, significantly improves the spectral fit, mostly at near fault distances.

3. Numerical Simulation of the 2009 Abruzzo Earthquake

The numerical simulations were carried out assuming the source model described in Sec. 2.1. The SPEED code run on the SCoPE cluster at University of Napoli Federico II on 128 cores, task for nodes 4 and number of threads equal to 2, with a total time of about 3 hours for a single run. In Fig. 3, the contours of E–W velocity component are

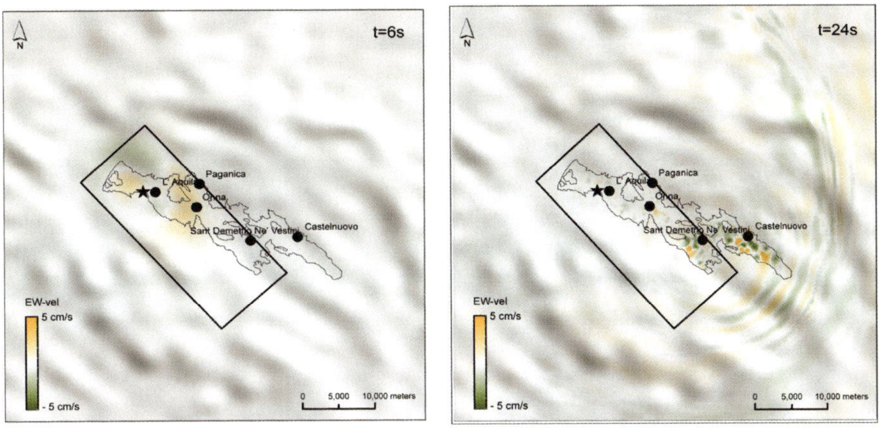

Fig. 3. Contours of E–W velocity component at 6 s and 24 s.

plotted at 6 s and 24 s, showing that the numerical model reproduces the up-dip rupture propagation and a wave focalization within the basin, particularly where the deepest depocenters are located.

The results of 3D analyses are shown as time histories of displacement monitored at selected stations of the Italian Accelerometric Network reported in Fig. 1. The numerical results are then compared with the records of the mainshock in terms of velocity, Fourier and normalized response spectra. To properly compare the observed and simulated data, the waveforms were processed with an acausal band-pass filter Butterworth between 0.1 and 2 Hz, according to the maximum propagated frequency.

Figure 4 shows the comparison between numerical and observed signals at the stations AQU and AQG, considering the 3 components (EW, NS and UP). The data clearly show that the energy content of the recorded signals is not consistently reproduced by numerical simulations while the frequency content is described coherently with the adopted source models, particularly at AQU (Fig. 4a). At the station AQG (Fig. 4b), instead, the simulations underestimate the amplitude of the signal. The shape of the recorded response spectra are well reproduced by the numerical simulations.

The comparison is also summarized in terms of Goodness-of-Fit scores, GoF.[2] For each monitored station, the GoF scores of peak

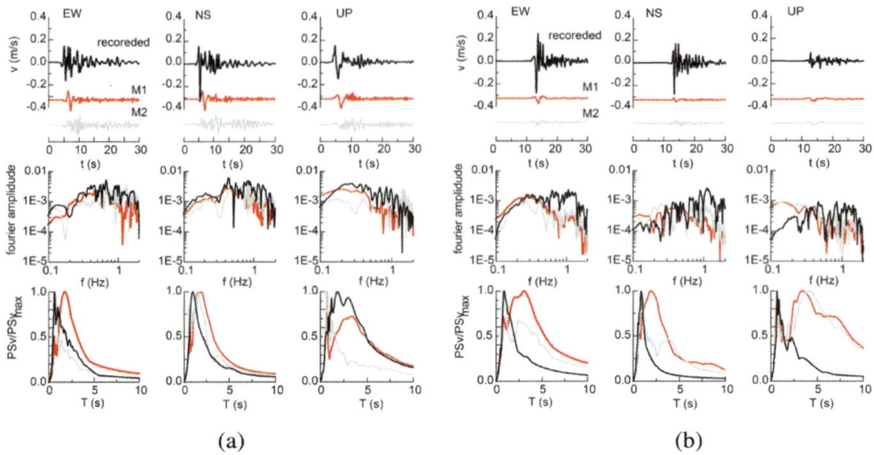

Fig. 4. Comparison of recorded and simulated signals at the RAN stations: (a) AQU and (b) AQG.

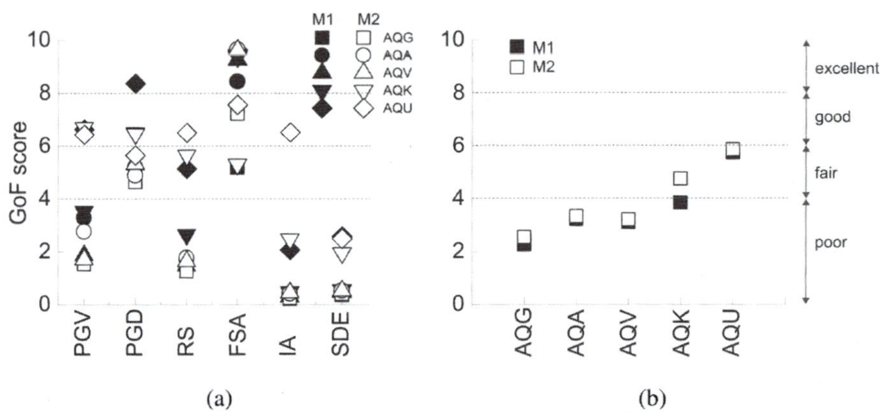

Fig. 5. Goodness-fit-scores in the frequency band between 0.1 and 2 Hz: (a) for the five monitored stations; (b) average values of the six metrics for each stations.

ground velocity (PGV), peak ground displacement (PGD), response spectral acceleration (RS), Fourier amplitude spectrum (FSA), Arias intensity (IA) and SDE (Energy density) are evaluated in the bandwidth 0.1–2 Hz (Figs. 5a and 5b). In Fig. 5, we also compare the results provided by the low frequency source model (M1) and by the coupled k^{-2} model (M2).

The scores of the 6 metrics are assessed as the average value of the 3-components for each station. The results clearly show a misfit between recorded and numerical waveforms, that cannot be ascribed to the simplified model adopted for the deposit filling the basin (see Sec. 3), but could be due to the lack of high-frequency energy content of the source model.

4. Conclusive Remarks and Future Activities

In this work, we performed numerical simulations of the *L'Aquila* earthquake with the code SPEED, based on the spectral element method.

We obtained an updated geological model of the Aterno river basin, suitable to devise a 3D numerical model for seismic wave propagation analyses. Then we included a kinematic description of the earthquake rupture based on the inversion of strong motion data. These ingredients were used to generate a numerical mesh of the region that was used by the code SPEED to simulate the ground motion during the earthquake, by exploiting the high performance computing (HPC) cluster SCoPE.

Several features of the near source ground motion were reproduced by the model, such as the low to intermediate frequency Fourier and response spectral amplitudes, peak ground velocities and accelerations. Nevertheless, loss of high frequency energy and un-modeled impulsive response at near source stations require a better description of the source and propagation. For this purpose, future research activities will be addressed to provide pre-processing tools to SPEED that, based on a relatively rough information on the fault geometry, hypocenter location and spatial distribution of the main fault asperities, with the goal of providing a numerical slip distribution consistent with the high-frequency radiation of seismic energy and to improve results of the numerical simulation of the *L'Aquila* earthquake by comparing several kinematic models available from literature.

Acknowledgments

This work was supported by the national project ReLuis 2014–2018 (Consorzio della Rete dei laboratori Universitari di ingegneria sismica) granted by the civil protection. The Authors would like to acknowledge Prof. Roberto Paolucci and Dr. Chiara Smerzini for the stimulating discussions and the precious help and suggestions in using the software SPEED.

References

1. G. Ameri, F. Gallovic and F. Pacor, Complexity of the Mw 6.3 2009 L'Aquila (central Italy) earthquake: 2. Broadband strong motion modeling, *J. Geophys. Res.*, **117**, p. B04308 (2012).
2. J. Anderson, Quantitative *Measure of the Hoodness-of-fit of Synthetic Seismograms*, Proc. of the 13th World Conference on Earthquake Engineering, Vancouver, Paper 243, (2004).
3. M. Causse, E. Chaljub, F. Cotton, C. Cornou and P. Y. Bard, New approach for coupling k-2 technique and Empirical Green's Functions: Application to the blind prediction of broadband ground-motion in the Grenoble basin, *Geophys. J. Int.*, **179**, pp. 1627–1644 (2009).
4. L. Chiaraluce, L. Valoroso, D. Piccinini, R. Di Stefano and P. De Gori, The anatomy of the 2009 L'Aquila normal fault system (central Italy) imaged by high-resolution foreshock and aftershock locations, *J. Gopehys. Res.*, **116**, p. B12311 (2011).
5. D. LaSalle and G. Karypis, *Multi-Threaded Graph Partitioning*, 27th IEEE International Parallel & Distributed Processing Symposium, (2013).
6. I. Mazzieri, M. Stupazzini, R. Guidotti and C. Smerzini, SPEED: SPectral Elements in Elastodynamics with Discontinuous Galerkin: A non-conforming approach for 3D multi-scale problems, *Int. J. Numer. Meth. Eng.* **95**(12), pp. 991–1010 (2013).
7. F. Santucci de Magistris, A. d'Onofrio, L. Evangelista, S. Foti, M. Maraschini, P. Monaco, S. Amoroso, G. Totani, G. Lanzo, A. Pagliaroli, C. Madiai, G. Simoni and F. Silvestri. Geotechnical characterisation of the Aterno valley for site response analyses. *RIVISTA ITALIANA DI GEOTECNICA*, 47(3), pp. 23–43, ISSN 0557-1405 (2013).
8. Working Group AQ-MS. Seismic microzonation for L'Aquila area reconstruction, 3 vol. + dvd. Regione Abruzzo (in Italian), (2010).

Chapter 31

The HPC ReCaS Infrastructure towards the Simulation of Subsurface Hydrological Processes

R. Campagna*,†,¶, G. Laccetti‡ and G. Severino§

** University of Naples Federico II, Center of Information Services, Italy*
† National Institute of Nuclear Physics, INFN seat of Naples, Italy
‡ Department of Mathematics and Applications,
University of Naples Federico II, Italy
§ University of Naples Federico II,
Division of Water Resources Management, Italy
¶ rosanna.campagna@unina.it

Well hydraulics concerns the process of flow toward/from wells, which occurs because of a vertical extraction or injection in the aquifer. The paper shows preliminary tests performed on the high performance computing (HPC) ReCaS infrastructure to compute simulations on the fluctuation of the water table in heterogeneous porous formations based on a stochastic parametrization of the transmissivity field of the aquifer for the horizontal flow around the pumping well. From a computational point of view, the more expensive aspects concern the storage of the samples and the numerical solution of the related boundary value problem. The numerical simulations were carried out by using packages/libraries included in the S.Co.P.E. toolkit v.3 available at the HPC ReCaS infrastructure.

1. Introduction

The extraction of water by a pumping well causes a cone of depression within the aquifer. Under stationary conditions, a Poisson-type equation is the mathematical model describing the so called hydraulic head.[1,2] The work focuses on the studies aimed to the validation of a solving approach. It involves aspects ranging from the model definition to the algorithmic description and implementation. Tests performed on the high performance computing (HPC) ReCaS infrastructure concern simulations of the fluctuation of the water table both in homogeneous and in heterogeneous porous formations. The numerical simulations of the scenarios were performed thanks to libraries and routines for Scientific Computing that are already included in the S.Co.P.E. Toolkit v.3,[3] available at the HPC ReCaS infrastructure for the user communities: PETSc[4] for the numerical solution of the PDE modeling the problem, Matlab[5]/Octave[6] environments for the output visualization and GSL (GNU Scientific Library[7]) for Special Functions, Statistics and Root-Finding routines. The complexity of the simulated problem, part of which is presented here, is not so high but it grows up when the size of the sampling domain increases or the spatial distance between samples decreases. However, even if our simulations don't solve large problems, an infrastructure of HPC and data manipulation in terms of storage and communications seems attractive to solve problems, related with real applications, with higher complexity. In Sec. 2, we describe the problem in short; in Secs. 3 and 4, we present the numerical approach, software details and results concerning the homogeneous case; Sec. 5 is about results in heterogeneous hypothesis. Section 6 refers to conclusions.

2. The Problem

In our model, the well is assumed pointwise and the pulse-like extraction of water from the well is modeled by the Dirac delta function; the solution describes the effects of this extraction on the piezometric height with respect to a reference level.

3. Homogeneous Simulations

3.1. *The Mathematical Model*

Firstly, we modeled the problem in Cartesian coordinates by assuming $\Omega \subset R^2$ be a rectangular domain extended around the well set in $P = (\alpha, \beta)$:

$$\Omega = [a,b] \times [c,d], \quad a,b,c,d \in R.$$

We denote with $d_{\alpha,\beta}(x, y)$ the distance to P from a point $(x, y) \in \Omega$. Let $\delta(d)$ be the Dirac delta function, we solve the Poisson's equation on the domain Ω, with boundary $\partial\Omega$ on which Dirichlet condition h^B is given:

$$\mathbf{M(P)}: \begin{cases} \nabla^2 h = \frac{Q}{T}\delta(d_{\alpha,\beta}(x,y)) & (x,y) \in \Omega \\ h(x,y) = h^B & (x,y) \in \partial\Omega. \end{cases} \tag{1}$$

A realistic hypothesis is $h^B \rightarrow 0$ when the distance from the well grows up. We solve (1) by a numerical approach based on a finite-difference scheme and a Dirac delta discretization properly defined on the problem.[8]

3.2. *The Numerical Approach*

Let Ω be covered by a regular grid:

$$\Omega_\Delta = \left\{ \begin{aligned} & x_i, y_j : x_i = a + i\Delta_x, \Delta_x = \frac{b-a}{n}, \\ & y_j = c + j\Delta_y, \Delta_y = \frac{d-c}{m} \end{aligned} \right\}_{\substack{i=0,\ldots,n, \\ j=0,\ldots,m}} \tag{2}$$

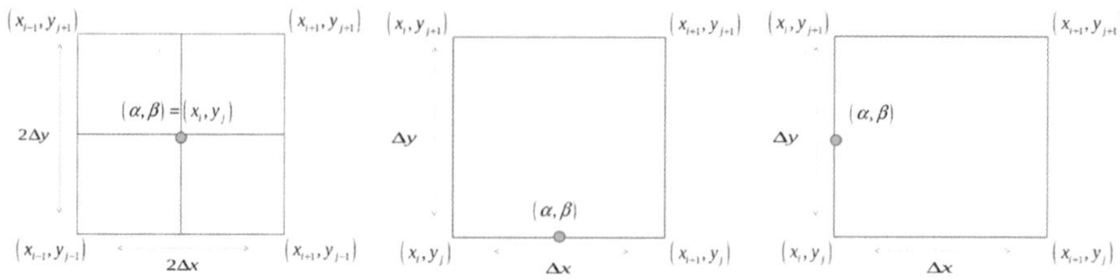

Fig. 1. Well at a grid spacing.

on which the h values are

$$h_{ij} = h(x_i, y_j), \quad i = 0,\ldots,n, \quad j = 0,\ldots,m.$$

In order to discretize the Dirac function, we distinguished[9,10] *irregular points*, those ones that have a coordinate one mesh spacing from the well point from regular ones. In matching to the cases in Fig. 1, it occurs:

*spb*1: the discrete delta function is nonzero only at P;
*spb*2: the discrete delta function is nonzero at the two grid points adjoining (a) P: (x_i, y_j) and (x_{i+1}, y_j) or (b) P: (x_i, y_j) and (x_i, y_{j+1}).

At the grid regular points of Ω_Δ, the five-point Laplacian scheme furnishes a second order approximation for derivatives of second order. A central difference scheme can be used to approximate first and second derivatives in (α, β) in the discrete delta definition. Let's consider *spb1* with $n = m$, $n = 2p$ and $p \in N$. The problem is symmetric with respect to $P \equiv (x_{n/2}, y_{n/2})$, so we solve only in a quarter of Ω_Δ (Fig. 2) the following:

$$\mathbf{M_h(P)}: \begin{cases} \dfrac{h_{i+1,j} - 2h_{i,j} + h_{i-1,j}}{\Delta^2} + \dfrac{h_{i,j+1} - 2h_{i,j} + h_{i,j-1}}{\Delta^2} = \tilde{\delta}_{ij}\dfrac{Q}{T} \\ i = 1,\ldots, n-1, \quad j = 1,\ldots, m-1 \\ h_{i,j} = h^B, \qquad j = 1,\ldots, m-1, \quad i \in \{0, m\} \\ h_{i,j} = h^B, \qquad i = 1,\ldots, n-1, \quad j \in \{0, m\} \end{cases} \tag{3}$$

where the discrete delta $\tilde{\delta}_{ij}$ is nonzero only at the irregular point P.

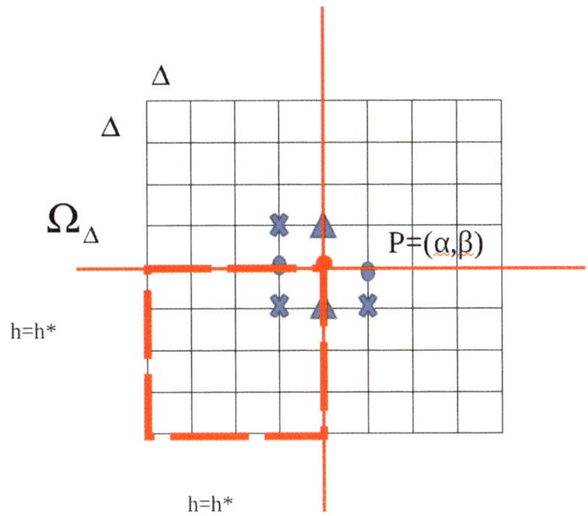

Fig. 2. Grid configuration: case *spb1*.

3.3. *Software Modules*

The numerical solution of (3) is reduced to solving a linear system

$$Ah = b \tag{4}$$

where A is $n/2 \times n/2$ block tridiagonal matrix

$$A = \begin{bmatrix} B_1 & I & 0 & 0 \\ I & B_1 & I & 0 \\ 0 & I & B_1 & I \\ 0 & 0 & B_2 & B_3 \end{bmatrix} \tag{5}$$

with B_1, B_2 and B_3 sparse matrices and the right-hand side b depending on the δ_{ij} definition. For the numerical solution of (4), we chose to use PETSc library, available at the HPC ReCaS infrastructure. PETSc is a sophisticated tool for the numerical solution of partial differential equations and related problems on high-performance computers. So we defined a software element based on the *scalable linear equations solver* KSP of PETSc, an abstract PETSc object that manages all Krylov methods, providing an easy-to-use interface to the combination of a Krylov subspace iterative method and a preconditioner. After creating A and b

```
1  ...
2  /* grid definition of size m x n ...    */
3  ...
4  /* Creation of the sparse matrix ...    */
5  ierr = MatCreateSeqAIJ(PETSC_COMM_SELF,N,N,5,0,&userctx->A);CHKERRQ(ierr);
6  ...
7  /* Creation of the right-hand side vector ...  */
8  ierr = VecCreateSeqWithArray(PETSC_COMM_SELF,1,N,NULL,&userctx->b);CHKERRQ(ierr);
9  ...
10 /* Create linear solver context ...  */
11 ierr = KSPCreate(PETSC_COMM_SELF,&userctx->ksp);CHKERRQ(ierr);
```

Fig. 3. PETSc modules for Vec and Mat object creation.

```
1  ...
2  /* Define the sparse matrix ... */
3  Ii = 0;
4   for (j=0; j<n; j++) {
5     for (i=0; i<m; i++) {
6       if (j>0) {
7         J    = Ii - m;
8         v    = ...;
9         ierr = MatSetValues(A,1,&Ii,1,&J,&v,INSERT_VALUES);CHKERRQ(ierr);
10      }
11      ...
12      Ii++;
13    }
14  }
15 /* Assemble matrix  */
16 ierr = MatAssemblyBegin(A,MAT_FINAL_ASSEMBLY);CHKERRQ(ierr);
17 ierr = MatAssemblyEnd(A,MAT_FINAL_ASSEMBLY);CHKERRQ(ierr);
18 ...
19 /* Set operators. Here the matrix that defines the linear system
20    also serves as the preconditioning matrix...  */
21 ierr = KSPSetOperators(userctx->ksp,A,A,SAME_NONZERO_PATTERN);CHKERRQ(ierr);
22 ...
23 /* Set runtime options...  */
24 ierr = KSPSetFromOptions(userctx->ksp);CHKERRQ(ierr);
25 ...
26 /* Solve the linear system */
27 ierr = KSPSolve(userctx->ksp,userctx->b,userctx->x);CHKERRQ(ierr);
```

Fig. 4. PETSc modules for system solution.

in (4), we have solved the system with the following commands (see line 11 of Fig. 3 and lines 21, 24 and 27 of Fig. 4):

```
KSPCreate(MPI_Comm comm,KSP *ksp);
KSPSetOperators(KSP ksp,Mat Amat, Mat Pmat);
KSPSetFromOptions(KSP ksp);
KSPSolve(KSP ksp,Vec b,Vec x);
```

```
...
/* use -binary option  ...*/
ierr = PetscOptionsGetBool(NULL,"-binary",&isbinary,NULL);CHKERRQ(ierr);
/* to save the output in a .mat file */
if (isbinary) {
  ierr = PetscPrintf(PETSC_COMM_SELF,"writing vector in binary to vector.dat...\n");
  ierr = PetscViewerBinaryOpen(PETSC_COMM_SELF,"vector.dat",FILE_MODE_WRITE,&viewer);
}
...
/* use -mat_view ::ascii_matlab option ...*/
ierr = PetscOptionsGetBool(NULL,"-mat_view ::ascii_matlab",&isascii,NULL);
/* to save the output in ASCII format  */
if (isascii) {
  ierr = PetscPrintf(PETSC_COMM_SELF,"writing vector in ASCII to mat.output...\n");
  ierr = PetscViewerASCIIOpen(PETSC_COMM_SELF,"mat.output",&viewerASCII);
}
```

Fig. 5. PETSc modules for output visualization.

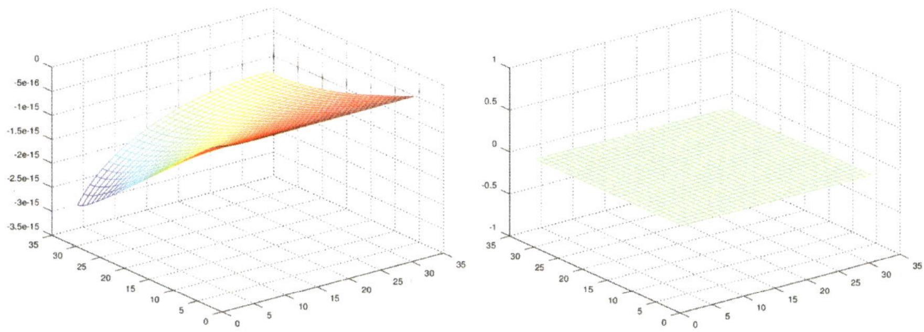

(a) (b)

Fig. 6. Homogeneous solution, (a) $h^B = -10^{-16}$, (b) $h^B = 0$.

The analysis of the results has been made using appropriate tools available in Matlab/Octave environments. Figures 3–5 describe the work flow of the PETSc core modules.

3.4. *Results*

The following results concern the solution computed in a quarter of grid, by input data $m = n = 61$, $T = 0.2$ cm^2/min, $Q = 60$ cm^2/min and $h^B = -10^{-16}$ or $h^B = 0$. Figure 6a highlights that the well presence at only (grid) point affects the values of the numerical solution elsewhere letting the solution values flat out towards the boundary condition h^B. So, in the limit (and realistic) hypothesis $h^B \to 0$, the solution doesn't highlight the well presence (see Fig. 6b). This limit, together with the relationship that we expect between the discrete model, describing the impulse function δ, and the head function values around the well has to be taken into account. Moreover, we observed that our $\tilde{\delta}_{ij}$ definition in (3) made the solution independent from the mesh size (i.e., from $\Delta_x = \Delta_y = \Delta$). In order to avoid this result, taking in mind the symmetry of the problem, we formulated the homogeneous model in polar coordinates so searching for a solution on a circular grid.

4. Homogeneous Simulations: Second Formulation

4.1. *The Mathematical Model*

We assume that $\Omega \subset R^2$ be a circular domain $\Omega = C(P,R)$, with center P and max ray $R \in R^+$, and we solve $\mathbf{M(P)}$ in polar coordinates:

$$\mathbf{M(P)}: \begin{cases} \frac{1}{r}\frac{\partial}{\partial r}\left(r\frac{\partial h}{\partial r}\right) + \frac{1}{r^2}\frac{\partial^2 h}{\partial \theta^2} = \frac{Q}{T}\delta(d_{\alpha,\beta}(r,\theta)) & (r,\theta) \in \Omega \\ h(r,\theta) = h^B & (r,\theta) \in \partial\Omega \end{cases}. \quad (6)$$

4.2. *The Numerical Approach*

Let Ω be covered by a grid Ω_Δ in the r–θ plane marked up by the points of intersection of the circles with center in P and radius less than R:

$$\Omega_\Delta = \left\{ r_i, \theta_j : r_i = i\Delta_r, \Delta_r = \frac{R}{m+1}, \theta_j = j\Delta_\theta, \Delta_\theta = \frac{2\pi}{n+1} \right\}_{\substack{i=0,\ldots,m+1, \\ j=0,\ldots,n+1}}. \quad (7)$$

The h values at the mesh points will be $h_{ij} = h(r_i, \theta_j)$, $i = 0,\ldots,m+1$, $j = 1,\ldots,n+1$; $h_{i,0} = h_{i,n+1}$

In homogeneous hypothesis, the h values on each ring will depend mainly on the distance from the well, while being constant with respect to θ. The discrete Dirac delta $\tilde{\delta}_i$ will be nonzero only at the irregular points, that have the first polar coordinate one mesh spacing from the well point, i.e., $(r_1, \theta_j) = (a, \theta_j)$, $j = 1,\ldots,n + 1$. We solve:

$$\mathbf{M_h(P)}:\begin{cases} \dfrac{h_{i+1,j} - 2h_{i,j} + h_{i-1,j}}{\Delta^2} + \dfrac{1}{i\Delta r}\dfrac{h_{i+1,j} - h_{i-1,j}}{\Delta^2} \\ \quad + \dfrac{1}{(i\Delta r)^2}\dfrac{h_{i,j+1} - 2h_{i,j-1}}{\Delta_\theta^2} = -\dfrac{Q}{T}\dfrac{1}{2\pi r_i}\tilde{\delta}_i, \\ i = 1,\ldots,m, \;\; j = 0,\ldots,\; n+1, \\ h_{1,j} = h_1^B \\ h_{m+1,j} = h_{m+1}^B \end{cases} \tag{8}$$

with $\tilde{\delta}_i$ and the boundary conditions suitably defined according to the problem.

4.3. *Software Modules*

The numerical solution of (8) requires to solve a linear system

$$Ah = F,$$

where A is a $m \times n$ pentadiagonal, block tridiagonal matrix:

$$A = \begin{bmatrix} B_1 & \left(1+\frac{1}{2}\right)I & & & \\ \left(1-\frac{1}{4}\right)I & B_2 & \left(1+\frac{1}{4}\right)I & & \\ & \left(1-\frac{1}{6}\right)I & B_3 & \left(1+\frac{1}{6}\right)I & \\ \cdots & \cdots & \cdots & \cdots & \\ & & \left(1-\frac{1}{2(m-1)}\right)I & B_{m-1} & \left(1-\frac{1}{2(m-1)}\right)I \\ & & & \left(1-\frac{1}{2m}\right)I & B_m \end{bmatrix},$$

where B_i, $i = 1,\ldots,m$ and I are $n \times n$ matrices, and the right-hand side matrix F depends on h_1^B and on the definition of the discrete Dirac delta function. For the numerical solution of (8), we chose to use the

```

      SUBROUTINE PWSPLR (INTL,A,B,M,MBDCND,BDA,BDB,C,D,N,NBDCND,BDC,
     1                   BDD,ELMBDA,F,IDIMF,PERTRB,IERROR,W)
C
C
C**********************************************************************
C
C          VERSION  2  OCTOBER 1976  INCLUDING ERRATA  OCTOBER 1976
C
C         DOCUMENTATION FOR THIS PROGRAM IS GIVEN IN
C
C        EFFICIENT FORTRAN SUBPROGRAMS FOR THE SOLUTION OF
C            ELLIPTIC PARTIAL DIFFERENTIAL EQUATIONS
C
C                              BY
C
C          PAUL SWARZTRAUBER   AND  ROLAND SWEET
C
C             TECHNICAL NOTE TN/IA-109    JULY 1975
C
C       NATIONAL CENTER FOR ATMOSPHERIC RESEARCH  BOULDER,COLORADO 80307
C
C        WHICH IS SPONSORED BY THE NATIONAL SCIENCE FOUNDATION
C
C**********************************************************************
C
C
C
C     SUBROUTINE PWSPLR SOLVES A FINITE DIFFERENCE APPROXIMATION TO THE
C     HELMHOLTZ EQUATION IN POLAR COORDINATES:
C
C          (1/R)(D/DR)(R(D/DR)U) + (1/R**2)(D/DTHETA)(D/DTHETA)U
C
C                                + LAMBDA*U = F(R,THETA).
C
...
```

Fig. 7. A scratch of PWSPLR routine in *ALGORITHM 541*.

FORTRAN routine PWSPLR, cited in Fig. 7, belonging to the software package *ALGORITHM 541* by Swarztrauber and Sweet,[11] specific for a finite difference solution of a Helmholtz-like equation in polar coordinates. The analysis of the results has been made using appropriate tools available in Matlab/Octave environments.

4.4. *Results*

We have compared the theoretical values of $\tilde{\delta}_1$ with the Thiem solution on the first circle, of radius a, for different grid sizes. The error behavior reveals that, a part from the errors at the points belonging to the first grid circle ($r = a$), according to the singularity due to the well presence, the number of correct digits increases when r grows up (Fig. 8).

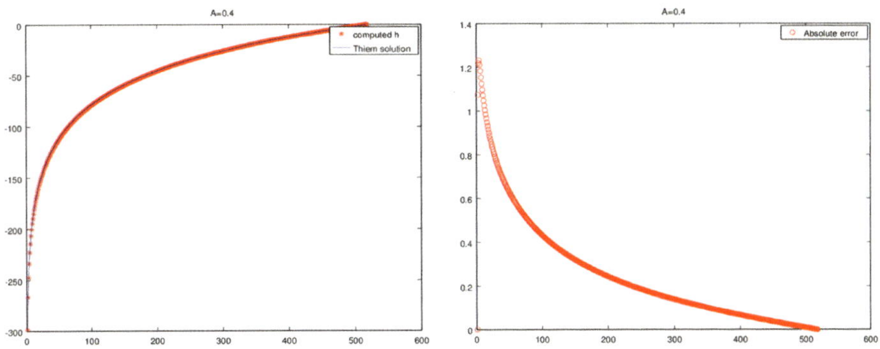

(a) (b)

Fig. 8. $\Delta_r = 0.4$, $m = 517$. (a) Computed solution. (b) Absolute error. The abscissae represent the grid r-values.

5. Heterogeneous Simulations

5.1. *The Mathematical Model*

The differential model describing the problem when the porous formation is heterogeneous, must take into account the local changes of the transmissivity T around the well. The mathematical model is

$$\mathbf{M}*(\mathbf{P}):\begin{cases}\frac{1}{r}\frac{\partial}{\partial r}\left(r\frac{\partial h^*}{\partial r}\right)+\frac{1}{r^2}\frac{\partial^2 h^*}{\partial\theta^2}=-\frac{1}{\bar{T}}\frac{1}{2\pi r}\delta(r) & (r,\theta)\in\Omega\\ h(r,\theta)=h^B & (r,\theta)\in\partial\Omega\end{cases} \tag{9}$$

with suitable boundary conditions. In (9) it is $\bar{T}=\frac{T}{\langle T\rangle}$, where $\langle T\rangle$ is the mean value of T; moreover $h = h^* Q^*$, with $Q^* = \frac{Q}{\langle T\rangle}$. The related numerical problem is:

$$\mathbf{M}_{\mathrm{h}}^*(\mathbf{P}):\begin{cases}\frac{h_{i+1,j}-2h_{i,j}+h_{i-1,j}}{\Delta^2}+\frac{1}{i\Delta r}\frac{h_{i+1,j}-h_{i-1,j}}{2\Delta r}\\ \quad+\frac{1}{(i\Delta r)^2}\frac{h_{i,j+1}-2h_{i,j}+h_{i,j-1}}{\Delta_\theta^2}=-\frac{1}{T_{ij}}\frac{1}{2\pi r_i}\tilde{\delta}_i,\\ i=1,\ldots,m,\ \ j=0,\ldots,n+1,\\ h_{1,j}=h_1^B\\ h_{m+1,j}=h_{m+1}^B\end{cases} \tag{10}$$

with $\tilde{\delta}_i$ and the boundary conditions suitably defined.

5.2. *Software Modules*

The $\mathbf{M_h^*(P)}$ definition is based on the assumption of a random variation of the transmissivity. So a random distribution generation has been assumed on data. A draft of the algorithm is in Fig. 9. The mathematical form of the function F_T in the algorithm depends on choices assumed about the data distribution. The numerical solution of (10) is still based on the FORTRAN routine PWSPLR, integrated with both statistical functions and root-finding routines, all available in GSL. Figures 10 and 11 describe the work flow in terms of GSL-routines.

5.3. *Results*

The results obtained in some case studies follow. The tests have been conducted for increasing values of *n*max, i.e., $n\text{max}_k = 100 \cdot k$, $k = 1,\ldots,12$. Set $a = 0.6$ and $\frac{Q}{\langle T\rangle} = 1$, Figs. 12 and 13 describe the computed solution for $\xi = 1$ and $\xi = 0.1$ respectively. Figures 14 and 15

for $SN = 1, \ldots, nmax$	
$\eta_j \in [0,1]$, $j = 1, \ldots, n+1$	*random gen.*
$\eta_j - \mathcal{F}_T(\overline{T}_j) = 0 \quad j = 1, \ldots, n+1$	*root finding*
${h^*_{i,j}}^{(SN)}$, $i = 1, \ldots, m+1$, $j = 1, \ldots, n+1$;	*(10) solution*
end for	
$E(h^*, nmax)_{ij} = \frac{1}{nmax} \sum_{SN=1}^{nmax} {h^*_{i,j}}^{(SN)}$	*mean value*
$\sigma^2(h^*, nmax)_{ij} = \frac{1}{nmax-1} \sum_{SN=1}^{nmax} ({h^*_{i,j}}^{(SN)} - E(h^*, nmax)_{ij})^2$	*variance*
$\xi* = \frac{\sigma(h^*, nmax)_{ij}}{E(h^*, nmax)_{ij}}$	*variation coeff.*
$\gamma^* = E\left[\left(\frac{({h^*_{i,j}}^{(SN)} - <{h^*_{i,j}}^{(SN)}>)}{\sigma(h^*, nmax)_{ij}}\right)^3\right]$	*skewness*

Fig. 9. Algorithm for heterogeneous simulation code.

show the behavior of the absolute and relative errors in 2-norm, between the mean values corresponding to two successive set of runs, when *nmax* increases.

```
...
for (i = 0; i < n+1; i++) {
/* use built-in functions for random number generation  ...*/
        eta = ((double)drand48());
   ...
/* gsl_roots.h: GSL routines for finding roots of one-dimensional functions */
        T = gsl_root_fsolver_brent;
        s = gsl_root_fsolver_alloc(T);

        gsl_root_fsolver_set(s, &F, x_lo, x_hi);
        do
        {
          iter++;
          status = gsl_root_fsolver_iterate(s);
          r = gsl_root_fsolver_root(s);
          x_lo = gsl_root_fsolver_x_lower(s);
          x_hi = gsl_root_fsolver_x_upper(s);
          status = gsl_root_test_interval(x_lo, x_hi, 0, 1.0e-8);
        } while (status == GSL_CONTINUE && iter < max_iter);
        ...
        gsl_root_fsolver_free(s);
}
...
```

Fig. 10. Block code for random number generation and finding roots by GSL functions.

```
...
for(j=0; j<m*n; j++){
  mean[j]      = gsl_stats_mean(data, 1, nmax);
  variance_m[j] = gsl_stats_variance_m(data, 1,nmax,mean[j]);
  sd_m[j]= gsl_stats_sd_m(data, 1, nmax,mean[j]);
  cv[j]=sd_m[j]/mean[j];
  skewness[j]=gsl_stats_skew_m_sd(data,1,nmax,mean[j],sd_m[j]);
}
...
```

Fig. 11. Block code for statistics by GSL functions.

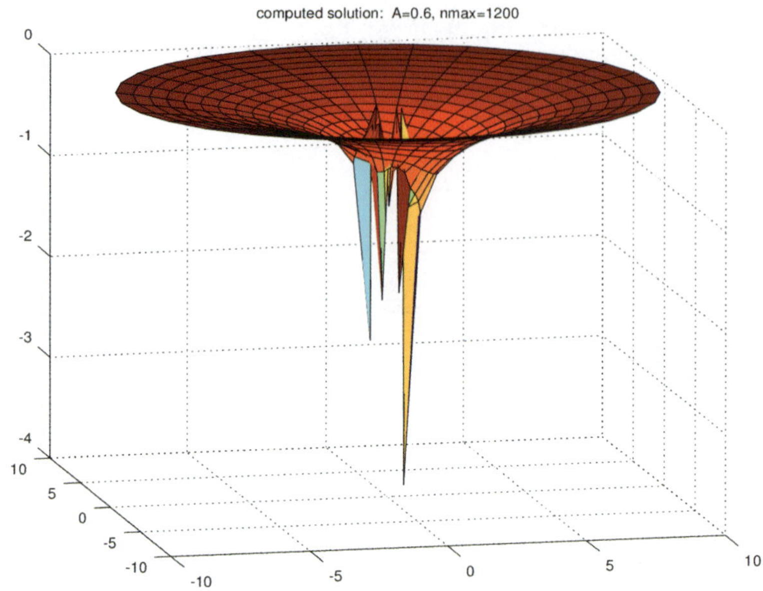

Fig. 12. Computed solution, $a = 0.6$, $Q/\langle T\rangle = 1$, $\xi = 1$ and $n\text{max} = 1200$.

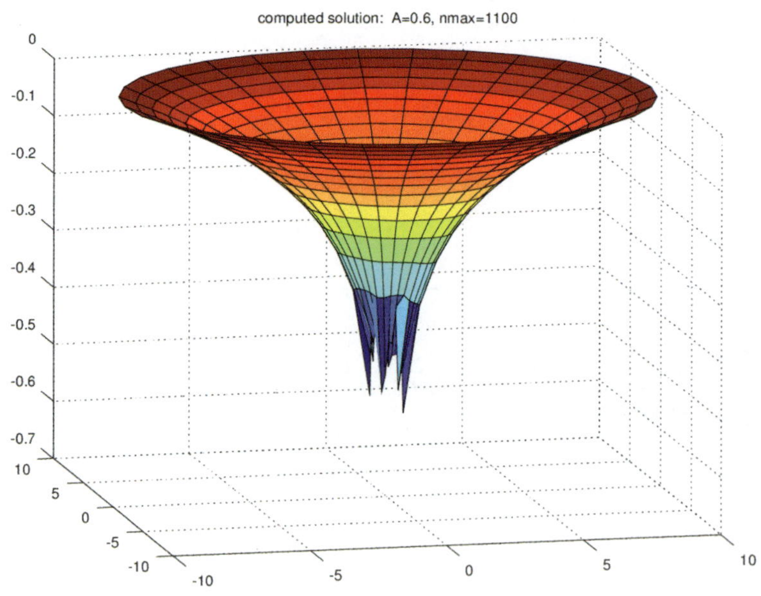

Fig. 13. Computed solution, $a = 0.6$, $Q/\langle T\rangle = 1$, $\xi = 0.1$ and $n\text{max} = 1100$.

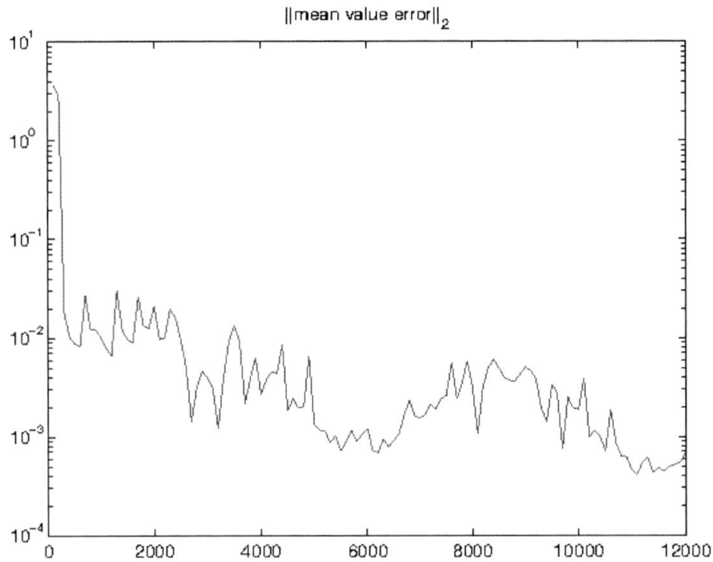

Fig. 14. E_2, *n*max = 100,...,12000, step 100.

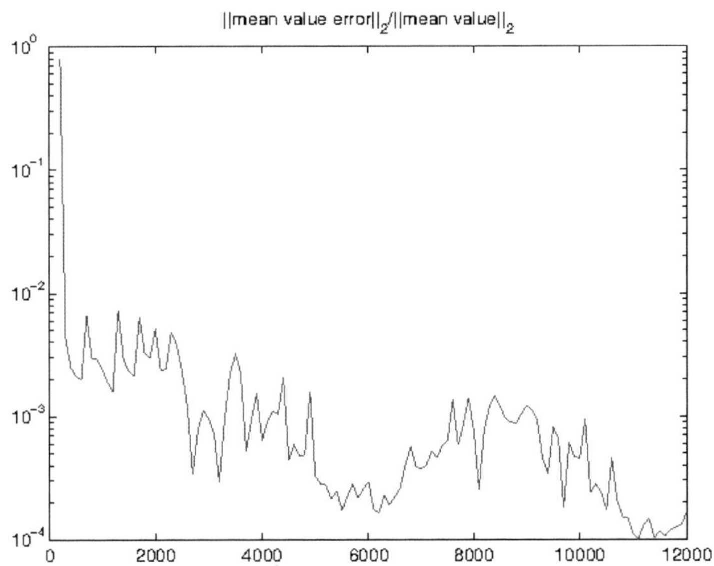

Fig. 15. E_2^{rel}, *n*max = 100,...,12000, step 100.

6. Conclusions

Pollution of environmental and agro-forestry resources is a widespread critical problem, representing the main obstacle to the development of the economy in a region due to the impact upon crop yields. Contamination of the environmental water resources, e.g., as a result of the widespread use of organic chemicals, has drawn the attention of hydrologists to the analysis of contamination versus decontamination phenomena. Well extraction and subsequent spread of groundwater is strictly connected to this problem. Reliable analysis and predictions are compulsory for engineering applications (such as decontamination procedures) as well as protection against long-term pollution of soils and groundwater. With this aim in mind, a HPC environment, such as the HPC ReCaS Infrastructure, is mandatory, also with the aim to deploy a software framework to model and predict e.g., contaminants transport in complex (both physically and chemically) environments, at different scales, including techniques of uncertainty and risk analysis.

Acknowledgments

This work has been realized thanks to the use of the HPC ReCaS/S. Co.P.E.computing infrastructure at the University of Naples Federico II.

References

1. G. Severino, A. Santini and A. Sommella, Steady flows driven by sources of random strength in heterogeneous aquifers with application to partially penetrating wells, *Stochastic Environmental Research and Risk Assessment*, **22,** pp. 567–582 (2008). Doi: 10.1007/s00477-007-0175-5.
2. G. Severino, Stochastic analysis of well-type flows in randomly heterogeneous porous formations, *Water Resources Research*, **47**, p. W03520 (2011). Doi: 10.1029/2010WR009840.
3. G. B. Barone, D. Bottalico, V. Boccia, L. Carracciuolo, S. Pardi, M. Scognamiglio and F. Serio, *Middleware applicativo: lo SCoPE Toolkit*, poster abstract at Italian National Conference "e-Science 2008", Naples, 2008 (in Italian).

4. Available at http://www.mcs.anl.gov/petsc/petsc-as/
5. Available at http://www.mathworks.com/
6. Available at http://www.gnu.org/software/octave/
7. Available at http://www.gnu.org/software/gsl/
8. G. Severino and P. Indelman, Analytical Solutions for Reactive Transport Under an Infiltration-Redistribution Cycle, *J. Contam. Hydrol.* **70**, pp. 89–115 (2004).
9. A. Mayo, The fast solution of Poisson's and the biharmonic equations on irregular regions, *SIAM J. Numer. Anal.* **21**(2), pp. 285–299 (1984).
10. P. Smereka, The numerical approximation of a delta function with application to level set methods, *J. Comput. Phys.* **211**(1), pp. 77–90 (2006).
11. P. N. Swarztrauber and R. A. Sweet, ALGORITHM 541: Efficient FORTRAN subprograms for the solution of separable elliptic partial differential equations, *ACM Trans. Math. Software* **5**, pp. 352–364 (1979).

Computer and Computational Science

Chapter 32

Scalability Analysis of Variational Data Assimilation Algorithms on Hybrid Architectures

R. Arcucci[*,†,¶,||], L. D' Amore[*,‡],
L. Carracciuolo[§] and A. Murli[‡,¶]

[*] *University of Naples Federico II, Naples (IT)*
[†] *Imperial College London, London (UK)*
[‡] *Euro Mediterranean Center on Climate Changes, Lecce (IT)*
[§] *National Research Council (CNR), Naples (IT)*
[¶] *SPACI s.r.l., Naples (IT)*
[||] *rossella.arcucci@unina.it*

Large-scale problems are computationally expensive and their solution requires designing of scalable approaches. Many factors contribute to scalability, including the architecture of the parallel computer and the parallel implementation of the algorithm. However, one important issue is the scalability of the algorithm itself. We have developed a scalable algorithm for solving large scale Data Assimilation (DA) problems: starting from a decomposition of the mathematical problems, it uses a partitioning of the solution and a modified regularization functional. Here, we briefly discuss some results.

1. The Data Assimilation Inverse Problem

A research collaboration between us and both CMCC (Euro Mediterranean Center on Climate Changes) and Imperial College London, give us the opportunity to work on Variational Data Assimilation (DA) in real Oceanographic Models for the Mediterranean Sea.[1,2] DA is an ill posed inverse problem and Variational approaches are essentially derived from the Tikhonov regularization formulation.[3]

We introduced[4] a decomposition of the global domain into subdomains. On these subdomains, we define local regularization functionals and we prove that the minimum of the global regularization functional can be obtained by collecting the minimum of each local functional. The (global) problem is decomposed into (local) sub problems in such a way. The resulted algorithm consists of several copies of a slightly modified sequential algorithm, each one requiring approximately the same amount of computations on each subdomain and an exchange of boundary conditions between adjacent subdomains. The data is flowing across the surfaces, as a result, the so-called surface-to-volume effect is produced.

Here, we report scalability results[4,5] obtained by running the parallel algorithm on two hybrid architectures (HAs). We note that heterogeneous and HAs are expected to play a preeminent role in the emerging and future computing architectures.[6]

The ReCaS project gave us the opportunity to use such kind of architectures: thanks to the obtained funding, the existing SCoPE computing infrastructure has been enriched with new multi-node computer systems with a high number of cores per node and, in some cases, also equipped with GP-GPU accelerators.

2. The Algorithm on Two Reference Hybrid Architectures (HAs)

We consider two HAs:

- HA1: a 288 CPU-multi-core architecture made of distributed memory blades each one with computing elements sharing the same local memory for a total of 3,456 cores.

- HA2: a GPU+CPU architecture made of the 512 threads NVIDIA Tesla connected to a quad-core CPU.

If *nproc* denotes the number of processing elements of the reference architectures, we have *nproc* = # cores-involved for HA1, and *nproc* = # threads-blocks, for HA2. We use the following correspondence between p, which is the number of subdomains and *nproc*:

$$p \leftrightarrow nproc.$$

We assume a 2D uniform decomposition of physical domain along the (x, y) axis, that is the x-axis is divided by s and the y-axis by q, then, the size of each subdomain is $r = \frac{NP}{nproc} = nloc_x \times nloc_y \times nloc_z$ where:

$$nloc_x = \frac{n_x}{s} + 2o_x,\ nloc_y = \frac{n_y}{q} + 2o_y,\ nloc_z = n_z. \tag{1}$$

These dimensions include the overlapping $(2o_x \times 2o_y)$.

Let $p_1, p_2 \in \aleph$ and $p_1 < p_2$. Let $T(\mathcal{A}(NP, p_i))$, $i = 1, 2$ denote the time complexity of algorithm $\mathcal{A}(NP, p_i)$, $i = 1, 2$, for all i different from j, running on p_i processing elements. We defined[4] the (relative) scale-up factor of $\mathcal{A}(NP, p_2)$, in going from p_1 to p_2, the following ratio:

$$S_{p_2,p_1}(NP) = \frac{T(\mathcal{A}(NP, p_1))}{(p_2/p_1)T(\mathcal{A}(NP, p_2))}.$$

We use the LBFGS method[7] for computing the minimum of the DA local problems. Then the time complexity of $\mathcal{A}(NP, p)$ is $T(NP) = O(f(NP))$ flops, on a problem of size NP, where $f(NP) \in \Pi_3$, and the scale-up factor of the algorithm $\mathcal{A}(NP, p)$ is

$$S_{p,1}(NP) = \alpha(r, p)p^2. \tag{2}$$

Remark: Let t_{flop} denote the unitary time required by one floating point operation. As a result, the execution time needed for algorithm $\mathcal{A}(NP)$ for performing $T(NP)$ floating point operations, is

$$T_{flop}(NP) = T(NP) \times t_{flop}.$$

Multiplying and dividing the (2) by t_{flop}, we get

$$S_{p,1}(NP) = \frac{T_{flop}(NP)}{pT_{flop}(NP/p)}. \tag{3}$$

In Ref. 8, authors define $T^{nproc}(NP)$, the parallel execution time of A(NP, p), as given by time for computation plus an overhead which is given by synchronization, memory accesses and communication time also.

$$T^{nproc}(NP) := T_{flop}^{nproc}(NP) + T_{ohnproc}(NP)$$

where

- HA1: $T_{flop}^{nproc}(NP)$ is computing time required for the execution of $T(NP)$ floating point operations; $T_{oh}^{nproc}(NP)$ is overhead time of $T(NP)$ data.
- HA2: $T_{flop}^{nproc}(NP) := T^{CPU}(NP) + T^{GPU}(NP)$, where
  - $T^{CPU}(NP)$ is the CPU execution time,
  - $T^{GPU}(NP)$ is the GPU execution time. Here we assume that

$$T^{GPU}(NP) := T_{flop}^{GPU}(NP) + T_{oh}^{GPU}(NP), \tag{4}$$

where $T_{oh}^{GPU}(NP)$ is the time for global and local memories transfers into the device (GPU) and $T_{flop}^{GPU}(NP)$ is the computing time required for execution of floating point operations.

Finally, $T_{oh}^{nproc}(NP) \equiv T_{oh}^{GPU}(NP)$ is the communication time between host (CPU) and device (GPU).

3. Discussion

In order to discuss results in Tables 1 and 2, in the following we report a brief discussion by considering, for simplicity of notations, the measured value of the scale up factor $S_{nproc,1}$. Let

$$SS_{nproc} = \frac{T_{flop}(NP/p)}{T_{flops}^{nproc}(NP/p)}$$

Table 1. Results on HA1.

NP	*nproc*	$T^{nproc}(NP)$	measured $S_{nproc,8}$	$S_{nproc,8}$
$O(10^6)$	8	2.0545e+02	1.0	1
	16	6.3316e+01	3.25	4
	32	2.0005e+01	10.27	16
	64	8.7835e+00	23.39	64
NP	***nproc***	$T^{nproc}(NP)$	**measured** $S_{nproc,16}$	$S_{nproc,16}$
$O(10^7)$	8	—	—	—
	16	3.9091e+03	1.0	1
	32	9.9952e+02	3.91	4
	64	2.7584e+02	14.17	16

denote the speed up of the algorithm running on *nproc* processing elements for solving the local DA problem, then we get:

measured $S_{nproc,1}$

$$S_{nproc,1} = \frac{T_{flop}(NP)}{\frac{pT_{flop}(NP/p)}{SS_{nproc}} + pT_{oh}(NP/p)} = \frac{SS_{nproc}\frac{T_{flop}(NP)}{pT_{flop}(NP/p)}}{1 + \frac{SS_{nproc}T_{oh}(NP/p)}{T_{flop}(NP/p)}}. \tag{5}$$

Let

$$\alpha = \frac{SS_{nproc}}{1 + \frac{SS_{nproc}T_{oh}(NP/p)}{T_{flop}(NP/p)}}$$

then from (5) we get

measured $S_{nproc,1}$

$$S_{nproc,1} = \alpha \frac{T_{flop}(NP)}{pT_{flop}(NP/p)} > S_{nproc,1} \Leftrightarrow \alpha > 1. \tag{6}$$

It holds that

- if $SS_{nproc} > 1$, then

$$\alpha > 1 \Leftrightarrow \frac{T_{oh}(NP/p)}{T_{flop}(NP/p)} = C, \qquad (C < 1/2),$$

where the condition

$$\frac{S}{V} := \frac{T_{oh}(NP/p)}{T_{flop}(NP/p)} < 1$$

guarantees that the so-called *surface-to-volume* effect on each local DA problem is produced.[9] So in this case, we get that

$$\frac{S}{V} < C \Leftrightarrow \text{measured } \; S_{nproc,1} > S_{nproc,1}$$

- if $SS_{nproc} = 1$ then $\alpha < 1$ and it comes out that

$$\text{measured } S_{nproc,1} < S_{nproc,1}.$$

In conclusion, since in our experiments it is $SS_{nproc} = 1$ for HA1, and $SS_{nproc} > 1$ for HA2, this analysis validates the experimental results reported in Table 1 for HA1 and in Table 2 for HA2, respectively.

Table 2. Results on HA2.

NP	P	$T_{flopGPU}(NP)$	measured $S_{nproc,2}$	$S_{nproc,2}$
$O(10^7)$	1	0.127	—	—
	2	0.027	4.7	4
	4	0.008	15.9	8
	8	0.007	18.1	16

Acknowledgments

This work has been realized thanks to the use of the SCoPE computing infrastructure at the University of Naples, also in the framework of PON *"Rete di Calcolo per SuperB e Altre Applicazioni"* (ReCaS) project.

References

1. L. D'Amore, R. Arcucci, L. Carracciuolo and A. Murli, DD-OceanVar: A domain decomposition fully parallel data assimilation software for the Mediterranean forecasting system, *Procedia Computer Science* **18**(0), pp. 1235–1244 (2013). ISSN 1877-0509. doi: http://dx.doi.org/10.1016/j.procs.2013.05.290. URL http://www.sciencedirect.com/science/article/pii/S187705091300433X. 2013 International Conference on Computational Science.
2. L. D'Amore, R. Arcucci, L. Marcellino and A. Murli, HPC computation issues of the incremental 3D variational data assimilation scheme in OceanVar software, *Journal of Numerical Analysis, Industrial and Applied Mathematics* 7(3–4), pp. 91–105 (2012). ISSN 17908140. Available at http://www.jnaiam.org/uploads/Volume7_Issues_3-4_Part_III.pdf.
3. E. Kalnay, *Atmospheric Modeling, Data Assimilation and Predictability*. Cambridge University Press (2003).
4. L. D'Amore, R. Arcucci, L. Carracciuolo and A. Murli, A scalable approach for variational data assimilation, *Journal of Scientific Computing* **61**(2), pp. 239–257 (2014). ISSN 0885-7474. doi: 10.1007/s10915-014-9824-2. Available at http: //dx.doi.org/10.1007/s10915-014-9824-2.
5. R. Arcucci, L. D'Amore, S. Celestino, G. Scotti and G. Laccetti, A parallel approach for 3D-variational data assimilation on GPUs in ocean circulation models, *International Journal of Computer, Electrical, Automation, Control and Information Engineering* **9**(5), (2015). ISSN 1307-6892. Available at http://waset.org/publications/10001417/a-parallel-approach-for-3d-variational-data-assimilation-on-gpus-in-ocean-circulation-models
6. J. Dongarra and P. Beckman *et al.*, The international exascale software roadmap, *International Journal of High Performance Computer Applications* **25**(1), (2011). ISSN: 1094-3420. Available at http://www.exascale.org/mediawiki/images/2/20/IESP-roadmap.pdf
7. C. Zhu, R. H. Byrd, P. Lu and J. Nocedal, Algorithm 778: L-BFGS-B: FORTRAN subroutines for large-scale bound-constrained optimization,

ACM Trans. Math. Softw. **23**(4), pp. 550–560 (1997). ISSN 0098-3500. Doi: 10.1145/279232.279236. Available at http://doi.acm.org/10.1145/279232.279236.

8. H. P. Flatt and K. Kennedy, Performance of parallel processors, *Parallel Computing* **12**(1), pp. 1–20 (1989). Available at http://dblp.uni-trier.de/db/journals/pc/pc12.html#FlattK89.
9. L. Carracciuolo, L. D'Amore and A. Murli, Towards a parallel component for imaging in PETSc programming environment: a case study in 3-D echocardiography, *Parallel Comput* **32**(1), pp. 67–83 (2006). ISSN 01678191. Doi: 10.1016/j.parco.2005.09.001. Available at http://dx.doi.org/10.1016/ j.parco.2005.09.001.

Chapter 33

Creating and Managing a Federated Cloud to Support Science Experiments

G. Andronico*,‡, M. Fargetta*, S. Monforte*, M. Paone* and M. Villari†

**Istituto Nazionale di Fisica Nucleare, Sezione di Catania, Italy*
†Università degli Studi di Messina, Italy
‡giuseppe.andronico@ct.infn.it

Cloud computing has evolved from a promising approach to the service provisioning to the reference model for all new data centers to build. Additionally, an increasing number of companies are choosing to migrate their business in the cloud "ecosystem" adopting the solutions developed by the biggest public Cloud Service Providers (CSPs). Smaller CSPs build their infrastructure on technologies available and to better support user activities and provide enough resources to their users, the federation could be a possible solution. In this work, we present different federation models, showing their strengths and weakness together with our considerations. Beside the highlighted existing federation models, we show the design of a new implementation under development at INFN (Italian National Institute for Nuclear Physics) aiming at maximizing the scalability and flexibility of small and/or hybrid clouds by the introduction of a federation manager. This new component will support a seamless resources renting on the base of acceptance of federation agreements among operators.

Additionally, we will discuss how the implementation of this model inside research institutes could help in the field of High Energy Physics with explicit reference at LHC experiments, digital humanities, life sciences and others.

1. Introduction

Small and medium clouds, to increase the adoption of their services, have to provide improvement for some cross cutting aspects, such as *availability*, *governance*, *interoperability*, *performance*, *portability*, *privacy*, *regulatory*, *security* and many others. These aspects are difficult to manage for single cloud operators and could be even worse when different operators trying to collaborate in order to increase the portfolio of resources and services made available. Therefore, implementing models for Intercloud Interoperability and Federation, and enabling the operator to facilitate their activities supporting the above aspects are the important roles in the current scenario.

Practical approaches to federation does not supply with any clearly defined real example leading to some sort of semantic clash on what federation means. In other words, some clouds declare to be federated because of a shared file-system or other distributed or replicated service. This is not true, and in order to understand the idea behind our approach, it is important to keep in mind the following assertion: federation and resource sharing are two distinct concept with different meaning. So what is federation and what we want federated clouds to act as? Let's start simply from a federation definition taken from a common dictionary:

> *Act of joining states or other groups with an agreement in common affairs they will be governed under one central authority.*

Translating this sentence into the cloud world is the idea underlain the proposed approach to federation; nothing is shared among federated clouds members; they have their own resources, users and autonomy but given the federation agreement they belongs to, each member supply the federation with its own resources.

Some standard organizations have started to recognize the need for a standardization in the area of federation cloud. This is very important because standard can facilitate the cooperation of different operators. Among these organization, an important role has the *IEEE Standard Association* which is working on Intercloud Interoperability and Federation with its project named *P2302 — Standard for Intercloud Interoperability and Federation (SIIF)*.[9] The SIIF project aims at developing standard methodologies for cloud-to-cloud interworking and it will be interesting to evaluate how its outcome will impact the federation of clouds but sometime is still needed.

In this work, we describe a model of cloud federation that are able to provide scalability and flexibility to a group of small clouds. Creating a federation among small cloud operators with heterogeneous and different administration domains and technologies raises many problems. However, it provides business benefits exceeding the drawbacks because the federation as a whole, and so each member, can compare with big cloud players, thanks to the possibility of accessing seamless resources according to federation agreements among the federated operators. The work is in a preliminary stage, but it represents a starting point for investigating and formalizing a model and able to consider all implications in accomplishing and managing Dynamic Cloud Federations. The model, aimed at small cloud operators, allows them to easily join and leave the federation minimizing all possible issues due to the evolving configurations. Moreover, the added-value of this work is in providing a concrete model that looks at heterogeneous cloud systems, in order to include in the federation different cloud middleware (e.g., OpenNebula, CloudStack, etc.).

The paper is organized as follows: Section 2 describes a brief survey on cloud federation models and useful positioning of our work with respect to the State of the Art. In Section 3, we make clearness on the concept of federation, distinguishing it from interoperability and orchestration, presenting the general idea of federation we are dealing with in this paper. Our model is presented in Sec. 4. Finally, some indication on the goal for INFN (Italian National Institute for

Nuclear Physics) and the aspected impact in the area of the physics research is discussed in Sec. 5. Section 6 concludes the work providing highlights for the future.

2. A Survey on Cloud Federation Models

Cloud federation refers to mesh of clouds that are interconnected by using agreements and protocols necessary to provide a universal decentralized computing environment. Introducing the federation concept is raising many challenges in different research fields on cloud computing (see Refs. 1, 8, 17 and 18). Most of the works in the field concerns the study of architectural models that are able to efficiently support the collaboration between different cloud providers focusing on various aspects of the federation.

The FP7 European founded project RESERVOIR[15], which operates at IaaS introducing an abstraction layer allowing to develop a set of high level management components that are not tied to any specific environment. Therefore, several sites can share physical infrastructure resources creating a kind of federation, with the condition that all the involved clouds have a homogeneous environment. The experience acquired in RESERVOIR leads up to the latest EU initiative known as FI-Ware.[6] In particular, the EC is encouraging a federated framework based on FI-Ware platform called XI-FI Federation.[7] Indeed, XI-FI federates homogeneous FI-Ware systems based on OpenStack framework. It is noteworthy that the work has been done in the area of formalization of federation cloud components, which are: *Federate Security*, *Federate Resources*, *Monitoring Resources* and *Define Scalability Rule.* However, XI-FI Federation maintains a static approach for making up the early phases of federation. XI-FI needs to formalize *a priori* agreements among the cloud parties interested in joining the federation.

In the work (see Ref. 2), the authors describe an architectural solution for federation by means of the Cross-Cloud Federation Manager (CCFM), a software component in charge of executing three main functionalities: (i) discovery, which allows to exchange information on federated Clouds, (ii) match-making, which performs the best choice of the provider that can loan its resources, and (iii) authentication, to create a secure communication channel among federated Clouds.

Despite the obvious advantages, the implementation of a federated environment is not trivial at all. Even the OpenStack framework[13] is looking at the possibility to federate two or more OpenStack clouds. In particular, OpenStack initiative, is investigating on *InterCloud Resource Federation Models* as described in Ref. 14, where the InterCloud Resource Federation Alliance is formalized. In brief, the idea presented is to give partners investing in a joint venture the opportunities to make a bigger cloud entity with massive resources capacity. OpenStack foundation realized that security is one of the main challenge in cloud federation, as from the first item within the list of issues to be overcome in the presented assessments:

> *Security: as Tokens management, Single Sign On features, Resource Access Across Clouds, Data Export Control, etc.*

Therefore, researchers are looking at the possibility to federate users and policies as presented in Cloud Infrastructures[10] exploiting Virtual Organization Membership Service (VOMS) originally conceived for GRID computing. It is also interesting to see works trying to federate Keystones, the Identity and Access Management systems of OpenStack like reported in Refs. 3 and 16. The complexity of Intercloud Architectures is well described in Ref. 4, where an architectural framework for cloud based infrastructure services provisioned on-demand is presented.

3. Reference Scenario

To face the issues concerning *cloud federation*, two aspects have been isolated and investigated separately by the scientific community. One aspect focuses on cloud interoperability, which mostly consist in the action of devising protocols that can access cloud services on different software systems (e.g., OpenStack, OpenNebula, EC2, etc.), thus the main effort is devoted to the design communication protocols and resources dissemination policies (e.g., the EGI federated cloud[5]). This activity has involved several standardization organization and produced standards like OCCI[11] and others. The other aspect deals with the definition of the entities operating in the cloud federation, and the actions these entities need to perform in order to manage the system (e.g., Federation joining, service negotiation, SLA monitoring, etc.).

In our opinion, the approach focused on interoperability requires a centralized entity which receives requests from a CC that have to access the federated cloud resources and translates them into requests to external cloud service providers (CSPs) (see Fig. 1). Actually, this model is not a cloud federation following the definition presented in this paper, since users are aware of the different CSPs and there is no cooperation among CSPs. Generally, this approach will imply that users need to adopt different software interfaces to access either their own internal resources or the external ones offered by "federated" sites. Hence, users are divided into *internal users* and *federated users*: the former access cloud services through native APIs, whereas the

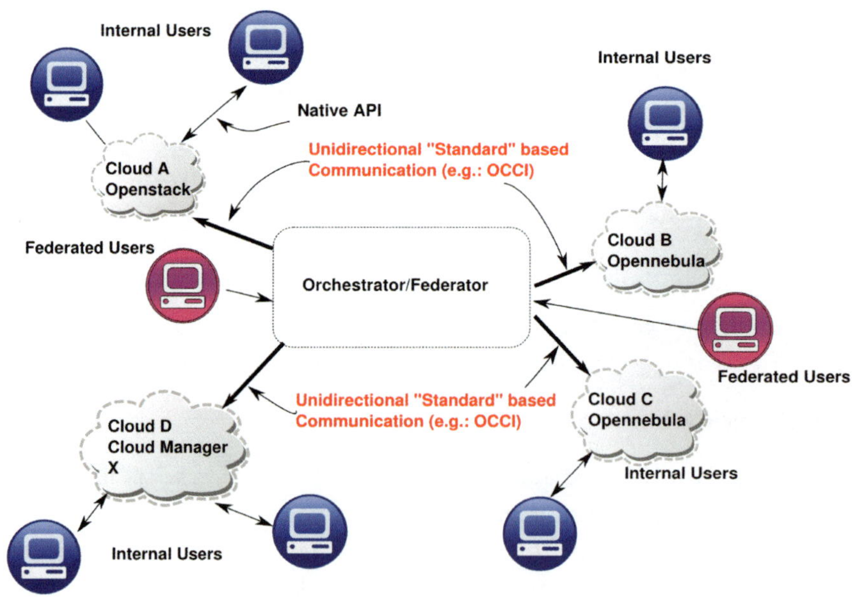

Fig. 1. Centralized approach to the federation: the users' requests are translated and forwarded to external CSPs by a central entity.

latter interact with the central entity through federation specific APIs. This simplification of federation presents some issues that can be critical in specific scenarios. Internal users cannot extend their cloud resources by taking advantage of federated CSPs, because they need another external software system that, in turn, will access resources not related to their own cloud. Additionally, internal users may have applications developed on cloud manager specific APIs, thus in order to exploit the federated resources, such applications have to be rebuilt. Nevertheless, each cloud may provide different services or interfaces (e.g., event notification or monitoring service), which cannot be available on the federation system.

Differently, the approach focused on the entities definition, as described above, allows to design a system which can transparently extend each cloud including external resources. Figure 2 depicts a simple scheme of such a system. This model implies no distinction

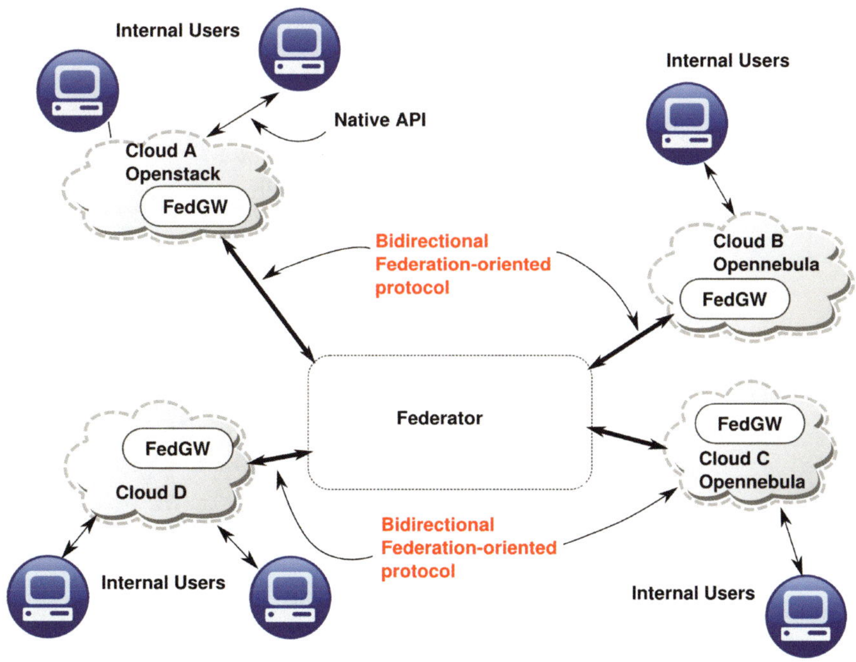

Fig. 2. An user-transparent approach to federation: each cloud extends its resource using external federated resources.

between internal and federated users: it defines only *cloud users*, which can access the resources offered by both their own CSP and external federated ones through cloud native interfaces. Most of the harmonization work among the federated CSPs is performed by software running on each site (represented by the FedGW graphical block in the figure). The entity Federator will carry out operations like resource discovery, marketplace of image templates, and so on.

To better understand the roles played by each actor, let us consider the scenario depicted in Fig. 3, where clouds A, B and C are small CSPs, whereas clouds D and E are big enough to internally address any request. CSP D is distributed around the world and its internal interconnection is depicted (link between D and D′).

Cloud brokers act as third part intermediary agents that make their business selecting the best solution satisfying both the CSPs and

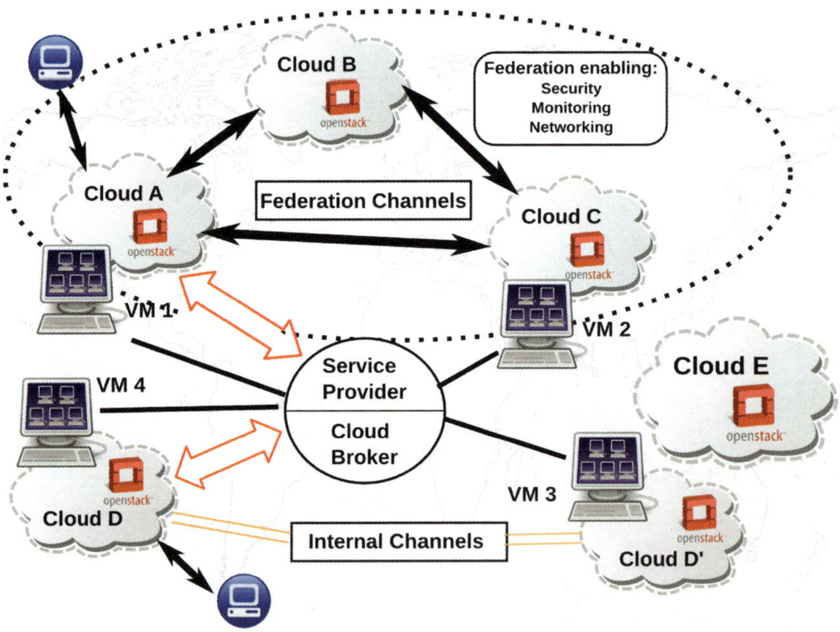

Fig. 3. IaaS: Scenario with stand-alone Clouds (D and E), Federated Clouds (A, B and C) and Cloud Brokers, Service Providers and Customers. Highlights of *In-Federation and Internal Channels*. Moreover, Cloud Customers interacting respectively with Cloud A and Cloud D.

SPs' requirements. Our interest is to provide clouds A, B and C the same type of business opportunities as for clouds D and E, in which neither brokers nor SPs might be aware of the capabilities each cloud operator supply with.

In this work, we analyze the steps required to define a federation agreement under which the cloud operators A, B and C can cooperate maintaining different administration domains. The model presented next treats all the solutions in a general way, hence they can be used also for different cloud middleware (e.g., OpenStack, OpenNebula, CloudStack, etc.).

Since the federation does not involve neither SPs nor brokers, it is necessary to setup a common federation framework where all rules and policies are respected. This transparency makes the task difficult considering issues such as compound SLAs (i.e., final SLAs towards SPs is made from a composition of more SLAs) or different network facilities. However, despite the complexity, such a federated environment allows CSPs to make new business leveraging their internal infrastructure, but also external renting resource. Thus, each CSP is able to satisfy their customers' demands and making profit of unemployed resources by providing them to other CSPs.

We remark the compelling work here is to investigate and formalize a model able to consider all implications required to accomplish all the goals of the federation described above. Section 4, "Proposed Cloud Federation Model", provides all highlights to overcome the problems discussed, trying to minimize all issues due to the evolving configurations of networks, security and monitoring parts and so on.

4. Proposed Cloud Federation Model

In Section 3, "Reference Scenario", several approaches to cloud federation have been presented, each and every one having different peculiarities. Nevertheless, none of them comply with our interpretation of the federation leading us to define our own reference model.

In accordance with our model, cloud federation lifecycle comprises of two distinct moments: join/exit and the resources access. The former is related to the activities performed by a CSP to create

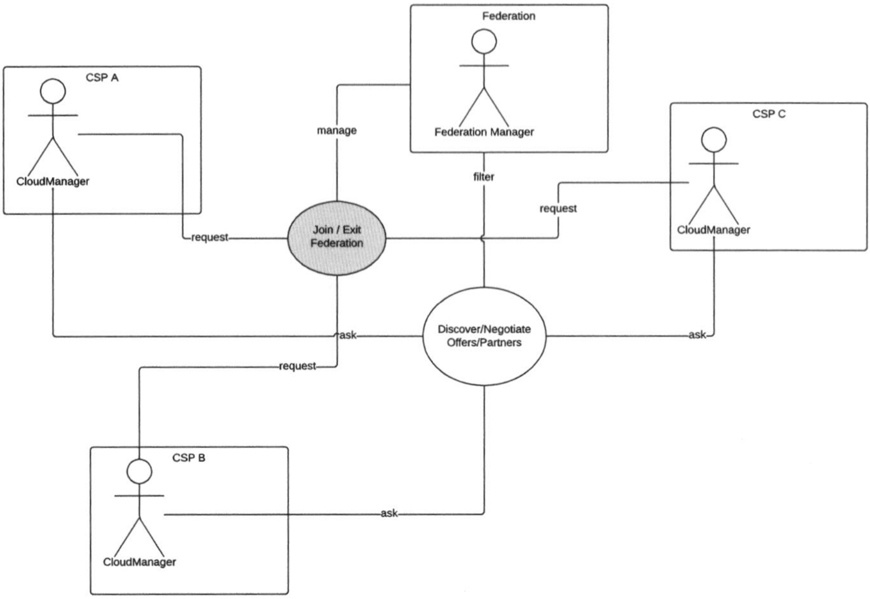

Fig. 4. Cloud federation model — use case.

or destroy the environment needed by the federation members to communicate each other. The latter is related to the discovery, negotiation and usage of federated resources. The relation between the two moments as well as the actors involved is shown in Fig. 4.

To join a federation, the CSP has to follow several steps as shown in the state diagram depicted in Fig. 5. In the first state, the joining CSP contacts the federation manager sending information about the resources (e.g., cores, storage, etc.) which might potentially be available to the federation parties as well as usage policies on those resources. Federated resources are not dedicated for exclusive use by the federation members but these are upper bounds of the resources available and its real usage depends on the actual request during federation lifecycle.

The federation manager, upon join request reception, checks whether the information provided about resources and policies matches with the federation rules or not. If the request is accepted, the just joining member is instructed to create a *tenant* with the

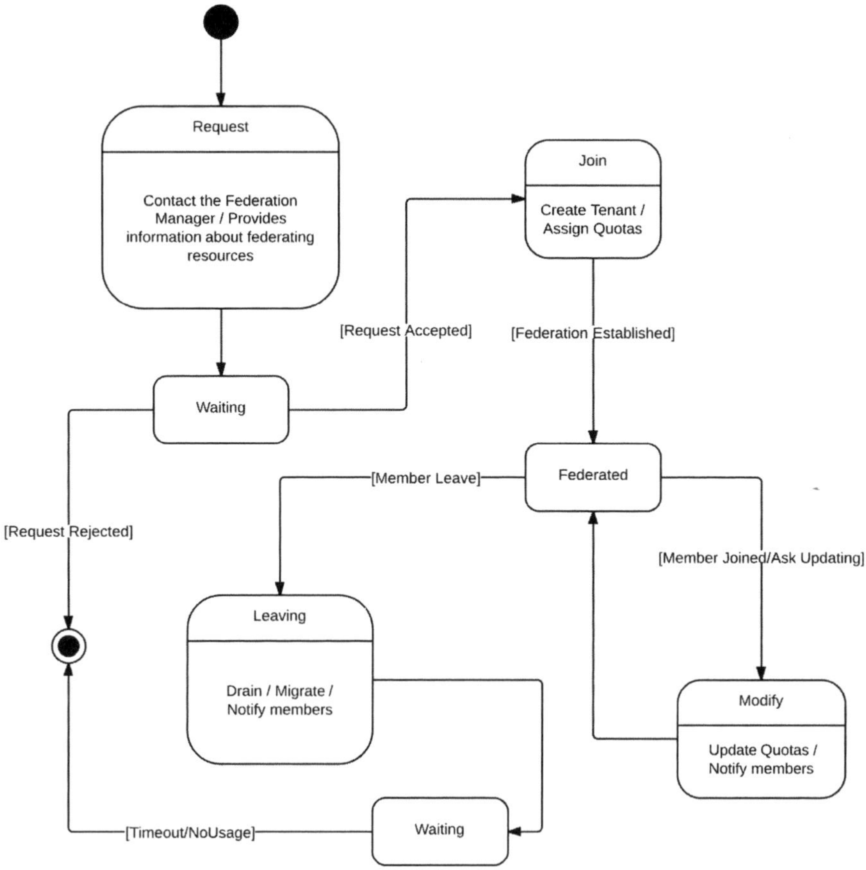

Fig. 5. Join and exit federation — State Diagram.

resources declared in the join request. At this point, the CSP can be considered as being federated and ready to fulfil requests from/to others federation members.

A CSP can modify the amount of resources committed to the federation and the policy in any moment but the changes must be notified in advance to the federation manager, who will propagate the information to all members. Obviously, during the information update, the federation can reject a member because it does not comply any more with the rules. A CSP can leave the federation, either for its or the federation manager decision. The federated CSP cannot leave immediately since some resources might be committed and still used, therefore the CSP enters in a leaving state. This state will terminate when all the resources are free or if the leaving period defined in the joint agreement has expired, in this case the resource will be forcibly released and remaining data or services discarded. The federation manager notifies all the members about the current disconnection of the CSP. The members have to release the resources the leaving CSP supply with before expiring of a given timeout period.

The federation defines the technical aspects in order to access remote resources and maintains a list of CSPs providing resources with both qualitative and quantitative information. Nevertheless, in order to access member resources, a new negotiation is requested between the two members, acting one as CSC and the other as CSP, with the supervision of the federation manager acting as CFA. Figure 6 shows the state diagram related with the discovery and negotiation of federated resources.

To access federated resources, the cloud manager has to send a request to the federation manager including the list of requested resources (e.g., number of cores, storage or other) and specify an optimization function used to pick the best fit among the possible results the CSPs supply with.[a] The optimization function contains constraints related to the resources, like performance, location, reliability, etc., as well as parameters describing QoS/SLA constrains. The federation manager, upon reception of the request, queries the members who are able to provide the relevant resources based on the information published in the federation about current availability and prices. The optimization function is then applied to select the best fit for the waiting CSC. This activity is performed automatically and unattended, so it does not require any human interaction.

The cloud manager can reject the offer selected by the federation manager and then could send a new request with a different optimization function. If the offer is accepted, an agreement has to be established before the CSC can access the resources. The agreement

[a]The request is an XML document based on WS-Agreement and include RDF elements for resource description.

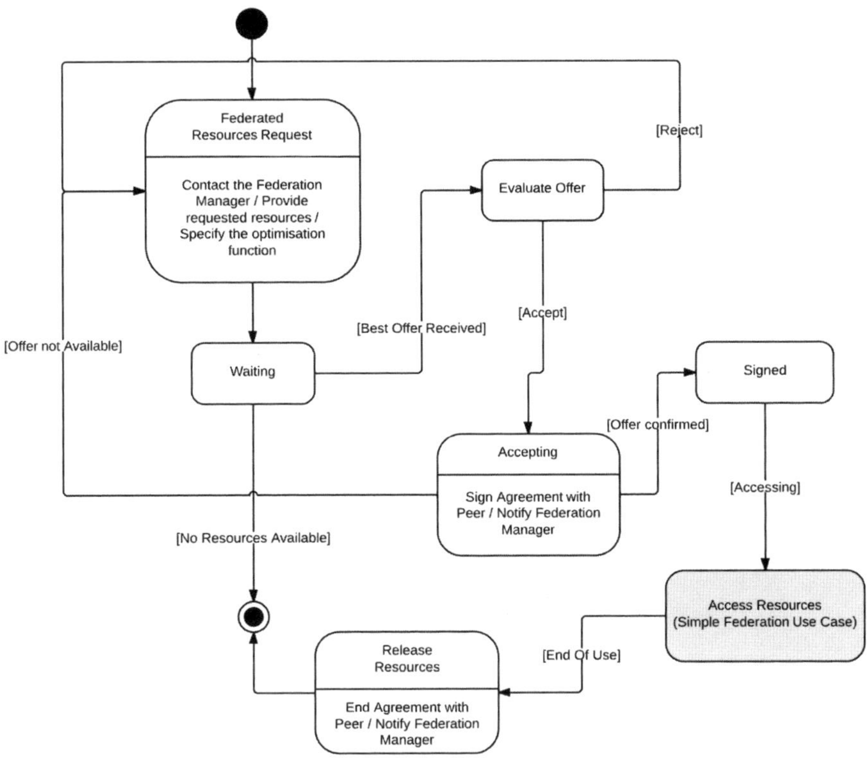

Fig. 6. Discovery and negotiation of federated resources — State Diagram.

is an XML document based on WS-Agreement,[12] which has to be signed by the two parties and the federation because it is responsible for all the relations among its members. Therefore, the federation manager is notified when the agreement takes place and is over, as well. This allows the federation manager to have full knowledge of the resources usage among the members and implements strategies for a better distribution and optimization of workload in the federation.

After the agreement is signed, the CSC can start deploying services on the resources of the CSP, upon user requests. The deploy and access to remote resources by the user is shown in Fig. 7 and described below. Resources under the agreement are reserved to the CSC and cannot be used by the owner.

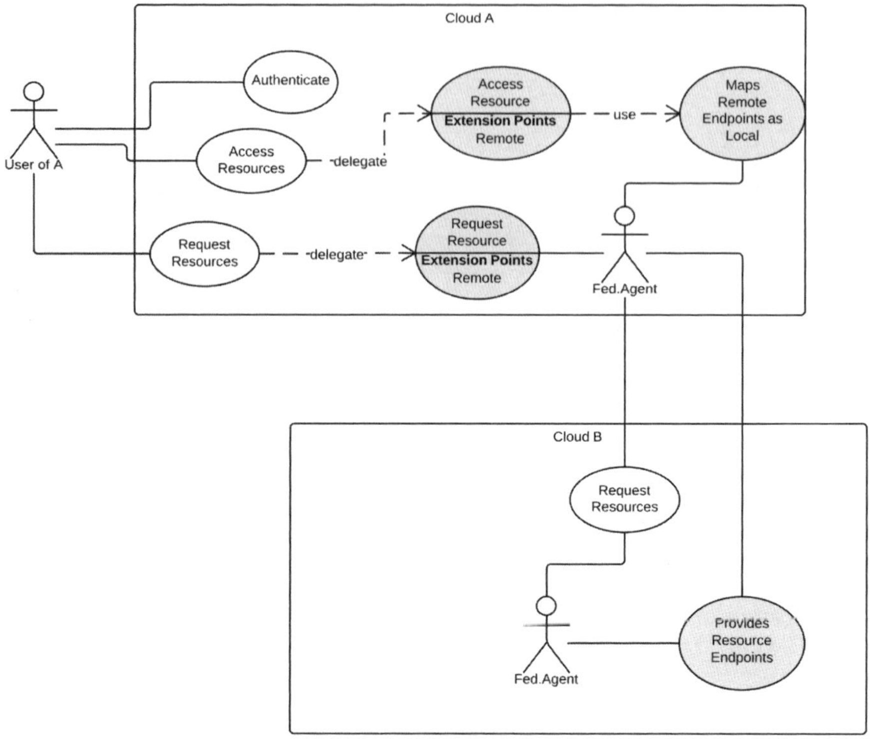

Fig. 7. Request and access federated resources — use case.

During normal operation, shown in Fig. 7, when a cloud user, or any CSC, requires new resources, the cloud manager discriminate whether these will be provided as internal resources or taken from the federation. In the former scenario, resources are managed by the CSP as usual. In case of federated resources, the cloud manager will become a CSC of the federation and will start the negotiation procedure described above. These operation are internally managed by a federation agent inside the cloud. Upon agreement establishing, the required resources are committed in the remote CSP and the relevant endpoints sent to the federation agent who will activate a mapping service to generate local endpoints for the users. The mapping is requested to masquerade the real location of the services. As a result, the user can access the services transparently as resources managed by the cloud itself, hiding to the user the real owner of the resources and their location, which is an important aspect of the federation.

Finally, agreements could be defined in advance, before users request new resources. Moreover, users might release requested resources before the actual expiration of the corresponding agreement, thus leaving them unused within the owning cloud. Hence, internal policies of federated resources usage should be defined and pursued by each member.

5. Cloud Federation for Research Institutes

Many research institutes, like *INFN*, where this work has been developed, have moved to cloud based solutions to support both their internal services (e.g., mail, web, etc.) and the computation needed to run experiments. In fact, computation and storage play a crucial role in the modern scientific investigation and cloud platform has simplified their access for the scientists removing the physical limitation, whereas cloud services are accessed from everywhere by everyone.

INFN manages some sites dislocated in different Italian cities and each site has its computing facilities for the local experiments and, at the same time, provides them to bigger experiments like ALICE, CMS and other.[b]

Many experiments require a well-defined amount of resources during its lifetime whereas other have some peak requests only in specific moment of the experiment. This makes difficult for a site to define the amount of physical resources to deploy to implement its cloud, avoiding waste of moneys and/or long waiting time to access the resources.

At INFN, the sites are following two different path: integration and federation. The integration aimed at joining small sites in a unique cloud so the sites can use resources from others. Although this help to mitigate the shortage of resources during peak requests, they

[b]ALICE and CMS are some of the experiments related with the Large Hadron Collider developed at CERN.

introduce problems in term of management. The administration will not be any more at site level and this will reduce the agility to support local experiments. Additionally, if peak requests increase resources, they can still be saturated with the result of an increased waiting time.

The other approach followed by INFN is the federation as described in this chapter. This should allow the creation of a federation among INFN sites and between these and external entities, either public or private. With this model, it should be easier to collaborate in order to run an experiment because the cloud of the partners can be federated and provide the resources needed. Nevertheless, including commercial cloud provider in the federation can create a big pool of resource to use on demand avoiding the problem of overplay the computing facilities.

6. Conclusion and Future Work

In this chapter, we have presented the idea of cloud cooperation among operators based on federation agreements. The challenge we want to address with the federation is to overcome all the problems raising in merging clouds with heterogeneous administration domains. Therefore, we introduced a high level model of cloud federation that are able to provide the scalability and flexibility needed by small clouds. The added-value of this work is in providing a high-level model not related to a specific technology which aims at federating different cloud infrastructures.

For the future, we are looking at a concrete implementation useful for testing the goodness of our model, but also for providing new features and solving real problems that may occur in cloud federation accomplishments.

References

1. S. Azodolmolky, P. Wieder and R. Yahyapour, Cloud computing networking: challenges and opportunities for innovations, *IEEE Communications Magazine* **51**(7), pp. 54–62 (2013).
2. A. Celesti, F. Tusa, M. Villari and A. Puliafito, *Three-phase cross-cloud federation model: The cloud SSO authentication*, In Proceedings of the

2010 Second International Conference on Advances in Future Internet, AFIN '10, pp. 94–101, Washington, DC, USA, 2010. IEEE Computer Society.

3. D. W. Chadwick, K. Siu, C. Lee, Y. Fouillat and D. Germonville, Adding federated identity management to openstack, ***Journal of Grid Computing*** **12**(1), pp. 3–27, (2014).
4. Y. Demchenko, C. Ngo, C. de Laat, M. X. Makkes and R. J. Strijkers, Intercloud architecture framework for heterogeneous multi-provider cloud based infrastructure services provisioning, ***International Journal of Next-Generation Computing (IJNGC)***, **4**(2), (2013).
5. European Grid Infrastructure. EGI federated cloud. Available at https://www.egi.eu/infrastructure/cloud
6. FI-WARE. Open APIs for Open Minds. http://www.fi-ware.org, 2014.
7. FI-XIFI. Joining The Federation Scenario Exploiting FI-Ware framework. Available at http://wiki.fi-xifi.eu/Public:Joining_the_Federation_scenario, 2014.
8. I. Goiri, J. Guitart and J. Torres, ***Characterizing cloud federation for enhancing providers' profit.*** In Cloud Computing (CLOUD), 2010 IEEE 3rd International Conference, pp. 123–130, (2010).
9. IEEE. P2302 — the IEEE standards association. Available at http://standards.ieee.org/develop/project/2302.html, 2014.
10. A. Lopez Garcia, E. Fernandez-del Castillo and M. Puel, ***Identity federation with VOMS in cloud infrastructures,*** In Cloud Computing Technology and Science (CloudCom), 2013 IEEE 5th International Conference, Vol. 1, pp. 42–48, (2013).
11. Open Grid Forum. An Open Community Leading Cloud. Available at http://occi-wg.org.
12. Open Grid Forum. Web Services Agreement Specification (WS-Agreement). Available at https://www.ogf.org/ogf/doku.php/documents/documents, 2007 (update 2011). GFD-R.192 (Obsoletes GFD.107).
13. The open source, open standards cloud, innovative, open source cloud computing software for building reliable cloud infrastructure. Available at http://openstack.org, Jan 2014.
14. OpenStack inter cloud resource federation. Available at https://wiki.openstack.org/wiki/Inter-Cloud-Resource-Federation, 2014.
15. B. Rochwerger, D. Breitgand, A. Epstein, D. Hadas, I. Loy, K. Nagin, J. Tordsson, C. Ragusa, M. Villari, S. Clayman, E. Levy, A. Maraschini, P. Massonet, H. Munoz and G. Toffetti, Reservoir — when one cloud is not enough, ***Computer*** **44**, pp. 44–51, (2011).

16. D. Sitaram, H. L. Phalachandra, A. Vishwanath, P. Ramesh, M. Prashanth, A. G. Joshi, A. R. Desai, C. R. Harikrishna Prabhu, Prafulla, R. Shwetha and A. Yashaswini, *Keystone federated security*, In Internet Technology and Secured Transactions (ICITST), 2013 8th International Conference, pp. 659–664, (2013).
17. F. Tusa, A. Celesti, M. Villari and A. Puliafito, ***How to enhance cloud architectures to enable cross-federation,*** In Proceedings of IEEE CLOUD '10, pp. 337–345, (2010).
18. G. Vernik, A. Shulman-Peleg, S. Dippl, C. Formisano, M. C. Jaeger, E. K. Kolodner and M. Villari, ***Data on-boarding in federated storage clouds,*** In Cloud Computing (CLOUD), 2013 IEEE Sixth International Conference, pp. 244–251, (2013).

Chapter 34

An Integrated Monitoring System, with Ganglia and Nagios, for the Cosenza ReCaS Site

N. Guarracino[*,†], V. Lavorini[†], A. Tarasio[†,‡] and E. Tassi[*,†]

[*]*Dipartimento di Fisica, Università della Calabria, Arcavacata di Rende, Italy*

[†]*Istituto Nazionale di Fisica Nucleare, Gruppo collegato di Cosenza, Arcavacata di Rende, Italy*

[‡]*alessandrotarasio@gmail.com*

We describe the monitoring architecture built at the ReCaS Cosenza Site. The system achieves high efficiency putting together many different tools such as: Nagios, Ganglia, customized plug-ins, Centreon and NagVis. A strong integration between Nagios and Ganglia is obtained by the extensive use of scripts querying, parsing and normalizing Ganglia data, later checked and managed by Nagios. Replacing Nagios check with Ganglia, we can see any typical Ganglia metrics (but also customized ones) being measured on each host (server, storage, network apparatus). Regarding equipment and sensors outside the computer hosts (chiller, rack, other sensors), monitoring was carried out by means of queries to Simple Network Management Protocol (SNMP) MIB tables.

1. Introduction

Administrators can detect and address problems before they affect application performance by monitoring system health proactively. Especially in multi-layer infrastructures management, monitoring systems can help in critical event managements that could happen at different architectural levels. In recent years, big efforts aimed to realize monitoring systems for distributed infrastructures, working at a single level or at multiple levels, based on standard software or made *ad hoc*. Between software working at one level, we can surely cite Ganglia[1] that is the most widely used to monitor the activity of the Local Resource Management System both for clusters and for more complex systems. Another well known, open source, powerful monitoring system available nowadays is Nagios.[2] It enables organizations to identify and resolve IT infrastructure problems before affecting critical business processes. Administrators typically assess the current state of network devices in one of the two ways. The pull method queries device instrumentation from a central monitoring console at specified intervals receiving a status or explicit data value in response. Alternatively, the push method constantly reports on device status by sending Simple Network Management Protocol (SNMP) traps, data, or both from device instrumentation to a central monitoring console. Both pull and push methods offer advantages. Although Nagios can receive SNMP information, its true strength is the ability to use any application or script to gather data. Data gathering programs, often called plug-ins, are placed and run on each client system being monitored.

This Chapter focuses on Nagios and Ganglia, two open-source monitoring tools that use the pull method and can integrate each other to help manage a moderately large group of servers and all data center equipment sensors. The multi-layered and distributed monitoring architecture built at the ReCaS Cosenza Site achieves high efficiency putting together many different tools like: Nagios, Ganglia, customized plug-ins, Centreon[3] and NagVis.[4] A strong integration between Nagios and Ganglia is realized by the extensive use of scripts querying, parsing and normalizing Ganglia data, later checked and managed by Nagios. Integrating Nagios and Ganglia is a good choice in order to achieve higher efficiency and scalability for the overall monitoring system architecture in a Cloud/Grid infrastructure. There are several ways for integrating Nagios and Ganglia. We choose the simple *check_ganglia_metric* python script modifying some of its more critical aspects as the cache management for large-scale systems. The rest of the Chapter is organized as follow: in Sec. 2, we review the ReCaS Cosenza site monitoring architecture describing the different components and how they achieve efficiency and scalability. Section 3 describes the script and the proposed solution to resolve the Nagios/Ganglia integration issues. Finally, in Section 4, we summarize our results.

2. The ReCaS Distributed Monitoring Architecture

The ReCaS Infrastructure is a multi-regional e-infrastructure for High Energy Physics and e-Science in South Italy. ReCaS is composed by four data centers distributed in the sites of Bari, Catania, Cosenza and Napoli. All the infrastructure components provide sensors for remote control and monitoring. ReCaS uses the Grid and Cloud paradigms to federate the four datacenter and to offer services to the final users. To monitor grids, in past years, many *ad hoc* monitoring systems have been used: [5] GSTAT and GridICE, or Service Availability Monitoring (SAM). Some of these are monitoring systems developed for controlling, at different level, services in an EGI distributed infrastructure. These monitoring systems cannot be used for other purposes, e.g., monitoring clusters or Grid/Cloud infrastructures based on middleware different from that chosen by EGI.

Therefore, new monitoring systems are needed to manage a Grid/Cloud infrastructure. Several and accurate monitoring activities are required to efficiently operate Cloud Computing platforms and to manage their increasing complexity and security requirements. Main improvements in terms of efficiency are expected for data management. In particular, more and more efficient algorithms and techniques are needed to manage, quickly and continuously, the large volume of monitoring data necessary to have a comprehensive view of

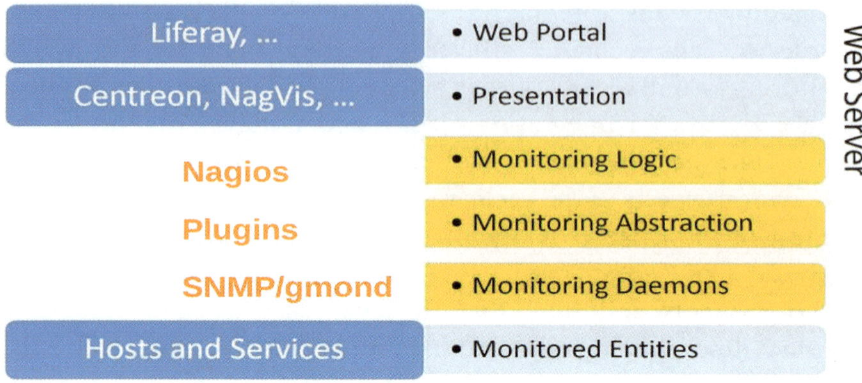

Fig. 1. ReCaS Cosenza site monitoring architecture.

the Cloud, without putting too much burden on the Cloud and monitoring infrastructures both in terms of computing and communication resources. The monitoring system should be therefore able to do several operations on data (collect, filter, aggregate, correlate, dissect, store, etc.) respecting strict requirements in terms of time, computational power and communication overhead. The multi-regional ReCaS monitoring architecture organize components into layers as shown in Fig. 1.

The software components of each layer use the information and capability provided by the lower level. At the lowest level of our architecture, we have the Monitored Entities that represent a wide range of monitoring resources present in the data center, that can be subdivided in two main categories: computer equipment (server, storage, network apparatus) and infrastructure elements (chiller, rack, other sensors). In the Monitoring Daemon layer, we collect all the components responsible to retrieve and publish information, including information providers and services like SNMP but also all the Ganglia metrics provided by the *Gmond* daemons running on each computer host. The Monitor logic layer provides advanced systems that are able to manage data and plug-ins. It is responsible for scheduling measurements, retrieving the real-time value, plotting graphs, managing the alarms and offering statistics. Nagios provides complete monitoring of Cloud Computing, web and storage services.

Nagios is capable of monitoring a variety of servers and operating systems — both physical and virtual. Moreover, it can also monitor all the infrastructure elements sensors querying every SNMP-device present in the data center. The Presentation level offers services to create maps and logical views of the monitoring information. It is used to improve the usability of the software component of the previous level. The final layer of the ReCaS Monitoring architecture is represented by the WebPortal, responsible to create an access point able to integrate different monitoring systems and presentation view.[6] Nagios, Ganglia, customized plug-ins, Centreon and NagVis work together to build our monitoring architecture as showed in Figs. 2 and 3. Each of the four sites has the same monitoring architecture as indicated in Fig. 1 but different software implementation. The ReCaS Cosenza Site's architecture implementation relies on two principal components: Nagios and Ganglia and on their strong integration as monitoring daemons.

Fig. 2. Nagvis shows some map metrics gathered from Nagios.

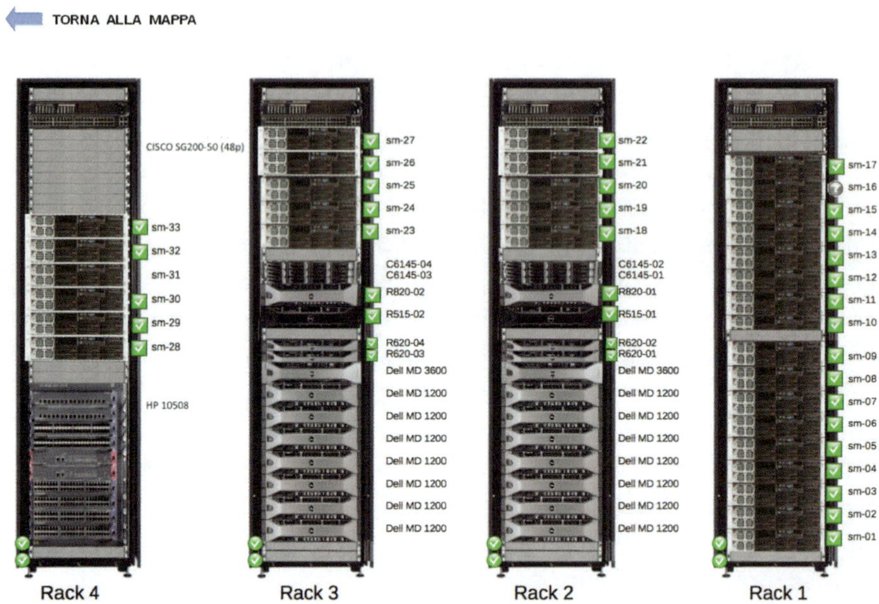

Fig. 3. Nagvis shows some room metrics gathered from the *check_ganglia* plugin.

2.1. *The Nagios Components*

Nagios is designed to quickly inform system administrators of problems. It uses Linux shell scripts and executables to retrieve and report status information. Nagios has a built-in notification system. Nagios, like Ganglia, is used heavily in HPC and other environments. While Nagios is more specifically an alerting mechanism, Ganglia is more focused on gathering and tracking metrics. Nagios previously only polled information from its target hosts, but has recently developed plug-ins that allows it to run agents on those hosts. Although Nagios can receive SNMP information, its true strength is the ability to use any application or script to gather data. Data gathering programs, the above mentioned plug-ins, are placed and run on each client system being monitored. Plug-ins return relevant data and one of the three states: "OK," "Warning," or "Critical." Administrators schedule plug-ins to run at specified polling intervals. Nagios takes the status

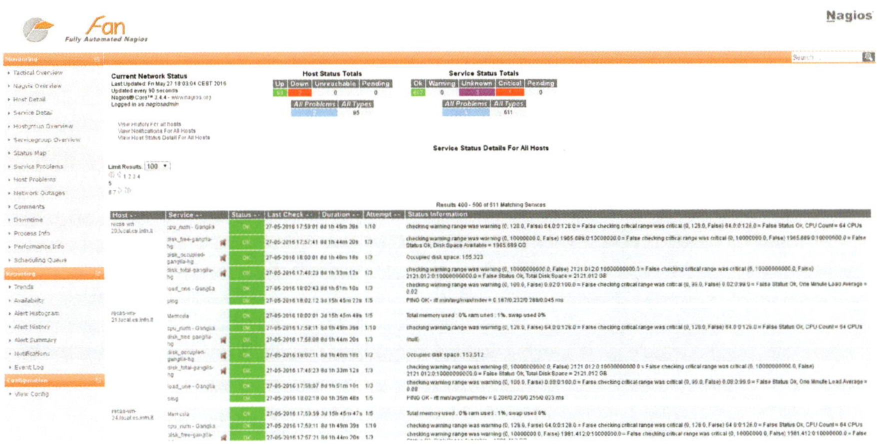

Fig. 4. An example of metrics gathered with the Nagios *check_ganglia_metric* plugin.

returned from a plugin, displays it on the Web console and stores its value in a database, enabling administrators to query the database for system status over specified time intervals. Nagios can also trigger alert actions either by sending an e-mail message or by launching a script based on the reported status. Figure 4 shows the Nagios Web console, which displays the output of the *check_ganglia_metric* plugin run on some of ReCaS Cosenza Site's servers.

Centreon, a network monitoring software based on the Nagios tool with its user-friendly interface makes it possible to view the state of the system, especially with graphics, regarding all hosts, sensors and network metrics. NagVis makes it possible to create functional views of monitoring because it can be paired with a network diagram in order to send the Nagios data to the diagram in real-time. As we can see in Fig. 5, many different graphs are being generated that can be analyzed over time.

2.2. *The Ganglia Component*

Ganglia is a real-time, agent-based monitoring tool that started at UC Berkeley. It can easily become a scalable distributed

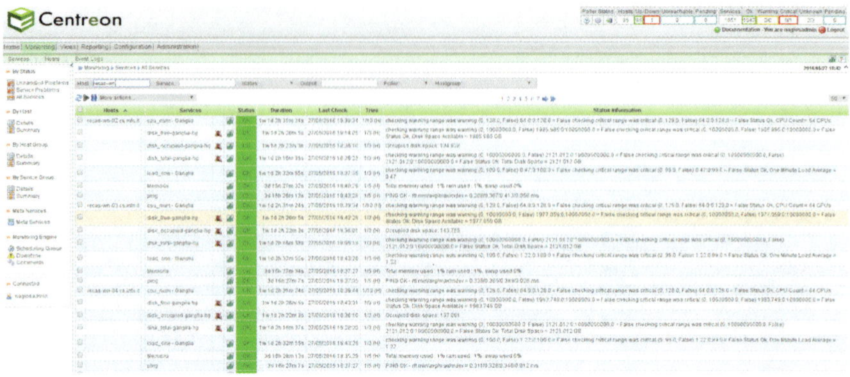

Fig. 5. The Centreon user-friendly interface.

monitoring system for high-performance computing (HPC) systems such as clusters and Grids. It uses carefully engineered data structures and algorithms to achieve very low per-node overheads and high concurrency. Each machine runs a daemon called *Gmond* which collects and sends the metrics (like processor speed, memory usage, etc.) it gleans from the operating system to a specified host. The host which receives all the metrics can display them and can pass them up a hierarchy on a condensed form. This hierarchical scheme is what allows Ganglia to scale so well.

Gmond has very little overhead which makes it a suitable element of code to run on every machine in the cluster without impacting user performance. No network traffic is exploited by running remote commands thanks to the benefits of Ganglia's scaling techniques that is able to see a large amount of output on the node from data that has been already collected on the nodes (provided Ganglia is up and running). *Gmond* is a multi-threaded daemon which runs on each cluster node that has to be monitored. *Gmond* has four roles: monitor changes in host state, multi-cast relevant changes, listen to the state of all other Ganglia nodes via a multi-cast channel and answer requests for an XML description of the cluster state. As described in Fig. 6, each *Gmond* transmits in two different ways: multi-casting host state in external data representation (XDR) or sending XML over a TCP connection. Many custom metrics can be sent on the Ganglia

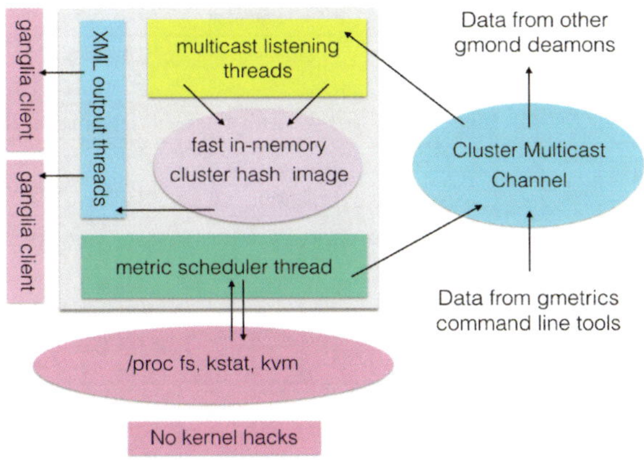

Fig. 6. Diagram of the Ganglia Monitoring Daemon (*Gmond*) with all its components.

multi-cast channel. To expand the list of metrics that can be monitored, the *Gmetric* tool is used.

Since a *Gmond* daemon can run on every machine, we choose to use Ganglia as the pathway for reporting performance data about all our IT equipment. This has several benefits as an easy integration for performance monitoring, a powerful analysis tools and a more simple and flexible integrated alerting system. The Ganglia Meta Daemon (*Gmetad*) works with Ganglia Monitoring Daemons (*Gmond*) to allow an easy cluster monitoring over unicast routes. While *Gmond* uses multi-cast channels in a peer-to-peer way, *Gmetad* pulls the XML description from Ganglia data sources (either *Gmond* or another *Gmetad*) via XML over unicast routes. *Gmetad* (a sort of backend for the Ganglia web frontend as we can see in Fig. 7) stores historical information to Round-Robin databases and exports summary XML which the web frontend uses to present useful snapshots and trends for all hosts monitored by Ganglia.

Our Ganglia infrastructure is strictly tied in with our Nagios alerting system. Once an alert is configured, an application can trigger an alert state by generating a specific metric with a sentinel value. This allows a centralized alert generation and event correlation. In Fig. 8, some useful Ganglia metrics are represented.

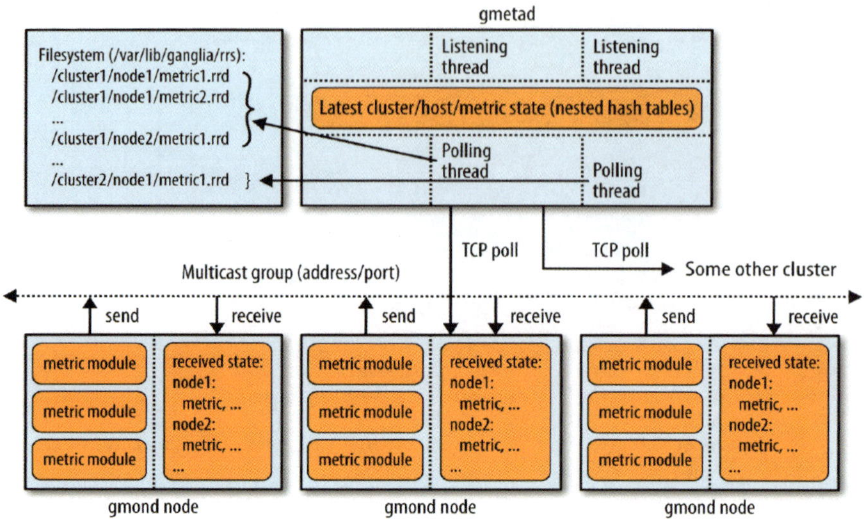

Fig. 7. A Ganglia Meta Daemon (*Gmetad*) example.

Fig. 8. The load one metric for a group of servers, as showed by the ganglia web GUI.

3. The ReCaS Cosenza Site Integrated Monitoring System

Because Nagios has no built-in means of polling data from remote hosts, Nagios users have historically employed various remote execution schemes to collect a litany of metrics with the goal of comparing them against static thresholds. These metrics, such as the available disk space or CPU utilization of a host, are usually collected by services like NSCA or NRPE, which execute scripts on the monitored systems at the Nagios server's behest, returning their results in the standard Nagios way. The metrics themselves, once returned, are usually discarded or in some cases fed into RRDs by the Nagios daemon. This arrangement is expensive, especially considering that most of the metrics administrators tend to collect with NRPE and NSCA, are collected by *Gmond* out of the box. If you're using Ganglia, it's much cheaper to point Nagios at Ganglia to collect these metrics. There are five Ganglia plug-ins currently available: (i) check heartbeat, (ii) check a single metric on a specific host, (iii) check multiple metrics on a specific host, (iv) check multiple metrics across a regex-defined range of hosts and (v) verify that one or more values is the same across a set of hosts. We choose to check a single metric on a specific host by a python script plugin.

The *check_ganglia_metric* plugin compares a single metric on a given host against a predefined threshold. *Check-ganglia-metric* is a simple Nagios plugin that collects ganglia data from *Gmetad* and transparently stores it in a local cache file.[7] This script requires some modification because as the number of Ganglia hosts grows, an heavy load appears on the Nagios server, due to the size of the cache file and the time it takes to refresh. Our customized plugin checks the cache and some services to implement Ganglia metrics monitoring with high availability. Then we add our plugin to the Nagios check command configuration so that a service check can be set for every host that require monitoring.

By replacing Nagios check with Ganglia, we can see the typical Ganglia metrics (but also customized ones) being measured on each host. Nice graphics are being generated that can be analyzed over time. Ganglia can afford problems of big scale problem using its high flexibility and scalability. For everything regarding equipment and sensors outside the computer hosts, monitoring was carried out by means of queries to SNMP MIB tables. There are different ways to integrate Ganglia and Nagios in distributed monitoring systems. The first one is ganglia-web/nagios[8] (PHP & bash based). Another technique is ganglia-nagios-bridge[9] (Python & cron based). The one we preferred is *check-ganglia-metric*[10] (Python). *Check_ganglia_metric* is basically a Nagios plugin that pulls data from Ganglia XML via sockets by a command using the following syntax:

```
check_ganglia_metric.py --gmetad_host=gmetad-server.example.com
    --metric_host=host.example.com --metric_name=cpu_idle
```

check_ganglia_metric is a Python script running on the same server where *Gmetad* is running (in our case, this is different from the management server where Nagios resides). So we only need to query the localhost on *Gmetad* port 8649. This is why there is no need to run remote commands wasting network traffic. This is one of the major benefit of Ganglia's scaling techniques.

4. Conclusions

The distributed monitoring architecture built at the ReCaS Cosenza Site achieves high efficiency putting together many different tools like: Nagios, Ganglia, customized plug-ins, Centreon and NagVis. A strong integration between Nagios and Ganglia is realized through the extensive use of scripts querying, parsing and normalizing Ganglia data, later checked and managed by Nagios. Ganglia is an incredibly powerful and flexible tool, but at its heart it is designed to collect performance monitoring data about the machines in a single HPC cluster and display them as a series of graphics in a web interface. Nagios is the most used monitoring tool, richly featured and easily integrable but definitely designed to gather all device instrumentation, especially the equipment and sensors outside the computer hosts. The ReCaS Cosenza Site Integrated Monitoring System has taken these two software and extended them across a complex infrastructure, made them work together taking advantage of their potential to reach major efficiency and scalability. Ganglia and Nagios combined together are our core monitoring solution to efficiently separate the two main data center items categorized as: IT equipment and sensors outside the computer hosts.

Acknowledgments

This work has been funded by PON 2007–2013, project code PONa3_00052.

References

1. The Ganglia Monitoring System, Available at http://ganglia.sourceforge.net/
2. Nagios, Available at http://www.nagios.org/
3. Centreon Documentation: Available at http://documentation.centreon.com/
4. NagVis project Documentation 1.4: Available at http://docs.nagvis.org/1.4/en_US/
5. P. Andreade *et al.*, Service Availability Monitoring Framework Based On Commodity Software, CHEP12, March 2012.
6. D. Del Prete, S. Pardi and G. Russo, *A Centralized, Extensible, Multilayer Monitoring System for Distributed Infrastructures: the Atlas Tier2-Naples experience*, In IEEE Proceeding of CCP2011, June 21–24, 2011, Italy. DOI 10.1109/CCP.2011.35
7. M. Massie *et al.*, *Monitoring with Ganglia*, O'REILLY (2012).
8. The ganglia-web/nagios plugin, documentation at https://github.com/ganglia/ganglia-web
9. The ganglia-nagios-bridge plugin, documentation at https://github.com/ganglia/ganglia-nagios-bridge
10. The check-ganglia-metric plugin, documentation at https://github.com/ganglia/ganglia_contrib

Chapter 35

Deploying a Stable Release of the SFINGE System Using the ReCaS Facility

M. Castellano*,†,‡, N. Corriero*, M. A. Tangaro,*
T. Del Vino† and G. Zaccheo†

**Politecnico di Bari,*
Dipartimento di Ingegneria Elettrica e dell'Informazione
†*Genesis Consulting s.r.l*
‡*marcello.castellano@poliba.it*

This Chapter describes a Solution Framework for Interoperable Network Grid E-learning (SFINGE) and its integration in the ReCaS facility through the PRISMA cloud services, to create a stable release of the platform during the prototype development.

1. Introduction

The Solution Framework for Interoperable Network in Grid E-learning (SFINGE)[1] proposes a collaborative approach for e-learning development[2–5] involving multiple organizations. The Virtual Organization (VO) workflow is developed in a controlled environment, using the resources shared according to the grid computing paradigm[6–9] to increase the competitiveness and the business capabilities of the involved subjects.

A collaboration with the INFN Bc^{2}S (Bari Computer Center for Science) is ongoing.

The ReCaS project[10] aims to increase the performances of the existing computing infrastructures of the involved institutions, including the Bc^{2}S, through the development of high level tools and services and the integration with national and international Grid/Cloud infrastructures. In this context, it is important that the interaction with the PRISMA cloud platform,[11] which foresee to develop a cloud platform for the management of different cloud services, supporting different use-cases, aiming to be considered the reference platform for public administration cloud services implementation. The PRISMA platform is developed using the hardware infrastructure (computing, networking, storage) of the ReCaS project.

We report here the main concepts and results of the SFINGE project and the integration of the system on the Open Stack platform of the PRISMA infrastructure.

2. The SFINGE Project

The e-learning process development requires a huge amount of computational resources (hardware and software), learning material and skills. Small and medium enterprises usually do not own all these resources. The Virtual Organization[12] structure, shown in Fig. 1, supports the collaborative work, aggregates hardware and software resources, individuals belonging to different institutions or enterprises, learning materials, documents and know-how, creating every time a new team to achieve common objectives.

The SFINGE system integrates different components to manage this set of heterogeneous resources:

- the web portal;
- the workflow engine;
- the grid infrastructure;
- the Learning Management System (LMS).

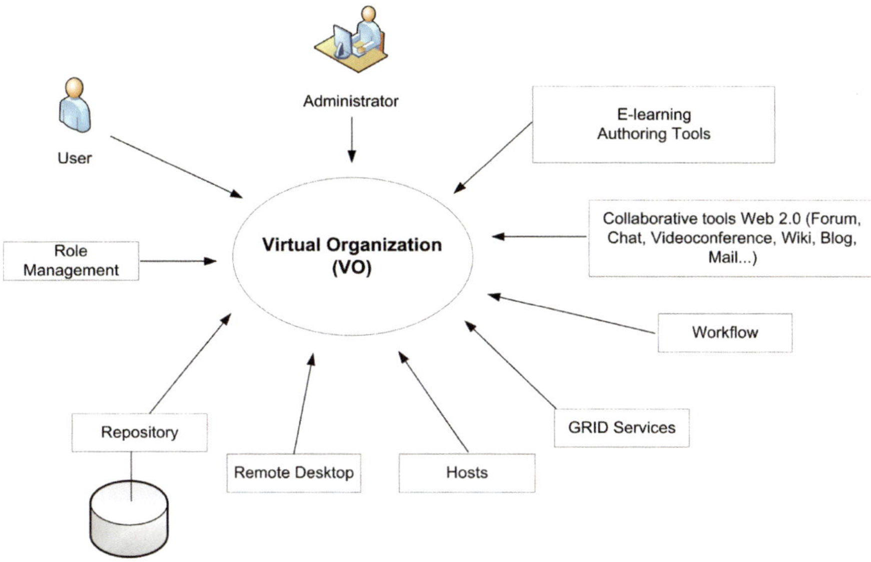

Fig. 1. Virtual Organization scheme.

The web portal is based on the Liferay[13] technology, a Java based open-source platform for portal and collaborative work management. It is the access point of the whole system and it is capable to create and manage different VOs. Moreover, since the e-learning development encompasses different phases and roles for the contents development (Fig. 2), the portal provides users and roles management. Each user has a digital identity to access, with appropriate permissions, to the portal services, to create or reuse e-learning contents. A digital identity consists of a set of digital characteristics, which will grant, through the implementation of shared policies, the identity and the right to access to the shared resources.

Some basic roles, for e-learning production,[2] have been created: project administrator, project manager, instructional designer, subject matter expert, LMS expert and evaluator. New roles can be added using the web portal interface.

To enhance and coordinate the collaborative work, the workflow technology has been exploited.[14] The workflow allows us the management of different activities for different roles, during the creation of a

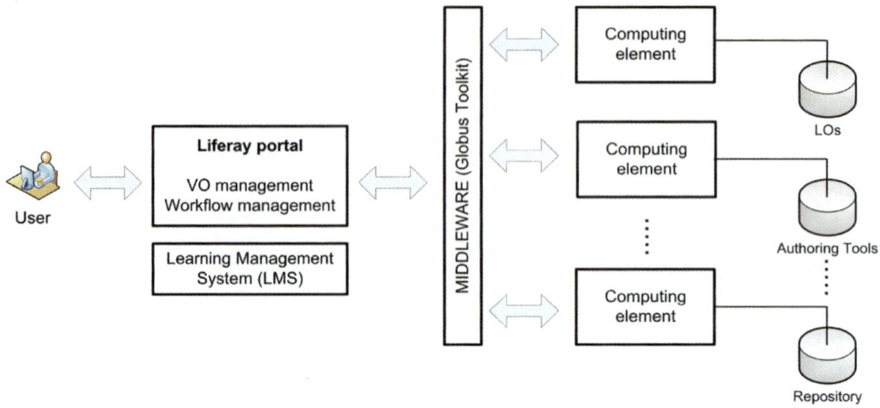

Fig. 2. The SFINGE system structure.

course, through team members' synchronization, task assignment and time control.

The workflow programming for each e-learning phase is done through the identification of the activities, the roles and the documents as depicted in Fig. 3. For instance, we briefly outline here the approval stages for each e-learning contents: when a document is created, it is sent to the users involved in the production cycle, for the review and the approval stage. If rejected, the document has to be resubmitted by its creator. If approved, it is available for the general consultation and to be used in the next phase.

Contents management, storage and delivery is usually done using Learning Objects (LOs), as unity of study, exercise or evaluation.

Software and resource sharing is done through the grid infrastructure, leading to a scalable system. New computing resources can be dynamically added, installing and configuring the grid client on new workstation or serves and easily removed. New organizations can join the VO, with its own licensed material, adding new grid nodes.

The grid services have been integrated in the Liferay portal. Learning Objects, authoring tools, learning materials and documents located on different grid nodes are accessible through the portal, providing a system capable to manage the development and sharing of

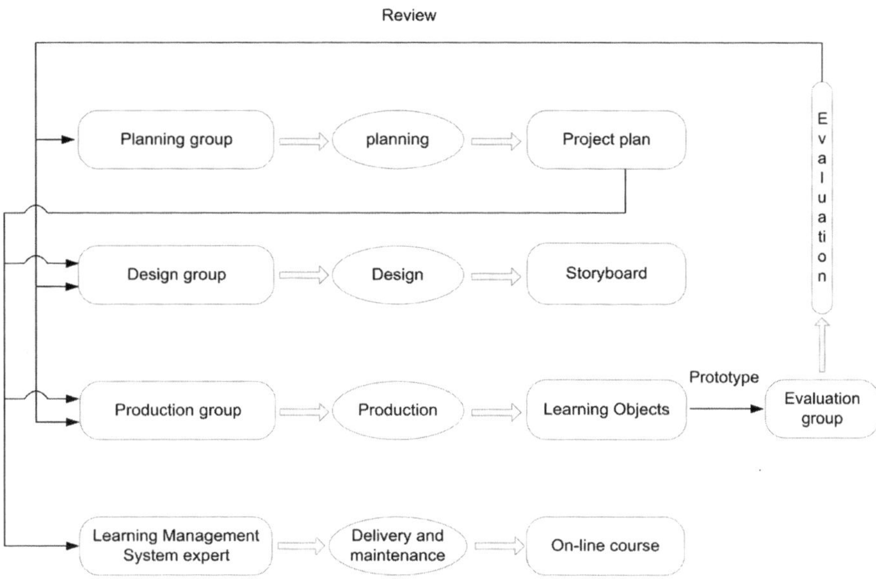

Fig. 3. E-learning production phases.

new contents and the reuse of the existing ones. The grid infrastructure has been developed using the Globus Toolkit[15] as middleware, which includes software, services and libraries for resource monitoring and management.

The security is a critical point in our grid system, since each VO member belongs to a different organization and the available resources may frequently change. Users are identified through a proxy certificate, released by a trusted Certification Authority (CA), allowing a Single-Sign-On authentication in the grid.[16] The CA specifies the user authorizations, providing the access only to the right resources and information.

Figure 4 shows the current home page of the SFINGE web portal. The grid status information are placed in the sidebar, while the contents are arranged in the biggest area.

Finally, the MOODLE platform (Modular Object-Oriented Dynamic Learning Environment) is used as LMS, for the delivery and the evaluation of the e-learning contents.

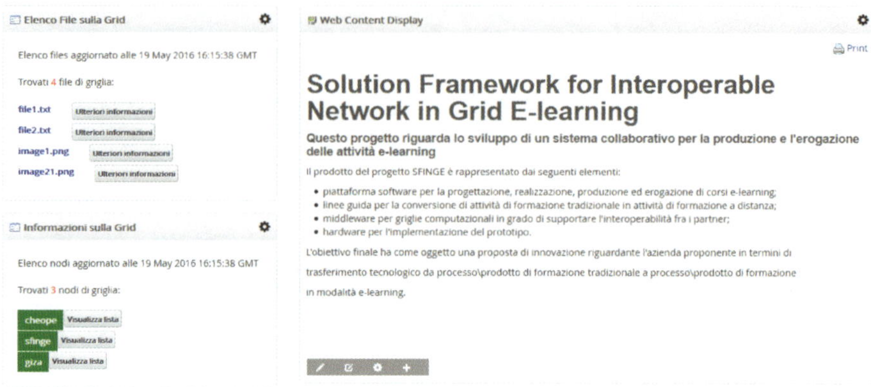

Fig. 4. SFINGE home page.[1]

3. The ReCaS and PRISMA Contribution

The aim of the integration of the SFINGE system in the ReCaS/PRISMA cloud infrastructure is to create a stable snapshot of our platform during the prototype development.

Open Stack is an open source cloud service, released under Apache license. It is composed by a set of different software that control processing, storage and networking resources and it is managed through a web-based dashboard, shown in Fig. 5. The OpenStack tenant includes:

- 1 Windows Virtual Machine (VM) with RDP protocol configured to access as remote desktop;
- 1 Ubuntu VM with remote desktop access enabled trough X2Go application[17];
- 10 Ubuntu VMs to provide the grid infrastructure;
- 1 OwnCloud[18] account, accessible via web interface or using the WebDAV protocol.[19]

We replicated the full SFINGE grid infrastructure on the cloud environment, using Virtual Machines (VMs) as grid nodes. The main

Fig. 5. OpenStack dashboard.

components of the Globus Toolkit have been installed in a VM named SFINGE: GRIDFTP, GRAM5, GSI, MyProxy, GSI-OpenSSH and SimpleCA. Also the Liferay portal and the MOODLE platform have been installed in a VM (Iside) and connected to the grid infrastructure.

Moreover, the PRISMA facility allows to extend the system capabilities through remote desktop services and cloud storage.

The remote desktop service allows to use a virtual machine configured with the authoring tools to create e-learning contents. Then these contents are shared using the Owncloud platform and are available directly on the SFINGE, the web portal.

4. Conclusions

In this Chapter, we described the SFINGE project, a grid based collaborative system for e-learning development, which aims to innovate the e-learning production process. The platform support the creation of VOs to increase the collaborative works between different institutions or enterprises and allows to create, deliver and maintain the contents.

The integration of a first stable version of our e-learning environment on the ReCaS/PRISMA cloud facility is ongoing, while the prototype is still under development.

This collaboration involves three fundamental research institutions: Politecnico di Bari, Bari University and INFN-Bari and represent an important test for the ReCaS and PRISMA infrastructure.

Acknowledgments

This contribution was made possible thanks to the grant of the Regione Puglia UNIONE EUROPEA — Fondo Europeo di Sviluppo Regionale MINISTERO SVILUPPO ECONOMICO REGIONE PUGLIA — Area Politiche: per lo Sviluppo Economico, il Lavoro e l'Innovazione Programma Operativo Regionale FESR 2007–2013 "Obiettivo Convergenza" Asse I: Promozione, valorizzazione e diffusione della ricerca e innovazione per la competitività Linea 1.2: Rafforzamento del potenziale scientifico-tecnologico della Regione a sostegno della domanda delle imprese Azione 1.2.4: Partenariati Regionali per l'Innovazione.

References

1. SFINGE project: Available at http://www.progettosfinge.it.
2. H. H. Adelsberger, B. Collis and J. M. Pawlowski, (eds.) *Handbook on Information Technologies for Education and Training*, Springer-Verlag, Berlin (2002).
3. M. Castellano, S. Marru and A. Pavone, E-services per griglie computazionali: un'applicazione all' automazione dei processi di e-learning, *Congresso MoodleMoot* (2010) (in Italian).
4. M. Castellano, L. Pisciotta, G. Tarricone, G. Amoruso and G. Stifini, Approccio flessibile e distribuito per applicazioni e-learning: Progetto "SFINGE", *Atti del Convegno MoodleMoot 2010, ISBN 978-88-8459-162-3, Convegno MoodleMoot* (2010) (in Italian).
5. M. Castellano and L. Pisciotta, *An Application of PVFs Technologies for Moodlebased e-learning*, ICERI2009 Conference, 16th–18th Nov 2009, pp. 1063–1069, Madrid, Spain (2009). ISBN: 978-84-613-2955-7.
6. F. Berman, G. Fox, T. Hey, (eds.) *Grid Computing: Making the Global Infrastructure a Reality*, John Wiley & Sons, Inc., New York (2003).

7. I. Foster and C. Kesselman, *The Grid: Blueprint for a New Computing Infrastructure*, Morgan-Kaufmann Publishers, San Francisco, CA (1998).
8. M. Castellano, G. Mastronardi, G. Piscitelli, A. Aprile, G. Di Giuseppe and V. Dicensi, Simulating a Computational Grid GESTS International Transaction on Computer Science and Engineering, *ISBN 89-953729-5-8 © GESTS* 40(1), pp. 9–20 (2007).
9. M. Castellano, G. Mastronardi, V. Bevilacqua, R. Bellotti and S. Tangaro, Distributed medical images analysis on a grid infrastructure Future Generation Computer System, *The Int. Journal for Grid Computing Theory Methods and Applications*, (2006). © 2006 Elsevier ISSN: 0167-739X 23: 4. Volume 23, Issue 3, pp. 475–484.
10. PON ReCaS, http://www.pon-recas.it.
11. PRISMA, http://www.ponsmatcities-prisma.it
12. I. Foster, C. Kesselman and S. Tuecke, The anatomy of the grid: enabling scalable virtual organizations, *Int. J. of Supercomputer Applications and High Performance Computing* 15(3), pp. 200–222 (2001).
13. Liferay open source portal: www.liferay.com
14. M. Castellano and C. Digregorio, *A Workflow Based Approach for Knowledge Grid Application*, In Proceedings of KMIS pp. 230–235 (2009).
15. Globus Toolkit 6: Available at https://www.globus.org/toolkit
16. J. Novotny, S. Tuecke and V. Welch, *An Online Credential Repository for the Grid: MyProxy*, Proc. of the 10th International Symposium on High Performance Distributed Computing (HPDC-10), Francisco, CA, pp. 104–111 (2001).
17. X2go: Available at http://wiki.x2go.org/doku.php
18. OwnCloud: Available at https://owncloud.org/
19. WebDAV: Available at http://www.webdav.org/

Chapter 36

Multi-Tiered Storage Based Cloud Environment in ReCaS DataCenter

S. Monforte*, G. Platania and G. Andronico

The Italian National Institute of Nuclear Physics (INFN), Division of Catania, Italy

**salvatore.monforte@ct.infn.it*

Tiered storage is an idea going back at least to 1990, when a three-tiered storage architecture was presented as a way of increasing storage efficiency. With the exponential growth of data handling in the last few years, the idea has gained new interest. Nevertheless, active tiering has been actually confined to homogeneous storage environments due to the lack of automated tools to manage the data. Tiered architectures are becoming more common, but usually they are passive, with all the data for a particular workload assigned to a specific tier to meet the application's requirements.

This Chapter illustrates successful deployment of multi-tier Cloud leveraging IBM General Parallel File System (GPFS) storage backend and OpenStack, for provisioning of different workloads on the ReCaS DataCenter at the INFN division of Catania. Additionally, the proposed solution provides end-to-end control over administration, automated storage optimization and other added enterprise class functionality such as Information Lifecycle Management (ILM).

1. Introduction

Studies show that data access follows the 80/20 rule: the most recent 20% of the data attracts 80% of the access. This means that after a certain period, data has aged to a level that the most activities performed are basically read operations. The extreme case is in financial trading, where the latest market data is extremely valuable and value drops precipitously literally in minutes.

Tiered storage has been an attractive theory since at least 1990 and due to exponential growth of data handling, in the last few years, the idea has gained new interest. Tiered architectures are becoming more common although, in practice, tiering has only been used in very limited storage environments due to the following critical issues:

- lack of automated data classification tools,
- lack of automated policy management tools,
- immaturity of fully virtualized environments.

As a result, while many data centers implemented their own kind of tiering, tiers are not integrated in any significant way. Data sets are assigned a tier based partly on the needs of the application and once written into a particular location stays there forever, that is, long enough after any real need for it.

A multi-tiered storage system with automated data movement provides the best solution for managing the data ReCaS infrastructure and is expected to handle in near future. The ReCaS Infrastructure is an initiative promoted by the University of Naples Federico II, The University of Bari Aldo Moro and the Italian National Institute of Nuclear Physics. It has been funded by the Italian Ministry for Education and Research with the goal of creating a multi-regional e-infrastructure for High Energy Physics and e-science in South of Italy.

ReCaS is composed by four datacenter distributed in the cities of Bari, Catania, Cosenza and Napoli and connected with 10 Gbit geographic links. Each site hosts a large number of racks, servers and storage for a final amount of about 50,000 cores and 10 PB of disk spaces.

2. IBM GPFS and OpenStack

The IBM General Parallel File System (GPFS) is a high-performance shared-disk file management solution that provides fast, reliable access to a common set of file data, online storage management. GPFS provides virtualization to file storage space, allowing multiple systems and applications to share common pools of storage, providing flexibility to transparently administer the infrastructure without disrupting applications and currently powers many of the world's largest scientific supercomputers and commercial applications that require high-speed access to large volumes of data. High-performance enterprise file management with GPFS provides high reliability and availability to eliminate production outages as well as policy-driven automation tools to ease information lifecycle management (ILM).

OpenStack is a promising open source software forming an easy solution for dynamic allocation of datacenter resources comprising of various components such as Nova, Glance, Swift, Neutron and Cinder forming a so called Software Defined Environment (SDE) technology supporting automation and a single window administration.

In this Chapter, we focus on the experience at the INFN ReCaS storage infrastructure deployment limited only to configuring Cinder component which interacts with multiple islands of storage resources and helps in provisioning of volumes to nova component. Challenges such as provisioning of flexible tradeoff (in terms of IOPS performance) tiered storage pools as well as intelligent ILM are successfully addressed by IBM GPFS Storage systems, and OpenStack cinder service is used in this solution as a cloud delivery platform for easy provisioning of volumes from multiple storage tiers.

The INFN division of Catania datacenter is equipped with heterogeneous storage systems consisting of:

- 320 TB raw on NL SAS disk controllers and 100 TB raw on IBM GSS storage system connected both via IB FDR to the HPC cluster[1] and 10 GbE to the standard computing farm.
- 200 TB raw on standard SAS disk controllers connected via FC 8 Gbps to the GPFS server and 10 GbE to the standard computing farm.

Therefore, the ReCas DataCenter setup at the INFN division of Catania assumes a setup containing a mixed combination of various storage devices and are grouped under respective NSDs (Network Shared Disks) comprising of three pools: NLSAS_NSD, NLSAS_GSS_NSD and SAS_NSD grouping Near-Line SAS, GSS JBOD NLSAS and SAS disks respectively.

Leveraging IBM GPFS storage backend and OpenStack these three pools form Tier-1 (Gold), Tier-2 (Silver) and Tier-3 (Bronze) storages respectively. OpenStack Cinder has been configured in multi-backend mode.[4] Follows the sample configuration changes in *cinder.conf* to support multi-backend defining the previously described three storage pools.

```
enabled_backends=nlsas_storage, nlsas_gss_storage, sas_
   storage

[nlsas_storage]
gpfs_mount_point_base=/nlsas_gpfs/volumes
gpfs_images_dir=/nlsas_gpfs/images
gpfs_images_share_mode=copy_on_write
gpfs_max_clone_depth=3
gpfs_sparse_volumes=true
volume_driver=cinder.volume.drivers.ibm.gpfs.GPFSDriver
volume_backend_name=GPFS_NLSAS

[nlsas_gss_storage]
gpfs_mount_point_base=/nlsas_gss_gpfs/volumes
gpfs_images_dir=/nlsas_gss_gpfs/images
gpfs_images_share_mode=copy_on_write
gpfs_max_clone_depth=3
gpfs_sparse_volumes=true
volume_driver=cinder.volume.drivers.ibm.gpfs.GPFSDriver
volume_backend_name=GPFS_NLSAS_GSS

[ssd_storage]
gpfs_mount_point_base=/sas_gpfs/volumes
gpfs_images_dir=/sas_gpfs/images
gpfs_images_share_mode=copy_on_write
gpfs_max_clone_depth=3
gpfs_sparse_volumes=true
volume_driver=cinder.volume.drivers.ibm.gpfs.GPFSDriver
volume_backend_name=GPFS_SAS
```

Moreover, three cinder volume types (Gold, Silver and Bronze) have been created as an OpenStack admin user and associated with values of key volume_backend_name as defined in *cinder.conf* configuration file. One can observe that volumes created by issuing cinder create command using types Gold, Silver and Bronze are stored in GPFS_NLSAS, GPFS_NLSAS_GSS and GPFS_SAS file systems path, respectively.

In OpenStack environment, this solution helps in effectively addressing the multi-tier environment where end-users want to position different workloads leveraging characteristics exhibited by various disk devices.

3. Multi-Tier Configuration for Value Added Features

The ReCaS DataCenter deployment at the INFN Division of Catania implements tiered storage management by ILM that helps better optimization and automated placement in Cloud/Software Defined environment.

IBM GPFS helps to achieve data lifecycle management through integrated policy-driven automation and tiered storage management. IBM GPFS provides ability to allocate physical storage hard drives to logical storage pools within the file system. With logical storage pools, it is possible to create tiers of storage by grouping physical disks based on performance, locality or reliability.

A new GPFS file system using multiple storage pools has been created consisting of disks belonging to different storage pools: *system* and *system_sas* grouping NLSAS_NSD and SAS_NSD disks, respectively. Cinder has been opportunely configured to handle a new backend pointing to this multi-tier file system and OpenStack nova instances created having their volumes on MULTI_TIER backend.[5] It is a common situation where users might not frequently use their volumes. In this case, migrating volumes to lower/cheaper tiered storage pools would be a cost effective action.

IBM GPFS enables the migration of volumes seamlessly and in an automated way by defining a custom policy for a given filesystem,[2] it follows the sample policy defined which provides a rule of migrating the files/volumes between pools if their last modification time is greater than 10 hours.

```
$ cat multi-tier.policy
RULE 'default' set pool 'system' RULE 'Migrate' MIGRATE
   FROM POOL
'system' TO POOL 'system_sas' WHERE ((CURRENT_TIMESTAMP) -
(MODIFICATION_TIME) > INTERVAL '10' HOURS)
```

4. Conclusion

In this Chapter, we presented a successful deployment of Multi-Tier Cloud leveraging IBM GPFS storage backend and OpenStack on the ReCaS DataCenter at the INFN Division of Catania, for provisioning of different workloads. Additionally, the proposed solution provides automated storage optimization configuring custom policies for data migration between storage tiers.

References

1. GPFS Concepts, Architecture, and Planning Guide. Available at http://www-01.ibm.com/support/docview.wss?uid=pub1ga76044100
2. Policy Templates. Available at http://publib.boulder.ibm.com/infocenter/tsminfo/v6/index.jsp?topic=%2Fcom.ibm.itsm.hsmul.doc%2Ft_mig_gpfs.html
3. OpenStack Installation Guide. Available at docs.openstack.org/trunk/install-guide/install/yum/content/
4. OpenStack Cinder Multi-Storage back-end. Available at wiki.openstack.org/wiki/Cinder-multi-backend
5. IBM GPFS Cinder driver Module. Available at docs.openstack.org/trunk/config-reference/content/GPFS-driver.html

Chapter 37

GaaS 2.0: The New Release Based on OpenStack with Features Implemented with OCCI

G. B. Barone*,¶, V. Boccia†, D. Bottalico*, L. Carracciuolo‡,||, G. Laccetti§, A. Solla†, B. Spisso† and A. Tebano†

** University of Naples Federico II, Naples, Italy*
† National Institute of Nuclear Physics, Naples, Italy
‡ National Research Council, Naples, Italy
§ University of Naples Federico II, Department of Mathematics and Applications "Renato Caccioppoli", Naples, Italy
¶ gbbarone@unina.it
|| luisa.carracciuolo@cnr.it

Cloud Computing has been widely adopted as an alternative paradigm with respect to Grid Computing to provide resources because of the advantages it brings to both users and providers.

GaaS (Grid as a Service) was born with the aim to enable the use of Cloud-provided resources with well-established Grid-like interfaces, avoiding the need for users to learn new resources access and use models. In previous experiences, we developed a prototype for GaaS, by using the most used Cloud open source technologies and frameworks available at that time (e.g., Eucalyptus, OpenNebula). Here, we present the final stage of our work on GaaS that is now based on several *de facto* and consolidated standards in Cloud, Virtualization and Grid worlds: OpenStack, KVM and EMI. Furthermore, we present both i) a production level version of GaaS and ii) GApp (GaaS APPlication) the mobile interface to the GaaS framework.

1. Introduction

An useful feature of the Cloud Computing is the chance to make available the users' resources on demand. Cloud paradigms can be also used to realize server consolidation and so to allow an efficient usage of resources in order to reduce the total number of servers or server locations that an organization requires. Moreover, as it happened for Grid infrastructures, the Clouds can be used efficiently by applications with a large amount of independent tasks, which can reduce their Total Time to Solution.

In the last decades, the need for computational power pushed the scientific community to deploy and sustain Grid Computing Infrastructures that are able to share huge amount of resources through a well-defined distributed infrastructure model, allowing the solution of large scale problems in a collaborative manner.

However, the Grid model is basically static: groups of users and organizations set up several Grid management services and computing resources in a multi-level structure which spread the management responsibilities among the organizations involved in the Grid. Users cannot define new Grid sites, add computational resources to existing ones and modify the resources aggregation scheme in accordance to their needs; it is also not possible to dynamically modify the resources number on the basis of the real system workload, in the name of saving energy and to achieve a more efficient and sustainable environment. But given the flexibility in resources management through the Cloud Computing paradigm, it seems a promising approach to provide flexible Grid Computing infrastructures through the combination of the Grid and Cloud paradigms.[3]

Hence the idea to develop GaaS[2] (Grid as a Service), a work started four years ago in order to find a solution to make distributed computing environments, based on Grid model, more "elastic" and more sustainable, exploiting the features of the Cloud model.

The GaaS model merges the two approaches combining the advantage of providing users with an usage model that is familiar to the traditional Grid users, with the chance of flexible management of computational resources in a *IaaS*-like fashion. Therefore, the GaaS paradigm can be classified as a *Platform-as-a-Service* due to the augmentation of Grid environments with elastic (i.e., virtual) resources.

We assume as Grid reference architecture the one implemented by the middleware EMI[1], developed in the context of EGI (European Grid Infrastructure). The EMI middleware provides a Grid infrastructure that is accessible to community members organized into Virtual Organizations (VO). The infrastructure provides users with high level services for scheduling and running computational jobs, accessing and moving data, and obtaining information on the infrastructure itself. Those provided are services for authentication/authorization (e.g., VOMS), resources allocation and discovery (e.g., LB/WMS) and infrastructure Information System (IS). Computing resources (WNs) are provided by means of CE (Computing Element) that is an endpoint with a set of queues handled by an LRMS (Local Resource Management System). User can access these services from a User Interface (UI).

GaaS consists of the following four services:

- to add new computing resources (Worker Node Service: GaaS WNS);
- to aggregate existing computing resource in a new queue (Queue Service: GaaS QS);
- to add a new grid site for an existing VO (Grid Site Service: GaaS GSS);
- to create a suited runtime environment for a set of applications on existing or new computing resources (Application Environment Service: GaaS AES).

The **GaaS WNS** service manages the integration of new WNs to an existing Computing Element queue. The system acquires cloud resources (public or private), configures resources as grid WN, select the queue where the WN has to be inserted and ask for a CE local resource management system configuration. This service is useful in these scenarios: to realize efficient datacenters management in term of energy consumption reduction and to provide users with more resources if they are not locally available.

The **GaaS QS** service manages the integration of a new queue into an existing Computing Element. In this case, the system ask for a CE configuration, specifying the set of underlying WNs and the queue policies. A possible use case for this service is a community who needs particular queue policy (i.e., queue priority for the community users, different values for max job execution time, etc.).

The **GaaS GSS** service manages the integration of a new grid site for an existing VO into an existing grid infrastructure. The system acquires cloud resources (public or private), ask for GaaS WNS and GaaS QS services, configures an Information System and select the grid infrastructure where new grid site has to be inserted into. A possible use case for this service is related to a community which has to share resources for the lifetime of a project. Another ideal scenario is related to the need to build on the scratch, and for a short time period, a tailored grid testbed.

The **GaaS AES** service performs the Application Environment Service. The system deploys a suitable combination of *GaaS WNS*, *GaaS QS* and *GaaS GSS* services and it provides, on allocated resource, all applicative software stack according to users' community requirements. The software stack can be built on the basis of a software portfolio. A possible use case for this service is related to a community which wishes add to the benefits of all the other listed above services also the chance of a proper "*applicative middleware*."[4]

So, by using GaaS, "*privileged*" grid users (e.g., the VO administrators) can "launch" new Grid sites, add computational resources to existing Grid sites and modify the resources aggregation scheme. Even if our approach is based on a *Grid-over-Cloud* model, in GaaS, resources are made available through their configuration in Grid abstractions, e.g., queues. Hence, provided resources can be reconfigured or differently aggregated on the basis of users' needs (in a way resembling the PaaS paradigm).

The Chapter is organized as follows. In Sec. 2, the previous GaaS implementation are presented, showing the development of the first prototype version and discussing about the insight gained from this early implementation. In Sec. 3, the current GaaS modular implementation based on the OpenStack cloud manager and on the Tomcat container as application engine is described. In Sec. 4, the user interface functions showing all the actions that an end user can perform to enquire and manage a GaaS service are discussed. Section 5 is devoted to an example of user interface designed for mobile devices, some screenshots from the app are displayed. Finally, in Sec. 6, conclusions are presented.

2. Previous GaaS Implementation

The first GaaS implementation (the GaaS 1.0 version) can be essentially considered a successful *proof-of-concept* of the approach in combining grid and cloud access paradigm.

All the GaaS services was developed, in a prototype version, into the context of the SCoPE Data Center at University of Naples Federico II. The SCoPE Data Center computational resources was, and are, integrated in a grid context that provides the user with high level services for scheduling and running jobs, accessing and moving data, and obtaining information on the infrastructure state.

This prototype was based on the EMI Grid middleware and on the OpenNebula Cloud management system. The modularity of OpenNebula allowed us for fast introduction of new features to the management system, hence, it allowed the easy integration of the Cloud-provided resources into EMI. Xen hypervisor was used to create virtual machines (VM), used to dynamically provide resources for the Grid infrastructure.

The main efforts in the prototype development were essentially the enabling of fast provisioning of grid resources.

To optimize the infrastructural resources used during the provisioning process, and to reduce the overall provisioning time, we took into account the peculiarities of the GaaS system. In particular, we made the following observations:

- all the VMs are mandated to host the same operating system and software configuration, i.e., there is just one VM template;

- the VMs operational destinations (e.g., WN, CE, etc.) are differentiated among them by a few configuration differences in respect to the VM template;
- most of the operations performed on the newly created VMs are *read*-type operations, since writes are usually performed on a dedicated network storage provided by the Grid Infrastructure.

Finally, some performance tests are performed to validate our approach in fast provisioning.

3. The GaaS 2.0 Implementation

Soon after the end of our activities on the development of the GaaS prototype, also thanks to the works of some students in the context of PON "*Rete di Calcolo per SuperB e Altre Applicazioni*" (ReCaS) project, we decided to build the *Production level* version of the GaaS services.

The architecture of the framework used to deploy GaaS services has been re-designed. The results of such efforts is the actual version of such framework which has a multiple levels and modular classic client/server structure (see Fig. 1).

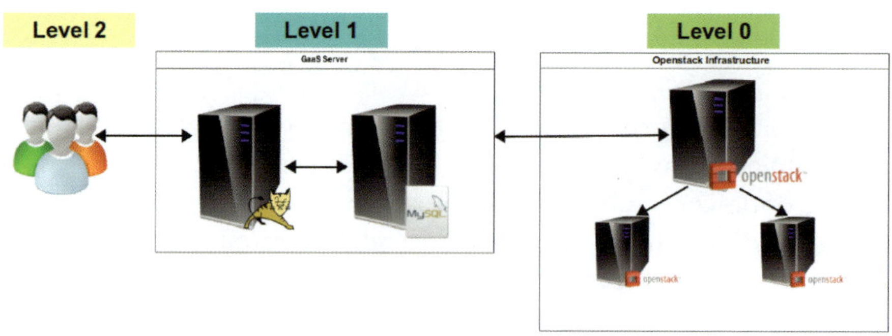

Fig. 1. The architecture of the GaaS framework.

At the lowest level (*Level 0* of this architecture), there is a Cloud infrastructure based on the OpenStack cloud manager consisting of

- a controller node that provides information on available resources and manage the virtual instances and
- some compute nodes that provide the creation of VM.

At a higher level (*Level 1*), there is the "GaaS server," which provides tools for interoperability between the OpenStack platform and user client: it acts as

- a client towards the OpenStack controller node, which sends requests related to the creation, deletion and display of the instances running on Cloud through appropriate REST-API,
- a server towards the user client (Level 2) responding to user requests.

Furthermore, the components of *Level 1* are:

- The Tomcat container which takes care of running the Java servlet that communicate directly with the users application;
- The Database (DB) system, based on the MySQL server, which stores the information about the GaaS users and instances running on OpenStack infrastructure.

Finally, at the highest level of abstraction (*Level 2*), there is the client, represented by an user application that, through a convenient graphical interface, gives to the user the possibility to create and manage a "*virtual*" GRID site and access it as if they were constituted by real physical machines completely abstracting the underlying Cloud Manager.

The choice of a such multi-level architectural scheme ensures an high level of transparency, indeed the end user is not absolutely aware of management methods and realization of the underlying VM, but will play purely the role of a user.

Moreover, the approach allows to delegate all expansive operations to the lower layers avoiding excessive workloads on the user client. This gives the possibility to develop client on small devices with low computational capacity (i.e., the first generation Android terminals). Finally, thanks to the characteristics of modularity of the architecture, each component can be easily replaced.

3.1. *GaaS MySQL Database*

The database system will store information about users and grid sites within a SQL database. This section will show briefly the structure of the database, describing the tables and their role within the system.

The USERS table is a key point for all those operations which provide the identification and profiling of the users. It contains the user's information gathered from the digital user credential (i.e., a X509 certificate) presented by the user when the registration to the GaaS infrastructure occurs. The table also contains the authorization informations related to the authorization procedures.

Among the fields in the USERS table, very important are:

- the *Common Name* field(CN),
- the TYPE and the APPROVED field.

The CN field contains the name of the user as stated on the certificate, the TYPE field is crucial for user profiling/authorization indeed this variable indicates to the system if the user in question is a normal user or an administrator one. The field APPROVED informs the system about the user grant to use the GaaS infrastructure.

The ELEMENTS table contains information regarding the nodes allocated in GaaS infrastructure. The main field of this table are:

- the ID field which is the unique identifier of the node,
- the TYPE field indicates the type of the node (i.e., playing the CE, Site BDII, WN or SE role),

- the SITE ID field that is the unique identifier of the grid site containing the node, and finally,
- the HOSTNAME which is the possible (if the role require it) public hostname of the node, it can be used to discover the public IP.

3.2. *A Sequence Diagram of a General Task*

In order to describe the interaction among the GaaS components, in this section, a basic sequence diagram is presented. In Fig. 2, the sequence diagram for a general task is shown. It highlights the temporal aspect of the interaction between the main components.

The time scan is from the left to the right and top to bottom for the black lines meanwhile the red lines flow in the opposite directions.

The sequence begins with an HTTP request from the client containing information about the task to be executed.

The Tomcat component of the GaaS server, after received the request, extract from it all the needed information such as the user and the type of request.

As soon as the user is known, MySQL DB is queried gathering the information about the users' authentication and authorization grants. If the user has the right credentials and authorization grants, the GaaS server proceed to the task execution: depending on the requested task, the Tomcat contact the OpenStack controller, the MySQL server or both.

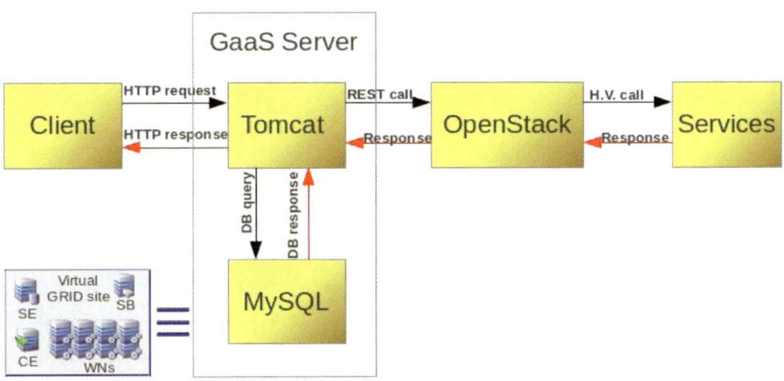

Fig. 2. Basic interactions between the GaaS components.

In the case of a "*discover request*," the Tomcat query the DB in order to collect the information about the services available to the user and then send this information back to the client.

Conversely, in the case of a "*managing task*," after gathering the information from the DB about the active service, the Tomcat sends through REST type calls all the commands required from the OpenStack controller for the execution of the task.

After receiving a command from the GaaS Server, the OpenStack controller contacts the hypervisors (H.V.) on the Compute nodes, which execute the command, and collect the response from the hypervisor. This response is propagated back to the GaaS server though the OpenStack controller.

At this point, the Tomcat provides update to the DB with the new information on the status of the services and, when the task is completed, communicates to the client about the outcome of the task. As will be discussed in next section, a set of functions are available which allow the client to create and manage a virtual GRID site, indeed for the GRID end user the entire system is equivalent to an ordinary site.

4. User Interface functions

There are at least 21 major user interface functions (corresponding to four different group of actions that the user might want to run) that were considered during the planning phase. For clarity, we have divided such function in classes (one for each type of actions) describing also when the function can be executed or not both by a normal or by an administrative user.

4.1. *User Authentication/Authorization Actions*

User registration: It is a preliminary phase as well as a fundamental in the use of the system. Each user, in order to use the services offered by the GaaS platform, needs a registration through the use of a certificate PKCS12 that contains all the information about the user. After recording, the user must wait until an administrator approves its registration (see use case Approve user). This action is available to both normal and administrative user.

User deletion: A user is able, if desired, to delete its GaaS account. The elimination of the account implies the subsequent removal of all sites and all instances connected to the user. This action is available to both normal and administrative user.

User login: It is the case of use prior to submission to any operation on the platform. The login is via sending their PKCS12 certificate extracted from the certificate information required for the recognition of the user. There exist two types of scenario:

- The user is not recognized by the system. In this case, the user can request to the administrator the association of its certificate to a new account (sub-use case "User Registration").
- The user is recognized by the system. The role of a user can be of two types: normal user and administrator.

This action is available to both normal and administrative user.

User credentials refresh: Upon registration, the system sends an e-mail to the address marked in the certificate containing an unlock code. The unlock code is used as an "emergency credential" and allows the user to update the user expired certificate. This action is available to both normal and administrative user.

User approval: Once the registration of a user is complete, the user must be approved by an administrator. So, the administrator enables the user the full use of the platform. This action is available to only the administrative user.

Revocation user approval: If necessary, an administrator withdraw may (temporarily or permanently) access the authorization of a user on the platform. This action is available only to the administrative user.

Raise an user to administrator: An administrator can raise the standard users to administrators. This action is available only to the administrative user.

Downgrade an administrator to a simple user: An administrator can revoke administrator rights to another user. This action is available only to the administrative user.

4.2. *Site Creation and Modification Actions*

Create site: The creation of a new virtual GRID site involves the creation and subsequent configuration of a site containing a minimal GRID Compute Element, a Site BDII and, if the user requests it, a Storage Element. In order to request the creation of a site, the user must specify the name of the new site, the maximum number of Worker Nodes associated with the site and the type of hardware of the CE, SBII and of the eventual SE. If desired, once the site is fully created, further elements can be added (see use case "Create element"). This action is available to both normal and administrative user.

Delete site: During this phase, the user, depending on its role, can eliminate a site from the GaaS platform, thereby eliminating all instances connected to it (EC, SB, SE and WN). This action is available to both normal and administrative user.

Create WN/SE: Once the creation and configuration of a site is complete, the user can add one or more elements in order to increase the computational capacity. The user has the possibility of adding a Storage Element as long as it is not been already added at the time of the creation of the site, or an arbitrary number of Worker Nodes provided they do not exceed the maximum number of Worker Nodes imposed at the time of the creation of the site. This action is available to both normal and administrative user.

Delete WN/SE: If no longer needed, the user (if he possesses the permission) can delete an item from a site Grid. This action is available to both normal and administrative user.

Fig. 3. The screen for login by a X.509 certificate.

4.3. *Management of Site/Elements Actions*

Start/Stop/Restart element: Being an element of a grid (even if virtual), the user has the capability to handle it in all its aspects, including the restart and shut-down device. This action is available to both normal and administrative user.

4.4. *Recovery of Site/Elements Information Actions*

List sites: The user has the option to request a list containing all the sites to which it has access (the administrators display all sites instantiated in the GaaS platform). This action is available to both normal and administrative user.

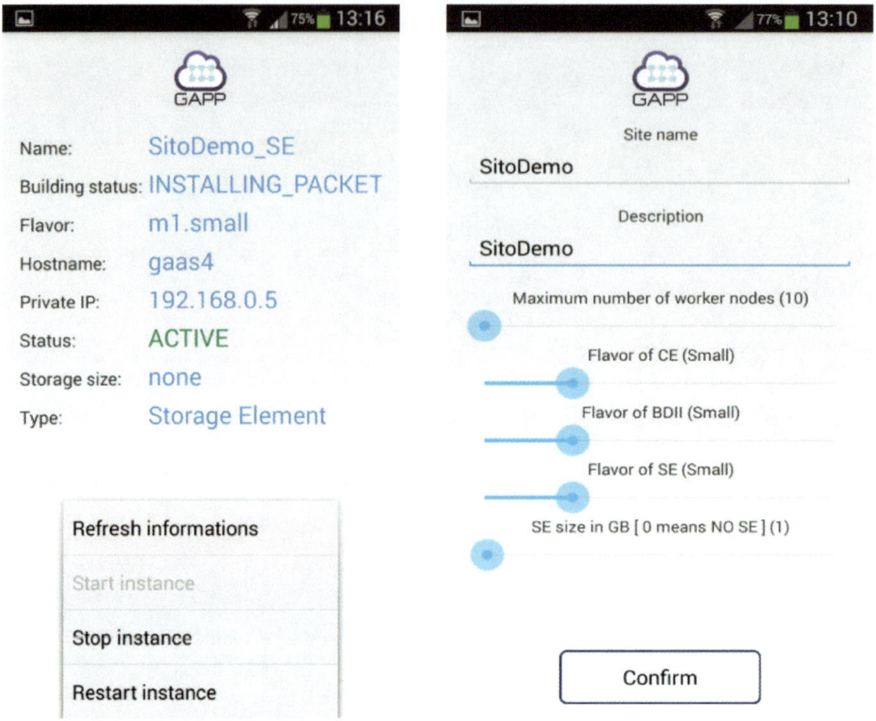

Fig. 4. The screens for the modification and management of an existing site components.

List elements: List the items for all sites which the user has access. The user is able to request a list containing the information overview of the elements of the site. This action is available to both normal and administrative user.

Site info: The user can view the complete information about a single site including basic information on the elements contained in the site and those related to the maximum number of WN instantiable. This action is available to both normal and administrative user.

Element info: The user can view the complete information about a single GRID element (node) obtaining information about its status, its hardware features (e.g., CPU, RAM and Storage) and the status of

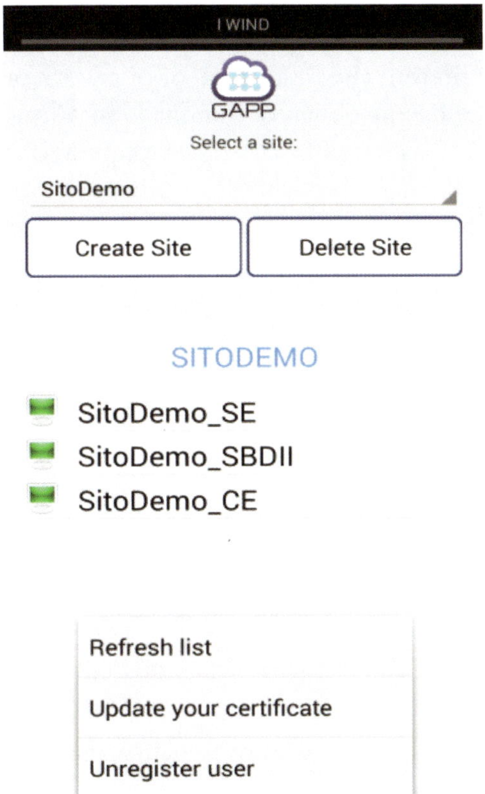

Fig. 5. The screens for the creation, modification and deletion of site.

its configuration. This action is available to both normal and administrative user.

5. An Example of User Interface: *GApp* the Mobile Interface to the GaaS Framework

As an example of user interface to the GaaS framework, we have developed GApp (GaaS APPlication): an App which is compatible with all Android mobile devices and, thanks to an appealing graphical interface, can be used to perform all the above presented user functions (see Figs. 3–5).

6. Conclusions

In this work, we present the final stage of our work on GaaS based on several *de facto* and consolidated standards in Cloud, Virtualization and Grid worlds. GaaS was born, some years ago, with the aim to enable the use of Cloud-provided resources with well-established Grid-like interfaces, avoiding the need for users to learn new resources access and use models. We describe some implementation details on (i) the developed *Production level* version of GaaS and on (ii) *GApp*, the mobile interface to the GaaS framework.

Acknowledgments

This work has been realized thanks to the use of the SCoPE computing infrastructure at the University of Naples, also in the framework of PON "*Rete di Calcolo per SuperB e Altre Applicazioni*" (ReCaS) project.

References

1. EMI — The European Middleware Initiative. Available at http://www.eu-emi.eu/documentation.
2. V. Boccia, G. B. Barone, R. Bifulco, D. Bottalico, L. Carracciuolo and R. Canonico, *GaaS: Customized Grids in the Clouds*, Lecture Notes in Computer Science, Proceedings of Euro-Par 2012: Parallel Processing Workshops, Vol. 7640, pp. 577–586, (2013). Available at http://dx.doi.org/10.1007/978-3-642-36949-0_67
3. M. Memon, Z. Nagy, E. Yen and O. Koeroo, *Virtualization and Cloud Computing Task Force Report V.0.7* (2012) Available at https://twiki.cern.ch/twiki/pub/EMI/EmiJra1T5TaskForceCloudandVirtualization/EMIVirtCloudReport-v0.7.doc
4. L. Carracciuolo, G. B. Barone, V. Boccia, D. Bottalico, S. Pardi, M. Scognamiglio and F. Serio, *Middleware applicativo: lo SCoPE toolkit,* In Conferenza Nazionale Italian e-Science IES08 — Sessione Poster (2008).

Chapter 38

Evaluation of HTCondor[A1], a Community Driven Local Resource Management System for the ReCaS DataCenter System

G. Donvito* and A. Italiano

INFN, Sezione Bari, Bari, Italy
**giacinto.donvito@ba.infn.it*

The ReCaS[2] DataCenter has to face a big challenge already at the start of its operation since it has to provide computing resources to a wide range of scientific communities and end users who will try to get the most out of the system also through an appropriate use of batch system. The way the resources will be allocated to the users is a key factor in the resource management process: in fact, the user committed resources have to be granted to them, while at the same time, one should avoid to waste even a single CPU cycle. Achieving these results at ReCaS DataCenter scale (more than 15K cores and 6 PB of storage) is challenging. The resource management process must be stable and reliable in order to efficiently allocate all the available resources. Last but not least, the computing resources hosted at ReCaS DataCenter must be accessible through different interfaces, in particular the GRID one, to enable end users spread around the world to access the resources in a transparent way. HTCondor[1] is emerging as a valid tool to manage computing resources in a stable, efficient and smart manner. The main purpose of this document is to try to demonstrate the sustainability of this tool as Local Resource Management System for ReCaS DataCenter. For this purpose, a virtual condor pool, with a size comparable to the ReCaS DataCenter one, has been deployed in order to simulate real use cases.

1. HTCondor Overview

The HTCondor architecture is described in Fig. 1, however for the purposes of this chapter, a brief overview of the HTCondor architecture is necessary. An HTCondor pool is defined by a collector, which serves as a registry, and distributed daemons in the pool. Each daemon is identified by a ClassAd record, that the daemon periodically sends to the collector to advertise its presence and its current status in the pool. Job execution nodes in the pool are represented by a startd, which is responsible for carrying out job execution requests. The startd divides the computing resources into one or more logical subdivisions called execution slots. A collection of jobs that have been submitted by users is maintained by a schedd. The schedd obtains a lease to run jobs on one execution slot. The lease is obtained from the negotiator, which is the daemon responsible for a pool-wide user priority management and for matchmaking. Matchmaking involves finding compatible execution slots for the resource requests (in the form of ClassAds) the schedd makes on behalf of the users.

2. The Architecture, a Key Point Toward the Sustainability

The HTCondor evaluation process could only start with a deep look at the architecture which is shown in Fig. 1. The figure only shows the main components but it highlights that they can be distributed. Actually you can execute each HTCondor component on a dedicated

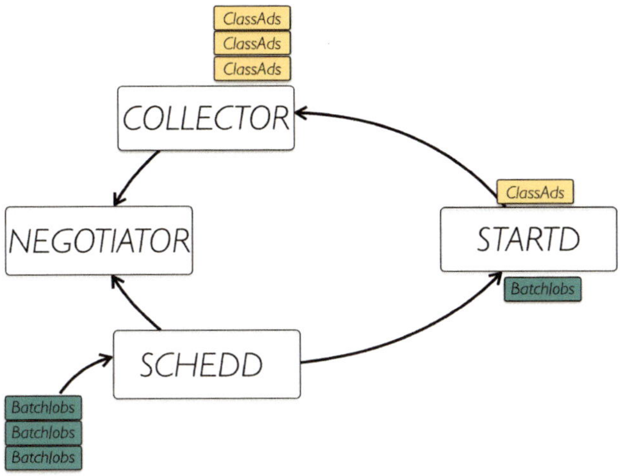

Fig. 1. HTCondor architecture.

host, so this means you do not have a monolithic service dealing with the entire pool but you can distribute load and functionalities among pool components. The negotiator will match batch job requirements with resource features, the collector will store and provide access to hosts information, the schedd will manage the batch job submission and the startd can even apply the pool policies. It is also possible to have multiple collector instances: in fact, it can just be considered as a keeper of the cluster status in terms of active clients ClassAds. Even better, it is strongly suggested to have more than one collector running at the same time. This is a mandatory requirement in order to deploy the negotiator in high availability. In the same way, it is possible to run multiple schedd instances which do not share the entire collection of batch jobs that have been submitted to the pool. Actually, each schedd handles only those batch jobs which have been directly submitted to it: as matter of fact, it is completely unaware about batch jobs submitted to another schedd. Note that this feature is not intended for scaling: each schedd instance can successfully manage more than one hundred thousands of batch jobs. From architecture point of view, HTCondor can be really considered a sustainable choice because there are no bottlenecks, it can easily sustain an increase in resources in the future.

3. Tests and Results

3.1. *Test-Bed Description*

Of course, it is necessary to evaluate all the declared HTCondor characteristics before definitely chose it as LRMS. For this reason, we have exploited the cloud computing feature to overcommit the available resources in order to deploy an HTCondor pool of the same size of the ReCaS DataCenter. Actually using OpenStack,[3] a cloud cluster in order to simulate an HTCondor pool with nearly 17K slots and 7K hosts has been deployed. There was the possibility to dynamically instantiate computing resources that automatically joined the HTCondor pool. In this way, a real simulation of concrete use cases was performed such as activating a brand new bunch a computing resources or destroying a bunch of virtual machines simulating, for instance, a power cut on a rack power line. The test-bed offered the possibility to measure how the pool reacts to these extreme circumstances and in the following paragraphs, we will describe the results.

3.2. *Reactiveness*

The cluster/pool status in terms of available resource is subject to change over the time. Let us suppose, for instance, that at t0, 100 cores area available, provided by five servers with 20 cores each. Sometime later, at t1, the pool status can change because one of the servers becomes unavailable; so the total of the available core will drop to 80. At the ReCaS DataCenter size, changes in the available resources are expected due to failures of the commodity hardware used. On the other hand, once the broken hardware has been repaired, it will join the pool again in production. Sometimes there will be even wide variation due, for instance, to a power cut on a rack power line or in the event to deploy a bunch of new resources. We assume that the state of the cluster will suffer variation over time. The way a LRMS reacts to a status change is an important factor which influences the LRMS main features such as the scheduling decisions which will be based upon cluster/pool status.

Response time to a basic HTCondor command provides a realistic measure of the time needed to access the pool status information so that high value in response time can be considered as an issue which can lead to poor cluster/pool reliability. Figure 2 shows the condor status command response time according to either small or wide variation in the pool status in terms of cores available. According to the measurements, the time needed to access the pool status grows linearly with the amount of data available. Anyway, it is important to highlight two remarkable aspects. The first one shows that a wide and sudden variation in the number of cores available generates a significant increment in the response time for a short period of time. Such condition can be considered expected because the pool is overloaded and it takes a lot of time to respond, however it does not become unresponsive which is the remarkable aspect. The second one shows as a slight variation in the pool status which can generate some spikes in terms of response time as the amount of that data representing the pool status increase. Anyway during these stress conditions, the HTCondor pool status has been always available.

3.3. *Stability*

We have also tried to understand whether the running batch jobs can negatively affects the pool stability so we have measured the condor

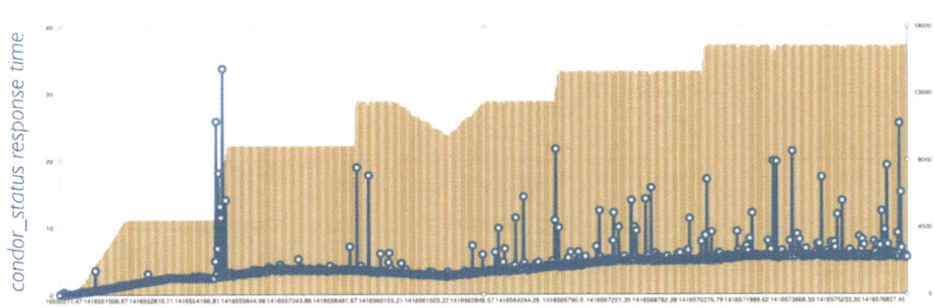

Fig. 2. Condor status command reports information about the hosts which have joined the HTCondor pool.

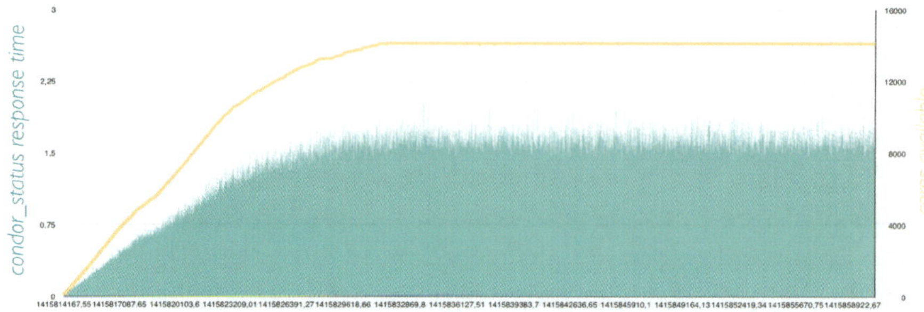

Fig. 3. Condor status response time while idle batch jobs get running as new slots become available.

status response time while idle batch jobs get running as new slots become available, see Fig. 3. We have realized that response time raises in a linear way with the number of hosts/cores, in other words, the information about the running jobs does not affect the pool status. Furthermore, in a high load condition with 14K running jobs, a low standard deviation of the response time confirms the stability of the batch system. We have experimentally measured the HTCondor stability, but at the same time, we have had the opportunity to demonstrate how the collected data reflects the HTCondor architecture distributed design. The cluster and the batch aspect of the LRMS do not influence each other. This feature provides a huge contribution to the stability of the LRMS.

In this Chapter, we have already mentioned that there seems to be a linear relationship between the registered cores in the pool and the response time to the condor status. It means that as the number of cores increases, the condor status will, of course, takes more time to respond to each query but the delta will not be high. In order to validate this sentence, we have applied a fitting process to the collected data and the results are shown in the Fig. 4. The plot confirms the linear relation between the number of cores available in the pool and the response time to the condor status command.

An other important factor which can influence the stability is the resource consumption by every HTCondor components mainly in

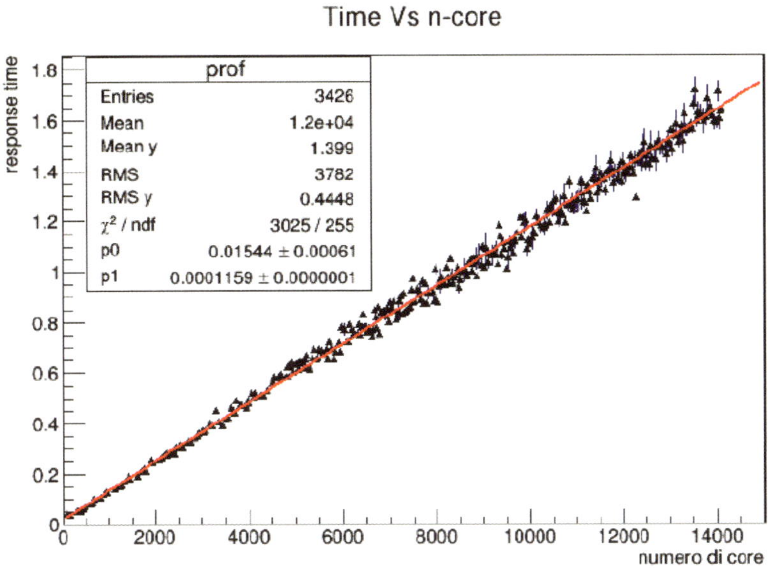

Fig. 4. Relationship between the cores/slots available in the pool and the response time to condor status.

terms of allocated memory. This cannot be considered a real issue because you can run each component on a dedicated server which will have enough room to run the service. Anyway we have measured how many resources a single component can request in order to work regularly. Figure 5 in particular shows how much Resident Set Size memory has been allocated to the HTCondor Collector process, while the number of the cores registered in the pool change over the time for a time period of four days.

- The allocated memory, less than 7 GB, is really restraint if you take into account that there were nearly 17K cores registered in the HTCondor pool.
- The ratio with the Virtual Set Size is 0.93.
- The memory is actively allocated or deallocated while the cores available change over the time.
- The memory allocation process seems really under control.

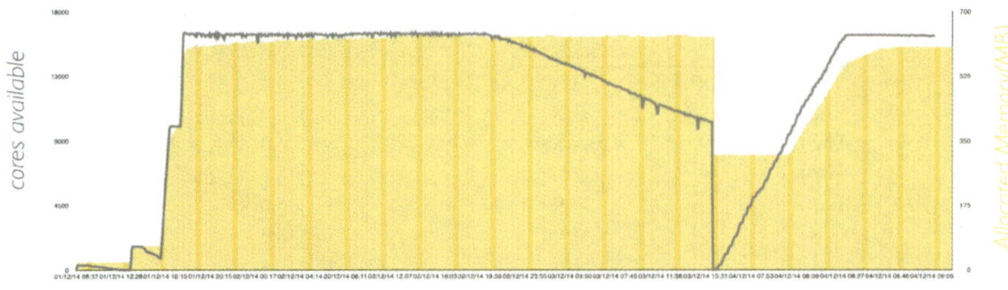

Fig. 5. Allocated Reside Set Size memory to the collector process while the cores available in the pool change over the time.

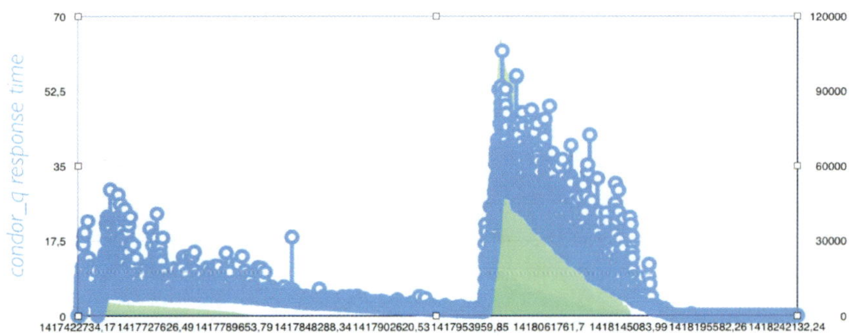

Fig. 6. Condor q command returns information about all the submitted jobs in this test managed by only one schedd.

3.4. *Scalability*

The entity which may vary from zero to a huge number is certainly the number of batch jobs submitted to the pool, because of that we have focused the scalability performance test on the batch jobs which can be managed by a single schedd. Figure 6 shows the results of this test and highlights the schedd scaling feature. We have measured such feature with the condor q command response time as the number of batch jobs managed by the schedd increase or decrease over the time. Significant variations of idle jobs do not negatively affect the LRMS functionalities. Response time has a direct relationship with the number of batch jobs managed by schedd and what is really important to highlight is the HTCondor schedd responsiveness, where actually nearly 120 K idle batch jobs have not killed the schedd, even better all of them have been correctly executed. You also have to take into account that there were a sustained number of running jobs about 16K.

4. Main Features

HTCondor LRMS is a tool rich of useful features and during this experimental activity, we have tried them taking into account those which could really have an important impact in the HTCondor deployment at ReCaS DataCenter.

- Priority: resources are allocated to users based upon a user's priority. A lower numerical value for user priority means higher priority, so a user with priority 5 will get more resources than a user with priority 50.
- Fairshare: it provides a fair access to the resources based upon user resources usage over the time and the user share assignment of the available resource.
- Security: this is a broad issue which implement different security levels. It provides what if needed to authenticate and authorize users and daemons. It implements the encryption and integrity checking over the network.
- Ranking: used to choose a match from among all machines that satisfy the jobs requirements and are available to the user, after accounting for the users' priority and the machines rank of the job. The rank expressions, simple or complex, define a numerical value that expresses preferences.
- Partitionable slots: allow to define a big slot that can be partitioned to accommodate actual needs by accepted jobs. It can address different CPU/memory requirements because it dynamically provides the needed resources.
- Policy: it provides a way to control resource access and limiting the resource usage over the time.
- GRID interface: even if not well supported, an HTCondor pool can be exploited through the GRID Computing.

5. Conclusions

We have evaluated a LRMS which can easily be defined stable, scalable and robust enough to be chosen as LRMS for ReCaS DataCenter. HTCondor can easily provide all the main functionalities in order that resources hosted at ReCaS DataCenter can be exploited through the batch job submission. Furthermore, HTCondor is a scientific community driven software licensed under the Apache License 2.0. The LRMS is a scientific community driven software. It is mainly developed by the University of Wisconsin with the contribution of other American Universities. Scientific community can influence the software development requesting new features and/or functionalities and at the same time, provides best effort support through two mailing lists and tutorials.

Reference

1. Available at https://research.cs.wisc.edu/htcondor/.

Chapter 39

The Catania Science Gateway Framework in the ReCaS Environment

R. Barbera[*,‡], R. Bruno[†],
M. Fargetta[†] and G. La Rocca[†]

[*]*Department of Physics and Astronomy, University of Catania, Italy*
[†]*The Italian National Institute of Nuclear Physics (INFN), Division of Catania, Italy*
[‡]*roberto.barbera@ct.infn.it*

Science Gateways are playing an important role in scientific research performed using e-Infrastructures. Through the highly collaborative environment of a Science Gateway, users spread around the world and belonging to various Virtual Research Communities can easily cooperate to reach common goals and exploit all the resources of the cyber-infrastructure they are entitled to use. One of the major tasks of a Science Gateway is to supervise the user access to the available services, denying the use to those people who are not authorized. In this work, we present a general framework to build Science Gateways and the customizations made in the ReCaS environment to meet the requirements of a couple of use cases coming from different scientific communities.

1. Introduction

By its most used definition, a Science Gateway (SG) is "a community-developed set of tools, applications, and data that are integrated via a portal or a suite of applications, usually in a graphical user interface, that is further customized to meet the needs of a specific community".[a] Science Gateways may provide access to a variety of capabilities including workflows, visualization, resource discovery, job execution and data management services.

In the last three years, the so called SG paradigm has grown both in terms of number of implementations and users.

At the beginning, most of the available SGs have been autonomously developed by specific research communities interested in facilitating the day-by-day life of their members. More recently, the European Commission and the US National Science Foundation have funded a few projects aimed at developing general frameworks (SCI-BUS)[b] or at coordinating the developments (Science Gateways RG[c]) and have strategic planning.

In the last couple of years, SGs have gotten an increased interest within EGI[d] at the point to be included in the "EGI Strategic Plan-Seeing New Horizons: EGI's Role in 2020."[e] Science Gateways, or Virtual Research Environments as they are called more in general, are presently quite a "hot topic" in the e-Infrastructure world especially when they are combined with the other important area of workflows. The first generation of single-project/single-community implementations is evolving into structured complex frameworks that are meant to make developers' life easier and application integration faster.

The power of SGs has been demonstrated in real cases and for real applications during workshops and other kind of events and in the framework of collaborations with some VRC-specific projects. Besides "in house" experience (the CSGF is developed at INFN Catania), another

---

[a]Available at www.xsede.org/gateways-overview.

[b]Available at www.sci-bus.eu.

[c]Available at www.sciencegateway.org.

[d]Available at www.egi.eu.

[e]Available at http://go.egi.eu/EGI2020.

important reason for the choice was the large number of standards supported by the CSGF:

- The JSR 168[f] and JSR 286[g] standards (also known as "portlet 1.0" and "portlet 2.0" standards) for the presentation layer;
- The OASIS[h] Security Assertion Markup Language (SAML)[i] standard and its Shibboleth[j] and SimpleSAMLphp[k] implementations for user authentication;
- The Lightweight Direct Access Protocol (LDAP) and its OpenLDAP[l] implementation for user authorization;
- The Cryptographic Token Interface Standard (PKCS#11)[m] standard and its Cryptoki implementation for X.509 robot certificate management;
- The Open Grid Forum (OGF)[n] Simple API for Grid Applications (SAGA)[o] standard and its JSAGA implementation for both local and Grid middleware interoperability;
- The Open Grid Forum (OGF) Open Cloud Computing Interface (OCCI)[p] standard and its rOCCI[q] implementation for Cloud middleware interoperability and cloud interoperation.

In particular, the support of the SAML standard allows SGs built with the CSGF to be configured as Service Providers (SPs) of Identity Federations (IdFs) in order to extend their potential user base to all individuals enrolled in Identity Providers (IdPs), owning federated credentials.

---

[f]Available at http://jcp.org/en/jsr/detail?id=168.

[g]Available at http://jcp.org/en/jsr/detail?id=286.

[h]Available at http://www.oasis-open.org/.

[i]Available at http://saml.xml.org/.

[j]Available at http://shibboleth.internet2.edu/.

[k]Available at http://simplesamlphp.org/.

[l]Available at http://www.openldap.org/.

[m]Available at https://docs.oracle.com/javase/7/docs/technotes/guides/security/p11guide.html.

[n]Available at http://www.ogf.org/.

[o]Available at http://www.ogf.org/documents/GFD.90.pdf.

[p]Available at http://occi-wg.org/.

[q]Available at http://occi-wg.org/2012/04/02/rocci-a-ruby-occi-framework/.

2. The SG for the ReCaS Project

This section describes the architecture and present status of the SG for the project. The SG is built within the Liferay web portal framework and portlet container[r] and it is fully compliant with the JSR 286 ("portlet 2.0") standard. Liferay is currently the most used framework to build SGs in the "Grid world" and ships with more than 60 portlets that can be easily combined (mashed-up) to build complex and appealing e-collaboration environments. Other 200+ portlets are available in the repository of the Liferay community.

Separate sub-sections are devoted to the various functional aspects of the SG.

2.1. *Authentication and Authorization*

The most important requirement of the ReCaS SG was to ease the access to the distributed computing and storage resources by the

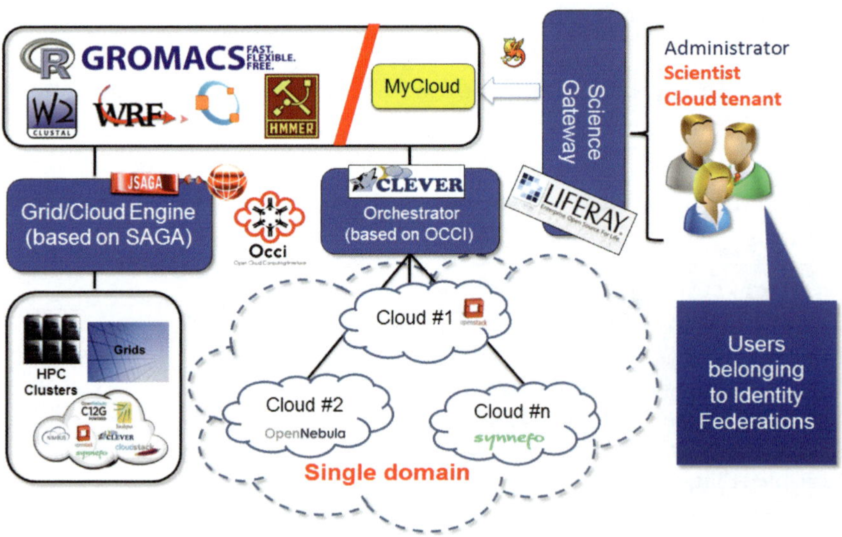

Fig. 1. The reference model of the ReCaS SG.

[r]Available at http://www.liferay.com/.

largest possible community of (non IT-expert) clinicians through a set of well-defined and domain specific applications. In order to meet this requirement, authentication and authorization mechanisms have been conceived to provide a smooth access to the applications yet preserving the security level requested by the distributed e-Infrastructure and the typology of the sensible information (clinical data) managed. Indeed, the neurological data stored in the SG have extra requirements in terms of security, anonymity and confidentiality. It must always be clearly defined who can access which images for his/her own analysis.

Therefore, several web and Grid technologies have been adopted and deployed to ensure that the authentication and authorization mechanisms fulfill the stringent requirements and implement the expected roles and the corresponding privileges. Such tools implement an authentication/authorization hierarchy moving from the web to the physical Grid resources in order for users to have to interact only with the top level tool, this taking care of transferring the credentials to lower levels.

The highest component in the authorization/authentication hierarchy has to be integrated in the SG and has to support a Single Sign On (SSO) mechanism across all services, given a user is entitled (i.e., has the right) to use, in order not to confuse non-experienced users with different sets of credentials.

Many web tools support SSO within a centralized or distributed authentication framework.

Nevertheless, in order to comply with currently adopted standards and support the most relevant IdFs in Education and Research, the ReCaS SG is compliant with the Security Assertion Markup Language (SAML) OASIS standard for credentials communication.

The Shibbolet implementation of SAML has been adopted in the ReCaS SG and a library has been developed to make Liferay manage the Shibboleth token.

In many countries, there is currently a big effort to create IdFs gathering all education and research institutions to simplify and ease the access to services for users working in different locations. Actually, they are generally managed by National Research and Education

Networks in EU countries and aim at the integration of networks, services and users. Therefore, it was important for ReCaS to follow this trend in order to allow its integration with other services and increase the number of potential users.

The use of Shibboleth allows an easy integration with IdFs and individual institutions (i.e., IdPs) wishing to include the ReCaS Science Gateway as one of the resources (i.e., SPs) for their users. When a user tries to use one of the ReCaS applications available on the SG, he/she is re-directed to a Discovery Service (DS) listing all the supported IdFs and IdPs among which he/she can select the one he/she is member of. The IdP identifies the user, generally through a pair of username and password. If the authentication by the IdP is successful, the control is returned to the SG where user authorization is checked.

The ReCaS SG[s] is a Service Provider of the GrIDP (Grid IDentity Pool)[t], a "catch-all" federation operated by INFN Catania and Consorzio COMETA to manage several SGs and generic web portals. This is the federation collecting the IdPs of institutions which are not members of any IdF and currently includes INFN (the Italian Institute for Nuclear Physics) IdPs.

Once a user is authenticated, the authorization system verifies his/her credentials. The Scientific Administrator grants authorizations, and a centralized LDAP-based registry connected to the ReCaS SG, has been created to store and manage roles and privileges. User roles are then mapped onto those performing Grid transactions using the Virtual Organization Membership Service (VOMS) functionalities.

2.2. *Robot Certificates*

The management of personal certificates to access e-Infrastructures has demonstrated to be difficult by non-expert users and represents a limiting factor to the rapid spread of this technology in new scientific domains where computer science is not a basic knowledge.

---

[s]Available at http://recasgateway.ba.infn.it/.

[t]Available at https://gridp.garr.it/.

Anotable step forward to make the access to Grid infrastructures as much transparent and smooth as possible, has recently been achieved with the introduction of robot certificates, also referred as portal certificates, and their integration inside traditional general-purpose portals[1] and SGs.[2]

The advantages introduced by this new kind of digital certificates are manifold and 10 Certification Authorities in Europe have already adopted them. For security sake, robot certificates are usually stored on board of tamper-resistant devices such as smartcards. This improves the security and avoids any fraudulent use of the private keys. A multi-threaded server, called eToken server, has been created and configured to manage a list of robot certificates (one certificate per application) stored in different USB eToken PRO 32/64 KB smart cards released by SafeNet.[u]

The eToken server provides the ReCaS SG (and other gateways developed at INFN Catania) with a 24 × 7 service and holds the web services to access the smart cards and interact both with the VOMS and the automatic proxy renewal (MyProxy) service. This crypto library developed,[2] combines different standards and programming native libraries with the latest release of robot certificates. A Java multi-platform client, configured for inter-service communication via HTTPS, completes the architecture.

The eToken server is built on top of the Apache Tomcat Application Server and configured to accept requests only from a set of authorized "clients" (i.e., the SGs). This ensures scalability and high performances especially when the server has to deal with huge numbers of requests. To further improve its performances and reduce the waiting time to get a proxy, the eToken server implements also a cache mechanism.

The usage workflow of the "light-weight" Grid crypto library used by the ReCaS SG is shown in Fig. 2.

Once the accredited IdP has successfully authenticated the user, his/her authorization rights on the local LDAP registry are checked and, if the verification is successful, he/she is logged into the ReCaS SG. Then, according to the information stored in the LDAP registry

[u]Available at http://www.safenet-inc.com/.

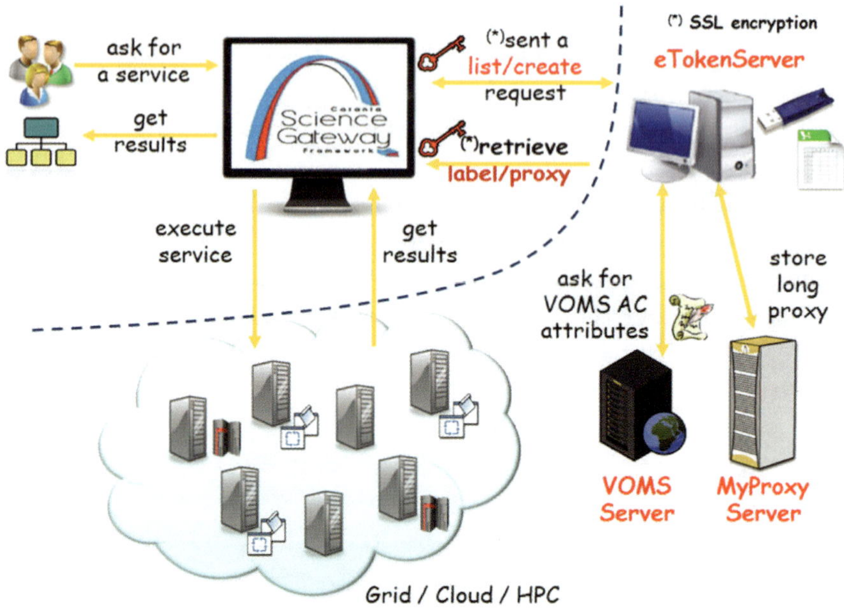

Fig. 2. The workflow for implementing Grid and Cloud authorization by means of the eToken server.

and the application(s) he/she wants to run from within the portal, the SG sends a requestID to the eToken server.

If the SG is authorized, and taking into account the information available in the HashMap, the eToken server sends back to the SG a new proxy certificate, if the requestID is not found or the lifetime of the old proxy is expired. Otherwise, it sends a cached proxy, if a valid proxy certificate is available in memory. This operation is completely transparent from the end-user point of view, and allows the user to execute the operations requiring the usage of the Grid services and get the results. The retrieval of a cached proxy takes only 20 ms while the creation of a new one requires about 5s.

2.3. *The Grid & Cloud Engine*

The ReCaS SG uses the Catania Grid & Cloud Engine to access different DCIs (Grid, Cloud and HPC clusters). The Catania Grid &

Cloud Engine is a generic software module able to interconnect the SG presentation layer with the underlying Grid infrastructures using standard technologies. It allows the quick creation of new SGs providing their developers with a simple interface and avoiding worry about middleware specificities.

In the design of the ReCaS SG, the adoption of international standards has been considered since the beginning as a mandatory practice, in order to protect the investment in the creation of this high-level user interface from middleware changes and lack of interoperability.

In this way, the SG can become a "gate" to a huge e-Infrastructure made of resources coming from different kinds of Grid infrastructures with different middleware deployed and connected via standard interfaces. As a consequence of the above considerations, the Catania Grid Engine adopts the Simple API for Grid Applications (SAGA) Core API,[3] a high level, application-oriented, software library for Grid applications specified by the Open Grid Forum (OGF), and its JSAGA implementation.[v] JSAGA allows creating a unique interface to different middleware stacks and makes SGs able to exploit resources coming from different Grid infrastructures.

The architecture of the Catania Grid Engine is sketched in Fig. 3.

The following modules are part of the Catania Grid & Cloud Engine:

- The SG Interface: an interface hiding all the complexity of the Grid world to the presentation layer providing additional capabilities suitably designed for the web world. It allows developers to create portlets for applications in a very short time (just a few days).
- The Job Engine: a module allowing users to fully exploit Grid job management services. It maps job operations to JSAGA functions;
- The Data Engine: a module allowing users to fully exploit Grid data management services. It maps data operations to JSAGA functions;

[v]Available at http://grid.in2p3.fr/jsaga/.

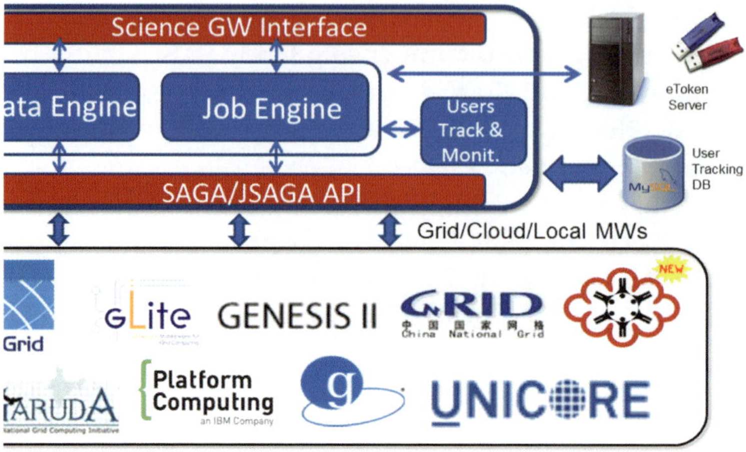

Fig. 3. The architecture of the Grid & Cloud Engine.

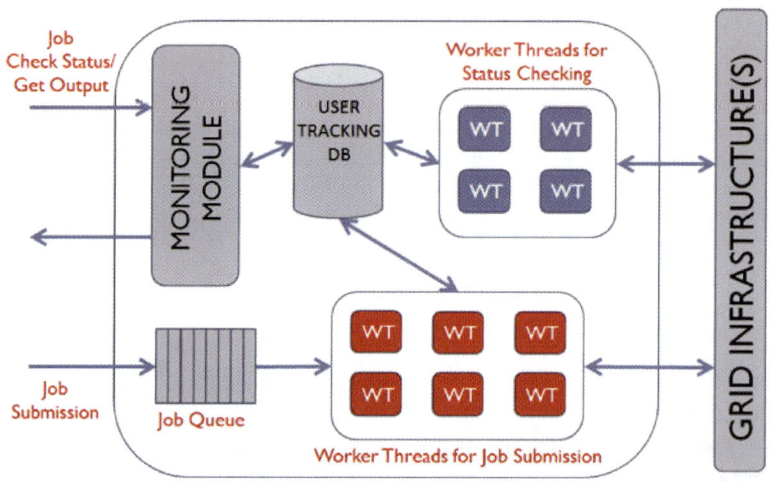

Fig. 4. Thread Pool usage in the Job Engine.

- The User Tracking and Monitoring DB: a user tracking and accounting tool able to store information on each Grid transaction in the Users' Tracking database. It is in charge of controlling the rate of Grid interactions (in total and per users) executed via the SG to implement flow control policies (e.g., max number of jobs submitted per users, max number of jobs submitted per SG, etc.).

It has been developed to make the SG complaint with the EGI VO Portal Policy[w] and the EGI Grid Security Traceability and Logging Policy.[x]

- The JSAGA API.

3. Summary and Conclusions

In this Chapter, we outlined the reference model of the ReCaS SG, a web-based technology to make easier the access and the exploitation of many Distributed Computing Infrastructures (DCIs). SGs have proven over the last years to be fertile ground for supporting technological and scientific collaboration across different e-Infrastructures operated in different regions and continents.

References

1. R. Barbera *et al.*, The GENIUS Grid Portal and robot certificates: A new tool for e-Science, *BMC Bioinformatics* **10**(Suppl. 6), S21 (2009).
2. G. La Rocca, R. Barbera, V. Ciaschini, A. Falzone and S. Monforte, *A new "lightweight" Crypto Library for supporting a new Advanced Grid Authentication Process with Smart Cards.* In Proceedings of Science (ISGC 2011 & OGF 31), pp. 29 (2001).
3. S. Jha, H. Kaiser, A. Merzky, O. Weidner *et al.*, *Grid Interoperability at the Application Level Using SAGA.* In Proceedings of IEEE International Conference on e-Science Grid Computing, pp. 584–591, Bangalore, India, 10–13 December (2007).

[w] Available at https://documents.egi.eu/public/ShowDocuments?docid=80.

[x] Available at https://documents.egi.eu/public/ShowDocuments?docid=81.

Chapter 40

VAF: A Virtual Analysis Facility Exploiting PRISMA OpenStack Infrastructure at Bari

F. Colamaria*,†, D. Colella*,†, G. Donvito†, D. Elia†, A. Franco†, G. Maggi†,‡, G. Miniello*,†,§ and G. Vino*,†

**Dipartimento Interateneo di Fisica, Università degli Studi di Bari, Via Edoardo Orabona 4, 70126 Bari, Italy*

*†INFN — Sezione di Bari
Via Edoardo Orabona 4, 70126 Bari, Italy*

*‡Politecnico di Bari, Dipartimento Interateneo di Fisica
Via Edoardo Orabona 4, 70126 Bari, Italy*

§giorgia.miniello@ba.infn.it

A Virtual Analysis Facility (VAF) for the ALICE experiment has been set up in Bari in the framework of the PRIN project n.20108T4XTM_004 "Sviluppo di Tecnologie per l'Ottimizzazione dell' Accesso ai dati di LHC (STOA)." The VAF uses the OpenStack Cloud Infrastructure built for the ReCaS PRISMA project. Elasticity and scalability features of Bari VAF have been studied, increasing progressively the number of PROOF workers available for running a test-analysis and evaluating both the time spent for raising them up and the wall time taken to perform a single analysis task parallelizing the jobs on the enabled workers.

1. The OpenStack Infrastructure at Bari

The Bari Computer Center for Science (Bc^2S) hosts **the Bari PRISMA**[1] **OpenStack Cloud Infrastructure**, which offers to users the availability of a maximum quota of Virtual Machines (VMs), providing *de facto* an elastic and highly scalable Cloud Computing service, the so called **IaaS** (**I**nfrastructure **a**s **a** **S**ervice).

The set-up of BARI-INFN (Istituto Nazionale di Fisica Nucleare) Cloud is shown in Fig. 1.

All the **core services** are provided in **High Availability** (H.A.). It means that replica sets of core services are available to provide an automatic failover: a secondary (backup) instance becomes a primary one if the primary instance is temporary unavailable.

The **MySQL Data Base** (MySQL DB) is on **Galera Cluster**. MySQL DB is also used by all OpenStack services as backend.

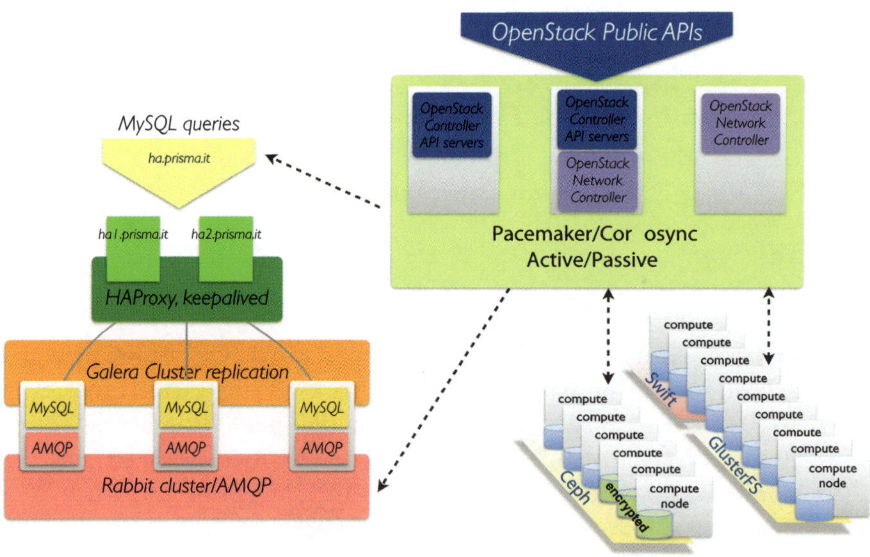

Fig. 1. The BARI-INFN Cloud set-up built for ReCas PRISMA project.

The **Messaging Server**, which allows communications among services is on **Rabbit Cluster**.

Both MySQL and Messaging Server Clusters are hosted on the same three physical instances, so if one instance is down, the other two become available.

The set-up includes two **Controller Node** instances and two **Network Node** instances in Active/Passive configuration on three physical hosts.

The software **Pacemaker Corosync,** which implements the H.A., orchestrates and manages this balance.

The distributed file system **Ceph** works as backend of the OpenStack **Block Storage Cinder**.

Finally, the current infrastructure includes nine **Compute Nodes**.

The **Gluster FS** file system is mounted on six of them, while the remaining three are based on **Swift**, which is the OpenStack **Object Storage**. The **Gluster FS** is used as **Nova** (OpenStack Compute) and **Glance** (OpenStack Image Service) **backend** to ensure **live migration**.

In the next section, a particular use case of the OpenStack IaaS at Bari will be discussed: the creation and deployment of a **VAF** for the ALICE experiment at Bari.

2. Virtual Analysis Facilities for ALICE Experiment

Virtualization and **Cloud Computing** are currently used to complement the GRID Computing, officially used for submit analysis jobs for all the LHC experiments, to achieve high-throughput tasks required by High Energy Physics.

A **VAF** for the ALICE experiment is a dedicated PROOF-based[2] cluster of VMs.

This project has been developed on the Bari **PRISMA** OpenStack Infrastructure to investigate more deeply such computing approach.

It is intended to be the evolution of previous Analysis Facility (AF) models which used a more static PROOF cluster configuration.

PROOF is an extension of the ROOT data analysis framework that allows interactive analysis on a local cluster.

In this context, GRID computing jobs are replaced by flexible VMs.

The flexibility is ensured by the possibility of creating and terminating the virtual instances just *on demand*, so allowing the cluster to be bigger (**scaling up**) when jobs are running on the PROOF workers and smaller if they are not busy (**scaling down**).

The VMs are provided and deleted *elastically* by the python tool called **elastiq**.[3]

The **elastiq** tool periodically checks the status of EC2[4] VMs belonging to HTCondor[5] cluster to find and terminate the idle ones.

The **VMs contextualization** and the **elastiq** configuration are provided using the web interface of **CernVM Online**.[6]

The working principles of **elastiq** is outlined in Fig. 2.[7]

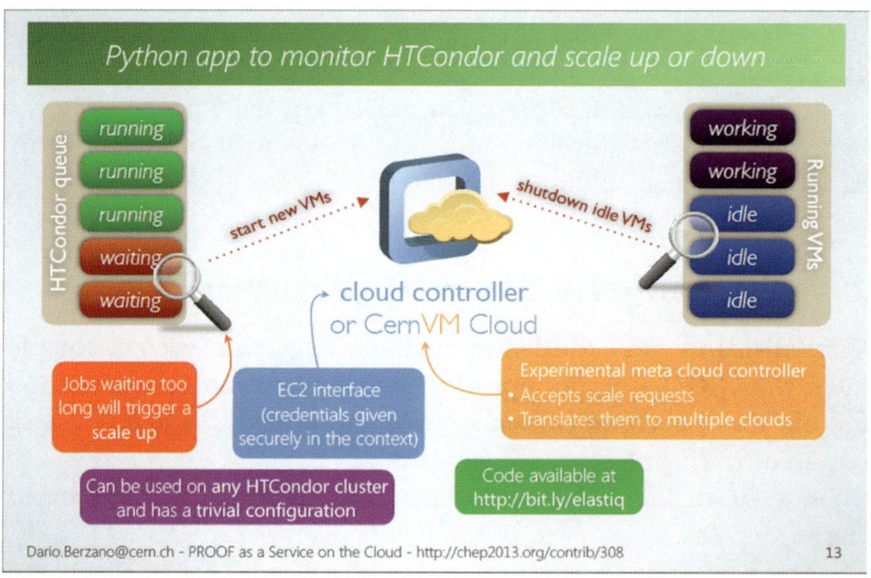

Fig. 2. A schematic representation of the operating principle of the python tool **elastiq**. This tool optimize *de facto* the performance of the infrastructure checking periodically the VMs status.

3. The VAF at Bari: Elasticity and Scalability Performances

The **latest VAF configuration at Bari** provides up to **56 PROOF workers**.

The **elasticity** of the cluster is clearly visible in Fig. 3 which represents the time taken to complete the whole boot of the VMs (pink markers) and the time taken by PROOF workers to join the cluster (blue markers) as a function of the number of available PROOF workers themselves (from 12 to 44).

As it can be observed, the cluster takes just seven minutes to complete the boot of the instances corresponding to a 44 PROOF worker configuration and the 44 PROOF workers are totally ready to join the cluster in just nine minutes.

As shown in Fig. 4, in the analysis used as a benchmark, the wall time of the task submitted to the VAF PROOF cluster has been

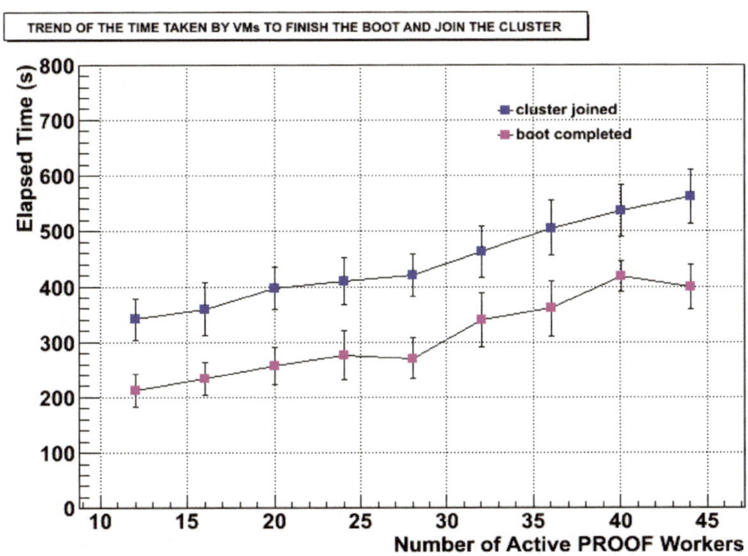

Fig. 3. Time elapsed for completing the VMs' boot (pink markers) and time elapsed to join the cluster (blue markers) versus the number of the available PROOF workers.

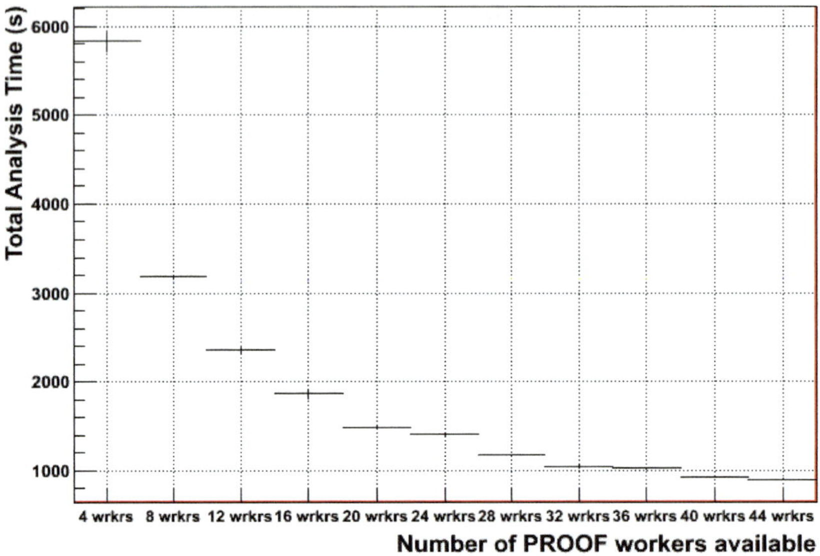

Fig. 4. The wall time of the analysis jobs versus the number of the PROOF workers available in the cluster.

reduced by a factor of 5, just passing from a 4- to a 28-active worker configuration with an analysis job wall time reduction greater than 80%.

Combining the elasticity information with the scalability analysis results, the user is able to determine the ideal cluster working conditions for his own specific data analysis.

4. Conclusions

A VAF for ALICE experiment exploiting PRISMA OpenStack Infrastructure has been setup and deployed in Bari. Elasticity and Scalability performances have been studied: more than 40 active PROOF workers can be made available in few minutes and a meaningful reduction of the analysis job wall time has been observed.

At present, the implementation of a Data Federation to preserve and share all the ALICE datasets needed to perform the analysis campaigns has been also studied and created at Bari, which currently hosts the unique national XRootD[8] redirector for all Italian VAF sites.

References

1. Available at http://recas.ba.infn.it/recas1/index.php/recas-prisma.
2. Available at http://root.cern.ch/drupal/content/proof.
3. Available at https://github.com/dberzano/elastiq.
4. Available at http://aws.amazon.com/ec2/.
5. Available at http://research.cs.wisc.edu/htcondor/.
6. Available at https://cernvm-online.cern.ch/.
7. D. Berzano, R. Meusel, G. Lestaris, I. Charalampidis, G. Ganis, P. Buncic and J. Blomer, *PROOF as a Service on the Clouda Virtual Analysis Facility based on the Cern VM Ecosystem*, Presentation at CHEP2013 — Amsterdam, 15th October, (2013).
8. Available at http://xrootd.org/.

Index